THE VENICE LAGOON ECOSYSTEM

Inputs and Interactions Between Land and Sea

MAN AND THE BIOSPHERE SERIES

Series Editor J.N.R. Jeffers

VOLUME 25

THE VENICE LAGOON ECOSYSTEM

Inputs and Interactions Between Land and Sea

Edited by

P. Lasserre

UNESCO, Paris, France

and

A. Marzollo

Universita degli Studi di Udine, Udine, Italy

PUBLISHED BY

PARIS

AND

The Parthenon Publishing Group

International Publishers in Science, Technology & Education

Published in 2000 by the United Nations Educational, Scientific and Cultural Organization,
7 Place de Fontenoy, 75700 Paris, France—UNESCO ISBN 92-3-103595-9
and
The Parthenon Publishing Group Inc.
One Blue Hill Plaza
PO Box 1564, Pearl River,
New York 10965, USA—ISBN 1-85070-083-4

and
The Parthenon Publishing Group Limited
Casterton Hall, Carnforth,
Lancs LA6 2LA, UK—ISBN 1-85070-083-4

British Library Cataloguing in Publication Data
The Venice lagoon ecosystem : inputs and interactions between land and sea. – (Man and the
biosphere series; v. 25)
  1. Lagoon ecology – Italy – Venice, Lagoon of 2. Ecosystem management – Italy – Venice,
Lagoon of 3. Nature – Effect of human beings on – Italy – Venice, Lagoon of
  I. Lasserre, Pierre II.Marzollo, Angelo
  577.7′385′027
  ISBN 1-85070-083-4

Library of Congress Cataloging-in-Publication Data
The Venice Lagoon ecosystem : inputs and interactions between land and sea / P. Lasserre and
A. Marzollo (editors).
        p.   cm.
    Includes bibliographical references and index.
    ISBN 1-85070-083-4
    1. Lagoon ecology – Italy – Venice. 2. Water—Pollution—Environmental Aspects—
Italy— Venice. I. Lasserre, Pierre. II. Marzollo, Angelo.
QH152.V46   1999
577.7′8385—dc21   99-31158
                                                                          CIP

Typeset by Speedlith Photolitho Ltd., Manchester, UK
Printed and bound by Bookcraft (Bath) Ltd., Midsomer Norton, UK

CONTENTS

Section 5   Modelling

PREFACE

UNESCO's Man and the Biosphere Programme

Improving scientific understanding of the natural and social processes relating to human interactions with the environment; providing information useful to decision-making on resource use; promoting the conservation of genetic diversity as an integral part of land management; enjoining the efforts of scientists, policy-makers, and local people in problem-solving ventures; mobilizing resources for field activities; strengthening of regional cooperative frameworks – these are some of the generic characteristics of UNESCO's Man and the Biosphere (MAB) Programme.

The MAB Programme was launched in the early 1970s. It is a nationally based, international programme of research, training, demonstration, and information diffusion. The overall aim is to contribute to efforts for providing the scientific basis and trained personnel needed to deal with problems of rational utilization and conservation of resources and resource systems and problems of human settlements. MAB emphasizes research for solving problems. It thus involves research by interdisciplinary teams on the interactions among ecological and social systems, field training, and application of a systems approach to understanding the relationships among the natural and human components of development and environmental management.

MAB is a decentralized programme with field projects and training activities in all regions of the world. These are carried out by scientists and technicians from universities, academies of sciences, national research laboratories, and other research and development institutions under the auspices of more than one hundred MAB National Committees. Activities are undertaken in cooperation with a range of international governmental and non-governmental organizations.

Man and the Biosphere Book Series

The Man and the Biosphere Book Series was launched to communicate some of the results generated by the MAB Programme and is aimed primarily at upper level university students, scientists, and resource managers, who are not necessarily specialists in ecology. The books are not normally suitable for undergraduate text books but rather provide additional resource material in the form of case studies based on primary data collection and written by the researchers involved; global and regional syntheses of comparative research conducted in several sites or countries; and state-of-the-art assessments of knowledge or methodological approaches used on scientific meetings, commissioned reports, or panels of experts.

The Venice Lagoon Ecosystem

On 4 November 1966, a tidal storm surge of exceptional height and duration caused the almost total flooding of the city of Venice. That day, the famous St Mark's Square was covered by more than a metre of water, an event that drew the attention of the whole world to the growing dangers threatening the City of the Doges. The Italian authorities immediately set up a Committee for the Safeguard of Venice, under the aegis of the Ministry of Public Works, and initiated a whole series of actions, including the approval of new laws and regulations. UNESCO, in turn, launched an international campaign for Venice and shortly thereafter published a first comprehensive report – entitled *Sauver Venise* – on the situation and on the various interrelated problems of Venice and its lagoon, a complex environment shaped over centuries by the Republic of Venice and its people.

If one main focus of public attention was on the restoration and preservation of important Venetian monuments, another parallel concern was that of developing an integrated scientific approach to the Venice Lagoon, as recommended by a group of experts jointly set up by the Italian government and UNESCO in 1981. Within such a context, plans took shape for a new scientific project on the Venice Lagoon System, under the aegis of the Italian Ministry for Universities and for Scientific and Technological Research. The project was launched in late 1990, as a co-operative initiative of the Italian National Research Council, the Universities of Padua and Venice, and UNESCO, with an agreement being signed between the Italian authorities and UNESCO for the execution of a project on the Venice Lagoon Ecosystem, and the co-ordination of the Italian and non-Italian research teams. From the UNESCO side, the primary focus for the project was provided at the time by the Major Interregional Project on Research and Training Leading to the Integrated Management of Coastal Systems (COMAR), in collaboration with UNESCO's other main undertakings in the environmental sciences field: the Intergovernmental Oceanographic Commission (IOC), the International Hydrological Programme (IHP), the International Geological Correlation Programme (IGCP) and MAB.

The main scientific results of the Venice Lagoon Ecosystem project are brought together in the present volume, edited by Pierre Lasserre and Angelo Marzollo. Pierre Lasserre acted as Scientific Co-ordinator of the project. He has been associated with scientific research initiatives in coastal areas throughout his professional career – as a marine biologist and oceanographer; as Director of the French CNRS Marine Biological Station at Roscoff in Brittany; as Professor at the University of Paris VI; from July 1993 to June 1999, as Director of UNESCO's Division of Ecological Sciences and Secretary of the International Co-ordinating Council for the MAB Programme; and most recently, from July 1999, as Director of UNESCO's Office in Venice.

Angelo Marzollo was the UNESCO officer in charge of the planning and execution of the project. He is an applied mathematician, full Professor of system theory from 1975. Previously visiting Professor at the University of California at Los Angeles, and at the University of Paris VII–Jussieu and research director at the Ecole Nationale des Mines (Paris), and he is now at the University of Udine. He is also a representative of the Italian Government on the Steering Committee of UNESCO's Venice Office and holds positions in several Venetian academies and associations such as Ateneo Veneto and the Forum for the Venice Lagoon.

In thanking the two editors and all contributors for seeing this co-operative undertaking into print, UNESCO hopes that the scientific results reported here will contribute distinctively to the understanding of one of the world's most renowned coastal lagoon ecosystems, as well as to decisions on the future development of Venice and its lagoon. UNESCO also hopes that the approaches and insights described in this volume may be of interest to scientists working on coastal lagoon ecosystems in other parts of the world.

The map featured on the cover represents the Lagoon of Venice and its hydrographic network, designed by Antonio Vestri in 1709; Archivio di Stato, sede di Venezia, S.E.A., disegni Diversi n. 109. The photograph on the cover is by Pierre Lasserre and the montage by Ivette Fabbri.

MAN AND THE BIOSPHERE SERIES

FOREWORD

The exceptional and dramatic 1966 flood drew world attention to the Venice Lagoon and to the physical damage to Venetian monumental and artistic treasures, as well as to its socio-economic decline. This situation was described in the comprehensive *Report of Venice* which UNESCO prepared and published in 1968. At the same time, UNESCO launched an International Campaign for Venice at the instigation of its Director General, René Maheu. Whereas most results of that campaign concerned the restoration and preservation of some important Venetian monuments, for various reasons, the attention of the international scientific community to the Lagoon was called upon, through UNESCO, only years afterwards.

The Venice Lagoon ecosystem is a unique case for developing and encouraging comprehensive interdisciplinary programmes of international scope. In 1981 an international group of experts was jointly appointed by UNESCO and Italy. This group considered it was important to develop an integrated scientific approach to the Venice Lagoon. The following aspects were taken into account: alteration of the Lagoon's morphology, due to human activities during the twentieth century; environmental degradation due to industrial, agricultural and urban pollution, as well as the proposed works aimed at controlling high water in the Lagoon. A report entitled *International Research Project Venice Lagoon* was prepared by the group of experts and published by UNESCO in 1983.

An important task of this overall project would have been 'to review previous works and to undertake a comprehensive study of the functioning and evolution of the lagoon ecosystem, including the exchanges at its interfaces'. The report underlined that this broad approach to the Lagoon ecosystem required international multidisciplinary co-operation among scientists, and recognised that the project fitted well into the framework of UNESCO's worldwide programme on coastal systems. Although the project was not realized in its entirety, it was used as a reference in the preparation of the Feasibility Study for the project *Venice Lagoon System* carried out on behalf of the Italian Ministry for Universities and for Scientific and Technological Research (MURST).

Following a decision by Minister A. Ruberti, this broad project was financed by MURST in late 1990, and executed by the Italian National

Research Council (CNR), the Universities of Padua and Venice, and UNESCO; each of these four Institutions being responsible for a part of the project. In November 1990 a Fund-in-Trust Agreement was signed between UNESCO and Italy for the execution of the project called *Venice Lagoon Ecosystem*, and whose results are presented in this volume of UNESCO's MAB Books Series. Its specific programme had been prepared by an international *ad hoc* expert panel chaired by Prof. Bruno Battaglia, from the University of Padua and co-chaired by Prof. Pierre Lasserre, from the University of Paris VI, and was approved by the MURST Co-ordinating Committee. Prof. Angelo Marzollo represented UNESCO in that Committee from 1990 to 1997, and also co-supervised with P. Lasserre (acting as scientific co-ordinator) the execution of the *Venice Lagoon Ecosystem* project. This included the practical co-ordination of the work of the multidisciplinary Italian and non-Italian research teams.

A scientific meeting, with the participation of selected senior research leaders, was held in July 1991 in Venice to revise the general aims of the project and elaborate detailed workplans. The first results were presented at the first Workshop of the entire MURST Project *Venice Lagoon System* held at the Istituto Veneto di Scienze, Lettere ed Arti, Venice, from 19 to 20 March 1992. The UNESCO teams made a general presentation of the objectives and workplans and described preliminary results. A second and third workshop were held in Venice at the same Istituto in 1993 and 1996 with presentations given by all team leaders, followed by discussion among project participants and representatives of the scientific community and Italian authorities. On the occasion of the third workshop, summaries of the essential results obtained by all groups were prepared, and printed, including those of the *Venice Lagoon Ecosystem* project, in a book belonging to the series of publications of the Istituto Veneto Lettere Scienze ed Arti (including abstracts in Italian).

Following the start of the Project good co-operation developed with the late Prof. A. Orio, Director of the Department of Environmental Sciences, University of Venice, and the Dutch scientists from Texel and Den Helder, involved in the European Community Project 'Benthos Eutrophication Studies'. This project was focusing on effects of organic material and nutrients on the benthic systems of the eastern part of the Venice Lagoon. Modelling and field experiments were tested by mesocosm studies at the Netherlands Institute of Sea Research, Texel. Significant results of the BEST Project have been incorporated in this volume (see Chapters 15–18 and 28).

The present volume, in the Man and the Biosphere Series, provides the scientific community with the extensive results of the whole UNESCO–MURST *Venice Lagoon Ecosystem* project. It contributes also to a multidisciplinary and synthesizing effort to understand the complex interactions between man-induced perturbations and natural biological phenomena so as to better understand their reciprocal effects. Hopefully, out of this interdisciplinary experiment a more enriching and useful vision of the major inputs and interactions between terrestrial and aquatic elements in the Venice Lagoon will emerge from the papers included in the four major sections: Introduction, Pelagic studies, Benthic studies, Biological effects of environmental pollution and Modelling.

We wish to acknowledge the tireless, voluntary efforts of all those who contributed to the project's conception, approval, financing and the complex practical implementation, including the Lagoon oceanographic campaigns. We take this opportunity to express our gratitude to the following talented and dedicated individuals: Prof. Bruno Battaglia, Chairman of the expert panel of the UNESCO–MURST Venice Lagoon Ecosystem Project; Minister A. Ruberti, without whom the whole initiative would not have materialized; G. Modena, who chaired with patient wisdom the MURST Coordinating Commission through recent years; and M. Steyaert, who represented UNESCO in the preliminary Coordinating Commission.

We also note that the field work and the present volume benefited greatly from the close and very friendly working relationship among the near 80 project participants, including scientists, technicians, and research fellows and, in particular, F. Carrera and D. Tagliapietra who took care of sampling campaigns, computerized data bases and maps; and the crew members of the research vessels *Umberto d'Ancona, Mysis* and *Orata*. We thank Prof. B. Battaglia and Dr P. Franco, former directors of the Istituto di Biologia del Mare of CNR, and Prof. O. Ravera director of the Department of Environmental Sciences of the University of Venice, for their help in providing excellent laboratory facilities. Undoubtedly there are others who helped this project whom we have unintentionally omitted. To them, too, we owe our appreciation. The editors are most grateful to the authors and the reviewers for their collaboration and for their involvement in the final form of contributed Chapters. We owe special thanks to M. Wafar for providing help in copy editing, and to our UNESCO colleague, P. Pypaert, whose help was essential in the final editing of this volume and to the persons who assisted him, V.E. Brando, E. Caramelli, and G. Dall'Olmo. We hope this book will provide insights, which will point the way for future interdisciplinary research on the Venice Lagoon.

Pierre Lasserre
Angelo Marzollo

CONTRIBUTORS

F. Acri
Istituto di Biologia del Mare
ICNR-IBM
Riva 7 Martiri, Castello 1364/A
30122 Venezia, Italy

Y. Aissouni
Institut National de la Santé et de
la Recherche Médicale
INSERM U 58
60 rue de Navacelles
34090 Montpellier, France

L. Alberighi
Istituto di Biologia del Mare
ICNR-IBM
Riva 7 Martiri, Castello 1364/A
30122 Venezia, Italy

M. Bastianini
Istituto di Biologia del Mare
ICNR-IBM
Riva 7 Martiri, Castello 1364/A
30122 Venezia, Italy

G. Bendoricchio
Dipartimento Processi Chimici
Ingegneria
Università degli Studi di Padova
Via Marzolo 9
35131 Padova, Italy

F. Bianchi
Istituto di Biologia del Mare
ICNR-IBM
Riva 7 Martiri, Castello 1364/A
30122 Venezia, Italy

M. Bocci
Dipartimento Processi Chimici
Ingegneria
Università degli Studi di Padova
Via Marzolo 9
35131 Padova, Italy

A. Boldrin
Istituto di Biologia del Mare
ICNR-IBM
Riva 7 Martiri, Castello 1364/A
30122 Venezia, Italy

G. Campesan
Istituto di Biologia del Mare
ICNR-IBM
Riva 7 Martiri 1364/A
30122 Venezia, Italy

G.M. Carrer
Dipartimento Processi Chimici
Ingegneria
Università degli Studi di Padova
Via Marzolo 9
35131 Padova, Italy

xix

G. Cauwet
Laboratoire Arago
Université P. & M. Curie and
  CNRS,
Banyuls-sur-mer, France

B. Cavalloni
Istituto di Biologia del Mare
ICNR-IBM
Riva 7 Martiri, Castello 1364/A
30122 Venezia, Italy

F. Cioce
Istituto di Biologia del Mare
ICNR-IBM
Riva 7 Martiri, Castello 1364/A
30122 Venezia, Italy

J.J. Cleary
Centre for Coastal and Marine
  Sciences
Plymouth Marine Laboratory
Prospect Place, West Hoe
Plymouth PL1 3DH, UK

G. Coffaro
Dipartimento Processi Chimici
  Ingegneria
Università degli Studi di Padova
Via Marzolo 9
35131 Padova, Italy

J.A. Coles
Centre for Coastal and Marine
  Sciences
Plymouth Marine Laboratory
Prospect Place, West Hoe
Plymouth PL1 3DH, UK

M-A. Coletti-Previero
Institut National de la Santé et de
  la Recherche Médicale
INSERM U 58
60 rue de Navacelles
34090 Montpellier, France

A. Comaschi
Istituto di Biologia del Mare
ICNR-IBM
Riva 7 Martiri, Castello 1364/A
30122 Venezia, Italy

L. Craboledda
Istituto di Biologia del Mare
ICNR-IBM
Riva 7 Martiri, Castello 1364/A
30122 Venezia, Italy

M. Dai
Department of Marine Chemistry
  and Geochemistry
Woods Hole Oceanographic
  Institution
Woods Hole
MA 02543, USA

N. Dankers
IBN-DLO, PO Box 167,
1790 AD An Den Burg (Texel),
The Netherlands

L. DaRos
Istituto di Biologia del Mare
ICNR-IBM
Riva 7 Martiri, Castello 1364/A
30122 Venezia, Italy

F. Dolci
Istituto di Biologia del Mare
ICNR-IBM
Riva 7 Martiri, Castello 1364/A
30122 Venezia, Italy

S. Foltran
Dipartimento di Scienze Ambientali
Università degli studi di Venezia
Calle Larga Santa Marta, 2137
30123 Venezia, Italy

V.U. Fossato
Istituto di Biologia del Mare
ICNR-IBM
Riva 7 Martiri, Castello 1364/A
30122 Venezia, Italy

J.M. Garnier
CEREGE
BP 80
13545 Aix en Provence, France

H. de Heij
Netherlands Institute for Sea
    Research
PO Box 59
1790 An Den Burg (Texel),
The Netherlands

W.W. Huang
Institut de Biogéochimie Marine –
    CNRS
Ecole Normale Supérieure (ENS)
1 rue M. Arnoux
92120 Montrouge, France

R.G. Jak
TNO Department for Ecological
    Risk Studies
PO Box 57
1780 AB Den Helder,
The Netherlands

P. Lasserre
Division of Ecological Sciences
MAB Programme
UNESCO
1 rue Miollis
75015 Paris, France

P. Le Corre
Laboratoire d'Océanographie
    Chimique
University of Bretagne
    Occidentale
Brest, France

S. Lemaire-Gony
Plymouth Marine Laboratory
Prospect Place, West Hoe
Plymouth PL1 3DH, UK

M. Lepesteur
Department of Water Engineering
School of Civil Engineering
University of New South Wales
Sydney
NSW 2052, Australia

D.R. Livingstone
Centre for Coastal and Marine
    Sciences
Plymouth Marine Laboratory
Citadel Hill
Plymouth PL1 2PB, UK

D.M. Lowe
Centre for Coastal and Marine
    Sciences
Plymouth Marine Laboratory
Citadel Hill
Plymouth PL1 2PB, UK

S.N. Lvov
Energy and Geo-Environmental
    Engineering
The Pennsylvania State University
110 Hosler Building
University Park
PA 16802-5000, USA

C. Macé
Observatoire Océanologique de
    Roscoff
Station biologique
CNRS and Université P. & M. Curie
29680 Roscoff, France

C. Madec
Laboratoire d'Océanographie
    Chimique
University of Bretagne Occidentale
Brest, France

A. Marcomini
Dipartimento di Scienze
  Ambientali
Università degli studi di Venezia
Calle Larga Santa Marta, 2137
30123 Venezia, Italy

J.M Martin
Institut de Biogéochimie Marine –
  CNRS
Ecole Normale Supérieure (ENS)
1 rue M. Arnoux
92120 Montrouge, France

A. Marzollo
Universita degli Studi di Udine
Dipartimento di Matematica e
  Informatica
Via delle Scienze, 206
33100 Udine, Italy

I.R.B. McFadzen
Centre for Coastal and Marine
  Sciences
Plymouth Marine Laboratory
Prospect Place, West Hoe
Plymouth PL1 3DH, UK

P. Morin
Observatoire Océanologique de
  Roscoff
Station biologique
CNRS and Université P. & M.
  Curie
29680 Roscoff, France

C. Nasci
Istituto di Biologia del Mare
ICNR-IBM
Riva 7 Martiri, Castello 1364/A
30122 Venezia, Italy

A.A. Orio
Dipartimento di Scienze Ambientali
Università degli studi di Venezia
Calle Larga Santa Marta, 2137
30123 Venezia, Italy

R. Pastres
Dipartimento di Scienze Ambientali
Università degli studi di Venezia
Calle Larga Santa Marta, 2137
30123 Venezia, Italy

B. Pavoni
Dipartimento di Scienze Ambientali
Università degli studi di Venezia
Calle Larga Santa Marta, 2137
30123 Venezia, Italy

C.J.M. Philippart
IBN-DLO
PO Box 59
1790 AB Den Burg (Texel)
The Netherlands

R.K. Pipe
Centre for Coastal and Marine
  Sciences
Plymouth Marine Laboratory
Prospect Place, West Hoe
Plymouth PL1 3DH, UK

A. Piva
Dipartimento di Scienze Ambientali
Università degli studi di Venezia
Calle Larga Santa Marta, 2137
30123 Venezia, Italy

A.L. Pulsford
Centre for Coastal and Marine
  Sciences
Plymouth Marine Laboratory
Prospect Place, West Hoe
Plymouth PL1 3DH, UK

S. Rabitti
Istituto di Biologia del Mare
ICNR-IBM
Riva 7 Martiri, Castello 1364/A
30122 Venezia, Italy

O. Ravera
Istituto Italiano di Idrobiologia
Largo Tonolli 50
28922 Verbania Pallanza, Italy

N. Riccardi
Istituto Italiano di Idrobiologia
Largo Tonolli 50
28922 Verbania Pallanza, Italy

N. Sabil
Institut National de la Santé et de
la Recherche Médicale
INSERM U 58
60 rue de Navacelles
34090 Montpellier, France

M.C.Th. Scholten
TNO Department for Ecological
Risk Studies
PO Box 57
1780 AB Den Helder
The Netherlands

A. Sfriso
Dipartimento di Scienze
Ambientali
Università degli studi di Venezia
Calle Larga Santa Marta, 2137
30123 Venezia, Italy

G. Socal
Istituto di Biologia del Mare
ICNR-IBM
Riva 7 Martiri, Castello 1364/A
30122 Venezia, Italy

G. Stocco
Istituto di Biologia del Mare
ICNR-IBM
Riva 7 Martiri, Castello 1364/A
30122 Venezia, Italy

D. Tagliapietra
Dipartimento di Scienze Ambientali
Università degli studi di Venezia
Calle Larga Santa Marta, 2137
30123 Venezia, Italy

M.E. Thomas
Centre for Coastal and Marine
Sciences
Plymouth Marine Laboratory
Prospect Place, West Hoe
Plymouth PL1 3DH, UK

G. Todesco
Istituto di Chimica Industriale
Università degli Studi di Padova
Via Marzolo 9
35131 Padova, Italy

M.M. Turchetto
Istituto di Biologia del Mare
ICNR-IBM
Riva 7 Martiri, Castello 1364/A
30122 Venezia, Italy

C. Turetta
Institut de Biogéochimie Marine –
CNRS
Ecole Normale Supérieure (ENS)
1 rue M. Arnoux
92120 Montrouge, France

R.M. Warwick
Centre for Coastal and Marine
Sciences
Plymouth Marine Laboratory
Prospect Place, West Hoe
Plymouth PL1 3DH, UK

N. Villano
Centre for Coastal and Marine
  Sciences
Plymouth Marine Laboratory
Prospect Place, West Hoe
Plymouth PL1 3DH, UK

J. Widdows
Centre for Coastal and Marine
  Sciences
Plymouth Marine Laboratory
Prospect Place , West Hoe
Plymouth PL1 3DH, UK

Y.Y. Yoon
Kwandong University
Department of Environmental
  Engineering
San 7
Yim Chun-Ri
Yang Yang Son
Kwangdong, Korea

Section 1

Introduction

CHAPTER 1

THE VENICE LAGOON ECOSYSTEM PROJECT: GENESIS, GOALS AND OVERVIEW

P. Lasserre and A. Marzollo

INTRODUCTION

The Venice Lagoon is unique: there is no other lagoon in the world where the mingling of human action and natural ecology has been so enduring, complete, complex, and profound. The Republic of Venice, which founded its life and development around this lagoon, constantly considered its integrity as essential, an absolute priority. Over the centuries, it adopted appropriate policies to preserve an equilibrium between the opposing aquatic and terrestrial elements symbolised in the engraving which appears in the book by Bernardo Trevisan, *Della Laguna di Venezia, Trattato*, published in 1715 (Figure 1.1). At that time this famous engineer was already concerned with the excessive sedimentation of the canals of Venice.

The Venice Lagoon is an open system and it is highly important to understand the exchanges through its boundaries, both with the open ocean and with the terrestrial systems, in order to understand its structure and function. The relationships between the biological, physical and chemical processes are typically very strong in this environment. The land–sea interface is dynamic. It is characterized by cyclic changes of different frequencies, such as tidal inundations, neap–spring tide cycles, and seasonal cycles in runoff, light, temperature or stratification, and by infrequent extreme events, such as high freshwater discharges and storms. The latter are unpredictable in the short term, but statistically predictable over longer time scales and therefore of evolutionary importance to living organisms.

Due to its position between the upland drainage and the sea, the Lagoon of Venice has been subject to important anthropogenic inputs of nutrients and pollutants, which have increased greatly with industrial and agricultural development. A shrinking of the Lagoon water space, the deepening of the inlets between the Adriatic Sea and the Lagoon for accommodating new types of sea traffic, some subsidence and soil erosion, as well as inadequate regulation of floodwaters from the mainland water catchment, are all considered as potential or actual causes of disturbance of the delicate Lagoon equilibrium that is vital for the survival of the city of Venice and other smaller inhabited Lagoon islands.

Figure 1.1 The opposition between terrestrial and aquatic elements in the Venice Lagoon. From Bernardo Trevisan, *Della Laguna di Venezia, Trattato*, 1715

THE VENICE LAGOON ECOSYSTEM PROJECT

One of the primary aims of the UNESCO Project was to provide a novel, quantitative, insight into the processes that determine the effects of eutrophication on bottom sediments and water masses, and which regulate the coupling of nutrient and energy fluxes between the pelagic and benthic components (Figure 1.2). Major disturbances may possibly result from the development of anoxic conditions and the collapse of biological communities. Although the broad outline of these interactions was already reasonably clear, the detailed mechanisms that define the rates of change, the vulnerability of different benthic systems and the coupling of benthic–pelagic processes, including biological, chemical and sedimentological aspects, were not fully understood. Moreover, the role of enzyme activity in the interaction between the sediment and the degradation of organic matter seemed to deserve pioneer study. Mathematical modelling of such a complex system was also desirable.

Rationale

Coastal lagoons act as natural traps for organic materials produced within their fertile waters as well as for materials entering from surrounding wetlands and terrestrial environments. The resulting organic-rich sediments are the sites of intense microbially mediated degradation processes, which control sedimentary geochemical distribution and recycle vital nutrient elements and other mobile chemical end-products to overlying waters. In order to understand and

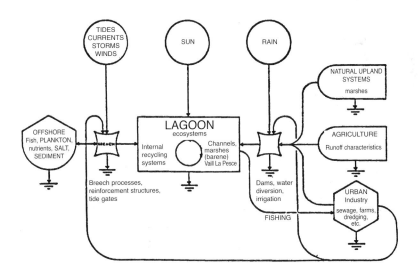

Figure 1.2 Conceptual diagram of the Lagoon of Venice and its connections with land, sea and atmosphere (modified from *International Research Project Venice Lagoon*, UNESCO, 1983)

to better control the seasonally variable eutrophication process, it is important to examine the mechanisms and rates of sediment–water recycling processes driven by the degradation reactions, leading to reducing conditions and anoxia. These processes are important factors in determining the distribution and species composition of the phytoplankton, macrobenthic plants (e.g. *Ulva*) and macrofaunal benthic populations. Quantitative measurements of such processes are still sparse so that those global estimates of coastal sources and sinks or of compounds of climatic and ecological importance remain uncertain.

Organisms living in the coastal zone have a suite of adaptation mechanisms ranging from biochemical and physiological adaptations to evolutionary change. These adaptations are integrated into the structure and function of ecological communities that have evolved in the changing and variable coastal environment.

The increased human influence on coastal systems leads to changes in the selective pressures on many biological and chemical processes. Such pressures arise from direct or indirect impacts, and also may be short term and visible, or long term and difficult to observe or predict. Process studies in the coastal zone cannot ignore selection and adaptation, but should consider them of primary interest when predicting the consequences of human impacts.

The interactions between human pressure, environmental constraints, the structure of the biological community and the functioning of the ecosystem as a whole also affect species composition of the coastal primary producers. For example, changes in phytoplankton composition result from changed inputs and ratios of nutrients in land runoff, changes in suspended matter concentrations, benthic–pelagic coupling and climatic change. The consequences of such changes in the hydrosphere may be far-reaching for the whole coastal ecosystem, since species differ widely in edibility, sinking rate or toxicity.

Of particular interest in this respect is the study of the link between the structure of ecological communities – for example in terms of species composition, functional groups, but also in terms of the genetic characteristics of the constituent species – and their functioning. Even the – relatively – low diversity communities in the Lagoon environment contain many species. However, several species may perform similar functions in the system and there is a certain redundancy in function which may permit reductions in diversity before system collapse occurs. System functioning may, therefore, in part be independent of system diversity. On the other hand, certain species can influence the structure of the entire ecosystem by their activity. Knowledge of their biology may significantly increase our ability to predict changes in coastal ecosystems. Obvious examples of such key species are vascular plants and macroalgae, the benthic primary producers at the land–sea interface. Their presence is characteristic of coastal lagoons and estuaries. They are a source of organic material that greatly increases productivity of coastal marine ecosystems and are becoming increasingly subject to stress.

A second feature of the coastal lagoon environment is the important function of benthic processes in the whole system. Even the zooplankton in the coastal zone are often largely composed of larvae of benthic invertebrates and therefore fundamentally differ from oceanic zooplankton. The impact of benthic reproduction on pelagic productivity is not known. The relationships between characteristics of benthic communities and biogeochemical processes, summarised as benthic–pelagic coupling, require further study in terms of quality and quantity of organic material deposited by pelagic processes, interactions between animals and bacteria and the structure of the benthic food web. Conversely, the structure of the benthos has a significant influence on the fluxes of nutrients from the sediment to the water column, on demersal fish production and on the stability of sediments.

Setting the objectives

One of the primary objectives of the *Venice Lagoon Ecosystem Project* was considered as providing a quantitative understanding of the major flows that comprise the coupling between the pelagic and benthic processes, and to relate this understanding to the increased eutrophication in the Lagoon of Venice. Other specific aims were as follows:

(1) To further understand the biological interactions within the benthic systems – particularly as these interactions respond to changes in the quantity, quality and timing of organic and inorganic inputs to the sediments; and to relate knowledge of these interactions to the key processes which regulate benthic/pelagic coupling.

(2) To develop an integrated approach involving multidisciplinary teams using controlled experimental ecosystems and process modelling.

(3) To undertake studies on enzyme systems in benthic flora and fauna and biodegradation of metabolites present in excess, as well as on biological buffering against pollutants.

(4) To develop mathematical models, in particular concerning trophic evolution and speciation.

Participant research terms, field surveys and sampling, and laboratory work

The intensive research campaigns integrating *in situ* and experimental studies were mainly focused on a well-defined salinity gradient in the northern basin of the Lagoon of Venice: Porto di Lido – canale di Treporti, Burano and Silone, and a study of the Palude della Rosa, taking account of the *Ulva* growth phenomenon and concentrating on eutrophication processes.

This effort required 78 persons including scientists, technicians and research fellows, of which 31 foreign scientists, from nine research groups and seven scientific institutions (from Italy, France and UK). All participants

Table 1.1 Venice Lagoon Ecosystem Project: Intensive field campaigns

1. **Intensive winter campaigns** ECOVENEZIA 1

    1A *25–30 November 1991*: First campaign organised by Istituto Biologia del Mare (IBM) of CNR (S. Rabitti, F. Bianchi *et al.*), to assess at the entrance of Palude della Rosa variations in hydrology, nutrients and plankton related to the short-term tidal variations, using discrete and continuous sampling

    IB *2–7 December 1991*: Second campaign with the joint participation of: Plymouth Marine Laboratory (D.M. Lowe *et al.*) and IBM/CNR (V. Fossato *et al.*): molecular and cellular pathology and toxicology,
    Plymouth Marine Laboratory (R. Warwick *et al.*) and Univ. of Venice at Celestia (O. Ravera *et al.*): benthos and *Ulva* (meiofauna and macrofauna)

    IC. *16–26 January 1992*: Third campaign with the joint participation of: Observatoire Océanologique de Roscoff group (P. Morin *et al.*) and IBM/CNR (F. Bianchi *et al.*): nitrogen and oxygen fluxes and primary production,
    Ecole Normale Supérieure, Montrouge (J.M. Martin *et al.*) and IBM/CNR (S. Rabitti *et al.*): particle size spectra, particle characterisation and chemical speciation

2. **Intensive spring campaigns** ECOVENEZIA 2

    2A. *30 March – 4 April 1992*: Campaign by:
    D.M. Lowe *et al.* and V. Fossato *et al.*: molecular and cellular pathology, toxicology and heavy metals,
    R. Warwick *et al.* and O. Ravera *et al.*: meiobenthos studies, macrobenthos and *Ulva* studies

    2B *25 April – 3 May 1992*:
    F. Bianchi *et al.* and P. Morin *et al.*: short-term nutrient dynamics at the entrance and in the middle of Palude della Rosa, and nutrient and oxygen dynamics along the salinity gradient,
    J.M. Martin *et al.* and S. Rabitti *et al.*: particle size spectra, particle characterisation and chemical speciation

3. **Intensive summer campaigns** ECOVENEZIA 3

    3A. *22–27 June 1992*:
    D.M. Lowe *et al.* and V. Fossato *et al.*: molecular and cellular pathology, toxicology and heavy metals,
    R. Warwick *et al.* and O. Ravera *et al.*: meiobenthos, macrobenthos and *Ulva* studies

    3B *22 July – 2 August 1992*:
    P. Morin *et al.* and F. Bianchi *et al*: short-term nutrient dynamics at the entrance and in the middle of Palude della Rosa, and nutrient and oxygen dynamics along the salinity gradient. Mesocosm experiments (entrance of Palude della Rosa)
    J.M. Martin *et al.* and S. Rabitti *et al*: particle size spectra, particle characterisation and chemical speciation

4. **Monthly campaigns**
    From 1992 to 1994, or in some cases weekly campaigns with a smaller number of participants on lighter flat-bottomed lagoon boats, by D. Tagliapietra *et al.*: sediment and benthos samplings on the Palude della Rosa, used for benthic studies (D. Tagliapietra *et al.* and R. Warwick *et al.*) and enzyme studies (Sabil *et al.*). Other Lagoon samplings, mainly for *Ulva* studies, were made by O. Ravera *et al.*

worked hard and enthusiastically, and a good team spirit developed allowing exchanges of ideas and the establishment of new personal links. This effort led to a further programme of common work well beyond the limits of the project. The list of project participants appears in Table 1.2, where the names of team co-ordinators appear in bold, and research fellows are marked with an asterisk.

Intensive sampling studies were organised along the salinity gradient Porto di Lido – Burano – Silone and in the Palude della Rosa. The CNR research vessels *Umberto d'Ancona*, *Mysis* and *Orata* and their crew members, were very actively involved in work developed on plankton, chemistry and geo-chemistry, and in toxicological studies. Other facilities included flat-bottomed boats for studies of the benthos of the Palude della Rosa. Laboratory facilities and research boats in Venice were provided by the Istituto di Biologia del Mare of CNR and by the Department of Environmental Sciences of the

Table 1.2   Venice Lagoon Ecosystem Project: Research teams

Istituto di Biologia del Mare del CNR, Venezia
F. Bianchi, *F. Acri, L. Alberighi, *M. Bastianini, A. Barbanti, A. Barillari, A. Boldrin, B. Bonora, A. Comaschi, *B. Cavalloni, A. Cesca, F. Cioce, P. Franco, M. Giorni, A. Locatelli, M. Marin, G. Penzo, S. Rabitti, F. Senigagliesi, G. Socal, *C. Turetta

Station Biologique, Roscoff, CNRS and University of Paris VI (France)
P. Lasserre, P. Le Corre, S. L'Helguen, E. Macé, C. Madec, A. Masson, P. Morin

Institut Biogéochimie Marine, CNRS, Montrouge (France)
J-M. Martin, C. Caillau, M. Dai, A. Fleury, J.M. Garnier, V. Huang, M. Le Pesteur, M. Rosset, Y.Y. Yoon

Istituto di Biologia del Mare del CNR, Venezia
V. Fossato, G. Campesan, C. Craboledda, L. Dalla Venezia, L. Da Ros, F. Dolci, *D. Liberalato, A. Menetto, C. Nasci, G. Stocco

Plymouth Marine Laboratory (UK)
D.M. Lowe, J. Holden, P. Lemaire, D.R. Livingstone, A. Matthews, M.N. Moore, R.K. Pipe, C. Porte

INSERM-CNRS, Montpellier (France)
A.A. Coletti-Previero, R.A. Boigegrain, A. Favel, A. Previero, M. Pugniere

Università di Venezia, Dipartimento di Scienze Ambientali alla Celestia
O. Ravera, S. Foltran, P. Pavan, A. Piva, N. Riccardi, *D. Tagliapietra, *C. Targa, C. Wagner

Plymouth Marine Laboratory (UK)
R. Warwick, *N. Villano, A. Rowden

Università di Padova, Istituto di Chimica Industriale
G. Bendoricchio, *M. Carrer, G. Coffaro

Università di Venezia, Dipartimento di Scienze Ambientali, Santa Marta
S.N. Lvov, A. Marcomini, R. Pastres, A. Sfriso

University of Venice (Celestia), and by the Military Authorities of the Venice Arsenal.

The laboratory facilities at IBM/CNR offered a very suitable base for pelagic studies (i.e. plankton, water chemistry, suspended matter, pollution and toxicology) undertaken by multidisciplinary teams from IBM/CNR, Plymouth (UK), Roscoff and Paris-Montrouge (France). Prior to the campaigns, specialised equipment not available in Venice (e.g. flow cytometer for particle analysis, polarographic system for trace metals, incubation chambers for ^{15}N analysis, multichannel oxymeters and specialised toxicology equipment was sent from the Roscoff, Paris and Plymouth laboratories and hosted at IBM/CNR. The laboratory space and facilities of the University of Venice at Celestia proved more than adequate for the multidisciplinary teams working on sediments, benthic communities, enzyme studies and aquatic flora. Important parts of the Project were also developed at the foreign participatory laboratories: the Institut National de la Santé et de la Recherche Médicale (INSERM) at Montpellier, for enzymatic studies based on samples of benthic sediments; and the Istituto di Chimica Industriale of the University of Padua, for mathematical modelling; as well as the Plymouth Marine Laboratory (biochemical indicators and meiofauna); the Roscoff Marine Station (primary production, nutrient and oxygen fluxes); and the Geochemistry Laboratory, Montrouge (flow cytometry, chemical speciation).

Training activities

Nine UNESCO research fellows (of Italian nationality) were trained for several months in the laboratories involved in the project, under the supervision of Italian and foreign researchers. N. Villano received training on meiobenthos at the Plymouth Marine Laboratory (UK) under the supervision of R. Warwick. D. Liberalato was trained at the same laboratory under the supervision of D. Lowe. F. Acri, M. Bastianini and B. Cavalloni were trained at the Roscoff Marine Station and at the Université de Bretagne Occidentale, Brest (France) on plankton productivity (pigments), nutrients analysis, ^{15}N techniques and oxygen uptake. C. Turetta was trained at the Institute of Biogeochemistry, Montrouge (France) on trace metal dynamics and flow cytometry. Short-term visits organised for C. Nasci and P. Tagliapietra to the Plymouth Marine Laboratory.

Databases and maps

In accordance with the recommendation of the MURST Co-ordinating Committee and to preserve an accurate record of the work carried out, the UNESCO Project initiated a computerised archive of both bibliographical and scientific data on subjects touched upon by the various research teams in the Project:

A bibliographical library and database   Collected reprints, mainly relating to the biological and geochemical aspects of the Venice Lagoon were gathered, and a database of papers and books was operational at the University of Venice (Celestia). The library of IBM-CNR was used as complementary facility.

A scientific database   A series of databases for the raw data collected by the various UNESCO research teams, both in the field and in the laboratory was established. Two reference manuals for collection of oceanographic data were utilised as guidelines, namely the ROSCOP (UNESCO) and the POEM programmes. A prototype relational database management system was developed. It includes, so far, 10 databases linked together into a unified system. Together, these databases constitute a modernised version of the above-referenced programme and contain items from field campaigns, sampling locations, parameters, laboratory studies and other specific topics.

Detailed maps were prepared to show and compare the sampling sites and are reproduced in this volume.

RESEARCH BACKGROUND

A holistic scientific approach was considered as a first priority for the Project. In most instances, a purely descriptive approach, usually consisting of the compilation of a long and costly series of inventories of the physical, chemical and biological parameters, does not permit objective evaluation of the production potential of an ecosystem, or control over production.The main task was therefore to collect and analyse comprehensive data on the structure and function of selected and representative parts of the Venice Lagoon ecosystem. This required the collaboration of scientists from the widest backgrounds – biologists, chemists, physicists, mathematicians – and their close co-operation to develop a comprehensive conceptual model reflecting the image of the workings of the Lagoon ecosystem and its interfaces (Figure 1.3). The Lagoon of Venice is ecologically, chemically and physically complex. The diversity of important forcing functions is high, i.e. sun, wind, tides, rivers, and is subject to complex coupling internally and with adjacent terrestrial systems and the Adriatic Sea. There is a high diversity of habitat types and primary producers, a highly complex food web, and a variety of life histories of estuarine organisms. There are also many behavioural adaptations and physiological tolerances within the system.

Due to the open nature of coastal lagoons, they are closely coupled with neighbouring ecosystems (Figure 1.2). The most obvious exchange is that of water. Passively moving within this aqueous medium are suspended sediments, organic and inorganic nutrients and organisms. Although the overall net water movement is generally seawards, there is a large movement of material in both directions. This exchange of materials and energy is the very substance of the lagoon environment. An important flow is the net sedimentation of organic and inorganic materials. During the last decades sedimentation of this type has been tremendously increased in the Venice Lagoon due to insufficient dredging of its canals, except for those excavated for new types of navigation, such as the so called 'petroleum canals' in the southern Lagoon basin. This latter type of excavation has unfortunately caused great soil erosion in the basin which has, therefore, partially lost its lagoon character.

Abundant sources of nutrients, efficient means of conservation and high rates of recycling characterize lagoon environments. Since lagoons are relatively shallow, processes at the sediment/water interface are very important. Generally, freshwater runoff carries a large load of suspended clays. Upon entering salt water, the charge structures breaks down and the clay particles clump. This flocculation and slower current allow the particles to settle. Therefore, the trapping of nutrients allows more time for incorporation into biogeochemical pathways. It is often difficult to measure dissolved inorganic nitrogen nutrients because the nutrient input from the river results in phytoplankton uptake and the nutrients are effectively and quickly used up.

The nutrients, as well as some of the contaminants, are decomposed toward the terminal acceptors, successively oxygen, nitrate, sulfate and carbonate, and distinct populations of microorganisms achieve this. The molecular events

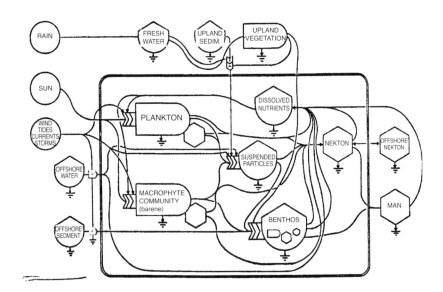

Figure 1.3 Conceptual diagram of the Lagoon of Venice and its connections with land, sea and atmosphere. From *International Research Project Venice Lagoon*, UNESCO, Venice, 1983; modified after P. Lasserre: Coastal Lagoons, Sanctuary Ecosystems, Cradles of Culture Targets for Economic Growth. *Nature and Resources*, 1979, **15**: 2–21

underlying this degradation by microorganisms through the actions of their catalytic systems (enzymes) provides the leading force in degradative transformations in the Lagoon. The enzyme activity within the sediments of the Venice Lagoon was extensively monitored and the biodegradative potential of the system was evaluated.

Another factor that is very important in lagoons is the presence of anaerobic zones, developing seasonally or permanently. Elements such as nitrogen, phosphorus, sulfur, iron and manganese exist in almost every possible chemical state. For example, nitrogen exists as NO_3, NO_2, N_2, NH_4 in the form of dissolved and particulate organic matter. Anaerobic conditions allow reactions such as solubilization of precipitated phosphorus, denitrification, and methane formation to take place. There are also large stores of nitrogen, phosphorus and other elements in anaerobic sediments that can be made available for plant uptake via mineralization.

Phytoplankton is not the only primary producer in coastal lagoons nor necessarily the most important. Marsh grasses, seagrasses and seaweeds (*Ulva*) decompose rapidly in temperate lagoons to release nutrients for further plant growth. Hence, natural fluctuations in photosynthetic production become sufficiently large as a result, for example, of excessive mineral or organic inputs originating from natural sources or urban wastes. When the ecosystem enters a period of crisis, or dystrophy, it is accompanied by such spectacular occurrences as the reddening or whitening of the water in shallow areas, and the release of hydrogen sulphide and even acetylene, due to the intense activity of anaerobic bacteria. These occurrences are accompanied by very marked reductions in the dissolved oxygen content of the water, followed by the wholesale death of the fauna and flora, especially fish, crustaceans and molluscs.

Interaction between the sediment and the degradation of organic matter

Studies on organic-rich coastal lagoon sediments have demonstrated that their biogeochemistry is dominated by a sequence of biological respiration and fermentation reactions which begin with aerobic respiration (utilizing dissolved oxygen as oxidant) and culminate in the use of alternative electron acceptors, such as dissolved NO_3^- and SO_4^{--}, by specialised bacteria under anaerobic conditions. When dissolved oxygen initially present in the interstitial waters of the sediments is exhausted, dissolved NO_3^-, solid phase manganese and iron oxides and dissolved sulfate are sequentially utilized by bacteria before methane production occurs. It has been shown that the rate of degradation of the metabolizable fraction of the total organic matter not only controls interstitial water chemical composition, but drives the recycling of mobile degradation products to overlying waters and the troposphere. Our ability to predict changes in the geochemistry of lagoon sediments is directly tied to our ability to quantitatively model the dynamics of degradation processes and their chemical end products as well as a variety of transport phenomena.

Hydrolytic activities of bacterial enzymes (phosphatase, cellulase and urease) have been measured in sediment samples from Palude della Rosa. The results have shown that the linkage between sediment–enzyme is qualitatively and quantitatively dependent on the inorganic composition of the sediment itself. Therefore, close attention should be paid to any interference that could effect the inorganic composition of the sediment. Preliminary studies showed that immobilized enzymes play a role in the biodegradation process of organic-rich sediments of the Venice Lagoon.

Major inputs and interactions (flows) between benthic and pelagic systems

The following three areas were selected:

- A study of nutrient, particulate matter and trace metal dynamics and their associated energy fluxes considered most appropriate to benthic/pelagic coupling;
- A study of the relevant features which characterise the structure and functioning of the benthic, phytal and faunal communities, and how these relate to key fluxes;
- An investigation of the responses of key species within the communities and how these responses contribute to observed community changes.

The approach was based on scenarios for ecosystem flow analysis (Figure 1.4). The programme took as its starting point the testable hypothesis that the quantity, quality and the frequency of organic inputs to the benthos determine the structural and functional features of the benthic communities. In addressing this hypothesis, two emphases were recognized, one related to the measurement of flows appropriate to benthic/pelagic coupling, the other to features of benthic community organisation. For both of these emphases, experimental approaches were suggested that would relate responses within the benthos to variability in the organic input from the water column. A protocol was suggested, based on the use of experimental mesocosms and modelling, to use knowledge gained from each approach to enrich the other. Both approaches were based on a scenario for ecosystem flow analysis (Figures 1.3 and 1.4).

Such compartmental flow diagrams were a useful way of summarizing the energy and/or material flows of an ecosystem. Existing information on nutrient inputs, primary and secondary production, respiration, etc. was used to budget the relative importance of different components of the community and to pinpoint weaknesses in existing knowledge. It was important to apply a 'top-down' approach, in which the major inflows and outflows, sources and sinks were identified, before detailed studies were made of internal community structure and function.

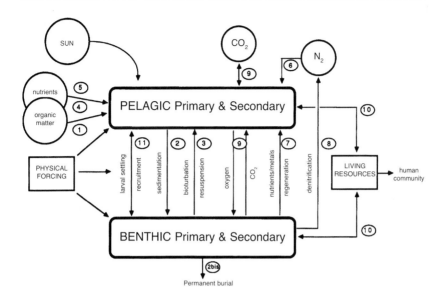

Figure 1.4 The major inputs and interactions (flows) between pelagic and benthic ecosystems in the Lagoon of Venice

The approach is illustrated by first describing the major forces (i.e. flows of energy or of limiting nutrients) affecting the pelagic and benthic systems. These flows can be separated into three major types (Figure 1.4):

A Organic flows

Allochthonous organic input from land (designated (1) in Figure 1.4) can constitute an important energy source in the Lagoon, complementing indigenous primary production. Sedimentation of organic matter (2) is the primary source that fuels the benthic system. This flow is primarily directed downwards, from the pelagos to the benthos, but an upward flux, by resuspension (3), may be important. The permanent burial of refractive organic matter in the sediments (2bis) represents a sink.

B Nutrient flows

Net primary production, occurring primarily in the pelagic system, is limited by isolation and by inputs of nutrients from land (4) and from the atmosphere (5), by nitrogen fixation (6) and by the upward flux of nutrient remineralised through organic matter decomposition in the benthos (7). Primary production in marine systems is predominantly nitrogen-limited. The major reason for this is the loss of nitrate through denitrification (NO_3 to N_2) in the sediment

13

(8). The oxygen supply to the benthos (9) controls redox conditions, nutrient cycling and faunal community structure in the benthos.

C Biological flows

Biomass produced within the ecosystem can be exported as fish food for species migrating between feeding and spawning areas, or as an exchange, mediated by fish feeding, between pelagic and benthic systems. (10). Fish populations entering the system will not only affect community structure and internal recycling rates, but may also generate inputs of material through spawning, defecation and excretion. Many benthic organisms have pelagic larvae, and some planktonic organisms have benthic resting stages. These flows between the pelagos and the benthos (11) are critical for recruitment, dispersal and survival.

Time scales

The consequences of variable organic and inorganic inputs on bottom living communities must be considered on various time scales.

Seasonal time scales

Many processes of biological interaction are most appropriately studied with seasonal changes in organic inputs as the major environmental variable. It is on this time scale that the most effective interaction between the two emphases taken in this proposal were anticipated; (1) budgeting of system attributes, and (2) the detailed biological structure of the key subsystems: (canals in the city of Venice, basins, barene and valli).

One of the most urgent requirements was to quantify nutrient and energy fluxes, when subjected to different levels of organic and inorganic inputs: (1) as properties of microbial processes in the sediment, and (2) as the seasonally variable bottom oxygen demand and anoxia due mainly to benthic macrophytes (*Ulva*, etc.). For example, it was suggested that microbial metabolism would play a larger role in community metabolism as the organic and inorganic inputs increase; this is amenable to measurements using benthic chambers in combination with direct enumeration of bacterial biomass. Modern technological developments in direct measurements of particle size (using flow cytometry), using probes for measuring profiles of oxygen tension, and direct calorimetry give data which can then be related to changes in the balance between various microbial and faunal components within the benthos and the water mass. Seasonal features, particularly the balance between organic input, temperature and metabolic demand, are paramount for differentiating the component processes which result in observed 'system level' nutrient and energy flux.

Events on this time scale are also most appropriate for experimental manipulation. It is for this reason that the use of experimental mesocosms in both challenging current conceptions regarding benthic pelagic coupling and in helping to formulate new concepts for testing by field observation, were identified and given prominence.

The annual time scale

Patterns of species abundance and biomass distribution in the Venice Lagoon ultimately depend on a complex of biotic interactions that may be studied on a time scale of years. Recruitment is important. More profound insights into recruitment phenomena will substantially improve understanding of the biological dynamics of the organisms that annually colonize the lagoon (notably larvae, juvenile and adult fish, molluscs and crustaceans).

To the manager, a better understanding of the scientific aspects of recruitment, the causal mechanisms that generate fluctuations in fish populations, will improve decision processes in traditional forms of fishery management as well as in the more futuristic management applications such as stock rebuilding strategies, development of pollution-control decisions, and early forecasting of stock collapses and explosions.

The processes (physical, chemical and biological) which control recruitment to benthic populations are still relatively unknown. Many invertebrate species recruit to adult stocks via a planktonic larval stage. An important question is: are the primary control processes those acting on the planktonic phase, or those in operation immediately following settlement from the pelagic to the benthic condition? The answer to this question is vital for a full understanding of benthic population structure, and the question is amenable to study via careful analysis of the spatial and temporal features of settlement and of post-settlement survival in communities exposed to different physical forcing functions and different levels of organic and inorganic inputs.

Recent developments in the theory of adaptive physiological strategies, based on premises of energy maximisation and the implications for genetic fitness, allow certain testable predictions to be made regarding the timing and the frequency of reproduction in species exposed to different environmental conditions. The results of such studies, when integrated with observations made on community-wide patterns of species abundance and biomass could provide a powerful theoretical and empirical framework for understanding the longer-term consequences of variable organic or inorganic loading to the benthos. One study designed to bring these various types of observation together in the context of different quantities and timing of inputs was proposed.

Experimental ecosystems methods (mesocosms and microcosms)

One of the most promising new methodologies for studying the fate, effects and transformations of materials in shallow waters and sediments involves the use of controlled experimental ecosystems. A mesocosm is a sea water and sediment enclosure of varying volume and surface area (according to the nature of the studied problem). These mesocosms (of volume ranging from 1 m^3 to 10,000 m^3) and microcosms (volume less than 1 m^3) are, therefore, relatively simplified ecosystems which have been enclosed in various ways (plastic bags, towers, etc.). These tools are extremely useful to verify specific rate terms in simulation models of marine ecosystems. They represent a scaling down of the ecosystem and are an intermediary between the chemostat or the aquarium and the highly complex natural system.

Experiments using laboratory and field experiments were done to verify specific rate terms in simulation models of marine ecosystems. Different systems were utilised, ranging from laboratory microcosms to *in situ* mesocosms.

Pollution and toxicology of Lagoon organisms

The many pollutants and the various kinds of pollution in the waters of the Lagoon have been studied and described in different ways. One of the most frequently used techniques is to identify and quantitatively assess the pollutant. This method can be used both for physical pollutants (e.g. thermal pollution), and chemical pollutants. The identification of the latter requires various often sophisticated techniques of qualitative and quantitative analysis. The most significant approach, however, consists in the study of the effects that water pollutants have on living organisms.

Conventional techniques allow a fairly accurate estimate of the short-term effects, certainly of the most evident of these, which involves the death of organisms. However, just as significant are these types of pollution or those concentrations of pollutants, the effects of which are felt only in the long term. In other words, traces of a pollutant may not cause immediate deaths (acute episodes), and therefore may not be considered dangerous by conventional techniques. Nevertheless, one cannot exclude the possibility that in the long run these quantities of pollutant, however small, could induce physiological or genetic stress capable of severely compromising or even extinguishing the population in question. Physiological and ecogenetical approaches to the problem can provide useful predictions and therefore constitute valid, even if unusual, methods for determining the quality of the environment; in particular in areas devoted to aquaculture.

Mathematical modelling

The experimental approach, advocated in the previous section can greatly strengthen mathematical models which address the impact of human activities in the lagoon. The complementarity of experimental approaches and mathematical models is evident from the different spatial scales which are inherent in these approaches.

The approach that was developed in the Project focused on models of trophic evolution of the Venice Lagoon (Istituto di Chimica Industriale, University of Padova). The results of the models developed for *Ulva* growth and seagrasses (*Zostera*), in a steady state condition reached after simulating the evolution over some years, showed good agreement with the few available field data.

A more theoretical study was also carried out in the project by applying some recent methods of far from equilibrium thermodynamics to the shallow Venice Lagoon basin.

CONCLUSIONS AND PERSPECTIVES

The UNESCO Venice Lagoon Ecosystem Project developed a multidisciplinary study emphasizing:

(1)　The major geochemical and biological inputs and interactions (flows) between pelagic and benthic systems in the Venice Lagoon, of which fundamental understanding is considered essential in order to progress the questions of pollution and eutrophication.

(2)　The understanding of mechanisms of productivity in order to be able to protect and further develop the fisheries and aquaculture in the Lagoon and outside in the near-shore waters.

(3)　The interactions between lagoon sediments and degradation of organic matter.

(4)　Mathematical models.

The Project produced an impressive number of results and facilitated data collection, interdisciplinary exchanges, and intensive training of young researchers. The results reported here meet the objectives set forth in the UNESCO–Italian Government Project Agreement, and provide a rough answer to at least some of the questions on the 'ecological state' of the Lagoon, described in particular, in the 1983 and 1987 proposed UNESCO projects on the Venice Lagoon. The computerised archive of both bibliographical and scientific data on subjects touched upon by the various research teams working on the UNESCO Venice Lagoon Ecosystem Project are available to interested scientists, engineers, managers and decision makers.

The Project contributed novel, quantitative and qualitative insights into the processes that determine the effects of eutrophication on bottom sediments and water masses, and which regulate the coupling of nutrient, oxygen and metal cycling. The different approaches were used in complementary research activities aimed at providing a basic understanding of the quantity, quality and frequency of organic and inorganic inputs from the water mass and the sediments that regulate the structural and functional features of planktonic and benthic communities.

Major damage may result from the development of anoxic conditions, due mainly to benthic macrophytes (notably *Ulva*), and temporary collapse of the biological communities. Although the broad outline of these interactions was reasonably clear, the detailed mechanisms that defined the rates of change were not understood. The vulnerability of different benthic compartments, and the physical, chemical, and geochemical coupling of benthic/pelagic coupling have been in part clarified in the Project. A most promising method of reducing the eutrophic condition of the Venice Lagoon is to decrease the input of phosphorus and nitrogen available to algae, by modifying agricultural fertilizer practices.

In addition to the seasonal (short-term) time scale, patterns of species abundance and biomass distribution in the Venice Lagoon ultimately depend on a complex of biotic and abiotic interactions which should be studied also on a time scale of years. The processes (physical, chemical and biological) which control recruitment to benthic populations are still poorly understood. Many invertebrate species recruit to adult stocks via a planktonic larval stage. An important question is: are the primary control processes those acting on the planktonic phase, or those in operation immediately following the settlement from the pelagic to the benthic condition? The answer to this question is important for a full understanding of benthic population structure (e.g. molluscs, crustaceans, annelids) and the question is amenable to study via careful analysis of the spatial and temporal features of settlement and of post-settlement survival in communities exposed to different physical forcing and different levels of organic and inorganic inputs. Some important elements of an answer appear in the Project results.

It was observed along the study gradient Porto di Lido – Canale di Burano – Palude della Rosa – Canale di Silone that concentration levels of free dissolved metals – cadmium, nickel, iron and lead – are too low to be acutely toxic to living organisms in this part of the Lagoon. Studies on the chemical speciation of copper and zinc show high accumulation into complexes by hydrophobic organic matter which is then absorbed by suspended matter and sediments, thus considerably lowering the toxicity of these metals. At a later stage, it would be important to determine if these metallic compounds in the sediments are irreversibly fixed, or whether the metal ions can return to solution through natural processes or human causes.

In spring 1992, increases in the concentrations and the dynamic properties of nitrogen compounds (nitrates, nitrites and ammonium) announced the development of *Ulva* near Palude della Rosa and the adjacent canals. However, the summer of 1992 was atypical in comparison with the 20 previous years: there was no *Ulva* growth that summer and, in parallel, there was a sharp phytoplankton bloom and balanced oxygen levels in the waters of the canals and in the Palude. This was the same as observed in the Venice Lagoon in the 1970s! During the summer of 1992 water oxygenation conditions were not limiting and compounds with ammonium and nitrite were present at higher levels than in spring, a sign of high regeneration of organic matter (studies using ^{15}N tracers). The summer phytoplankton bloom, therefore, seems to use the degraded organic matter from the spring *Ulva* stocks.

This work provides a large quantity of data which points the way towards a return to more balanced chemical and biological water conditions, at least along the Porto di Lido–Burano–Silone gradient. Additional studies in spring and, if possible, in summer would be very useful for verifying whether this return to equilibrium is in fact taking place.

The results of this project, though experimentally and conceptually important, obviously leave many concrete questions concerning Venice Lagoon open. They point to the need for a framework of general knowledge integrating the various scientific data and results and including socioeconomic, historical, cultural, artistic and tourism aspects. In this connection, it may be recalled that the Venice Lagoon Ecosystem Project also promoted cultural activities. It resulted in the recent publication of the CIERRE–UNESCO volume *La Laguna di Venezia*, which deals with numerous aspects of the Venice Lagoon, including geographical, physical, geological, historical, ecological, socioeconomic and juridical, as well as an impressive series of maps describing its evolution over the centuries.

In a more global perspective, the project fits into current studies on coastal lagoons, which provide a striking example of the importance, on a worldwide scale, of the interface between land and sea. Coastal lagoons comprise 13% of the world's coastline and are found from tropical zones to the poles. They are enriched by oceanic and continental inputs and are among the most productive ecosystems in the biosphere. In many parts of the world the human conquest of lagoons is very ancient and has given rise to a wide variety of cultures and activities, extending from subsistence economies, small-scale fisheries and extensive aquaculture, to massive concentrations of port facilities, urban centres and tourist infrastructure. The accelerating pollution of lagoons, the increasing over-exploitation of their resources, and rising pressure from engineering projects in lagoons around the world have greatly stimulated our efforts to understand these environments and to manage them in a responsible and effective way.

Section 2

Pelagic Studies

CHAPTER 2

DISSOLVED TRACE METALS IN THE VENICE LAGOON

J.M. Martin, W.W. Huang and Y.Y. Yoon

SUMMARY

Concentrations of both particulate and dissolved trace metals in the Venice Lagoon are comparable to those measured in pristine rivers and uncontaminated seawater. In the mixing zone between the Silone channel water and the Adriatic seawater, lead, zinc, nickel, iron and cadmium behaved conservatively, indicating the absence of any significant particulate–dissolved exchange processes. Copper was released in significant amounts from the particulate to the dissolved form, leading to non-conservative behaviour.

Mobilization of copper is associated with the intense production of macroalgae and phytoplankton in the Lagoon. The C-18 Sep-Pak column extraction method was used to isolate the hydrophobic organic fraction of dissolved trace metals. This fraction increases in the following order: Cd, Pb < Fe < Zn < Ni < Cu. A first estimate of total dissolved trace metal input to the Venice Lagoon shows that, for all trace metals studied, the atmospheric flux was more important than the riverine flux.

INTRODUCTION

The Venice Lagoon is a semi-enclosed body of water (area – 549 km^2; mean depth – 1 m; volume – c. 0.5 km^3) connected to the Adriatic sea through the Lido, Malamocco and Chioggia inlets. The water exchange with the Adriatic Sea averages 140 km^3/y and the mean renewal time, hence, is close to 1 day, although it may be much longer in some confined areas. The Lagoon receives domestic sewage, agricultural drainage and various wastes from the surrounding industrial areas (Battiston et al., 1988). In spring–summer, a bloom of the macroalga *Ulva rigida* occurs over vast areas of the Lagoon. With its collapse in late summer, the nutrients assimilated by the bloom are regenerated almost completely in superficial sediments, a process that continues throughout the winter (Sfriso et al., 1988). Seasonal cycling of the organic matter is the most important process driving the nutrient exchange at the sediment–water interface (Degobbis et al., 1986). The seasonal growth and decay of algae are likely to affect the extent and nature of trace metal complexation. In fact, dissolved trace metals in waters can exist in different forms: free hydrated ions, inorganic complexes and various organic complexes. As different species intervene in different biogeochemical cycles, the speciation of trace metals influences both their toxicity and bioavailability to aquatic organisms. Thus, both the form and concentration of trace metals are important in evaluating their interactions with plankton (Bruland *et al.*, 1991). For example, the availability of zinc and iron and the toxicity of zinc, copper and cadmium are controlled by their free ionic activities in aquatic systems (Sunda and Guillard, 1976; Anderson and Morel, 1978; Foster and Morel, 1982). Over the range of their dissolved free ion concentrations in natural waters, some trace metals such as copper may be both biolimiting and toxic.

It has also been shown that different species of algae can regulate trace metal cycling. For example, the depletion of ambient manganese by *Phaeocystis* colonies in the North Sea can cause a growth limitation of other phytoplankton species. This might explain why the occurrence of *Phaeocystis* blooms are often associated with a low abundance of other species (Lancelot *et al.*, 1991). Thus, it is crucial to determine whether a similar depletion occurs during the massive *Ulva* bloom, rendering other phytoplankton species trace metal-limited.

The aims of this study were:

(1) To determine the levels and fate (total concentration and speciation) of dissolved and particulate trace metals in the Venice Lagoon; and
(2) To relate the total concentration and speciation of dissolved trace metals to the biological cycle of *Ulva* in terms of complexation, uptake and mobilization, and limitation of phytoplankton growth.

Sampling and pretreatment

The winter survey was done from 16th to 26th January, 1992 and the spring survey was from 27th April to 2nd May, 1992. Seven stations were occupied in winter and nine stations, in spring (Figure 2.1). The stations were located along a salinity gradient from the Porto di Lido inlet to the Silone channel, a small tributary river flowing into the Lagoon (average flow rate: 7 m³/sec). R/V *Orata* and *Mysis*, belonging to IBM-CNR, Venice, were used for the collections.

Surface samples were taken by hand in precleaned polyethylene bottles from a small rubber boat 30 m away from the mother boat, in order to avoid contamination. Subsurface water samples were collected with a precleaned teflon pump. The samples were filtered under nitrogen pressure through precleaned 0.4 μm Nuclepore filters. The filtrate was divided into three parts so as to determine the labile and hydrophobic fractions and the total trace metal concentrations (see below). One 20 cm-long sediment core (CT2A) sample was taken at station 3 (Palude della Rosa) during the spring survey, using a plastic corer. All samples were processed under a laminar air flow clean bench to minimize risks of contamination.

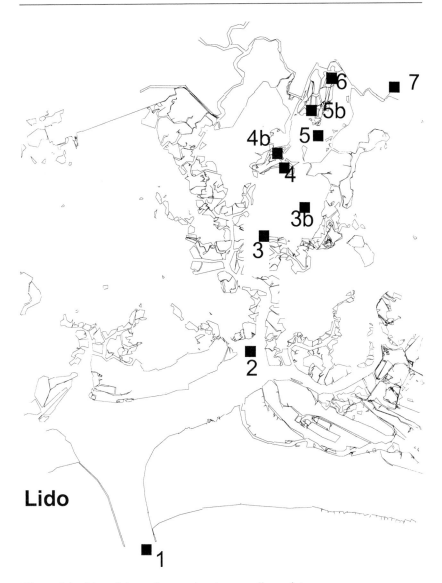

Figure 2.1   Map of the study area showing sampling points

ANALYSIS

Total dissolved trace metals

The filtrate was concentrated following the APDC/DDDC (ammonium pyrolidine dithiocarbamate/diethylammonium diethyldithiocarbamate) freon extraction procedure (Danielsson *et al.*, 1982) and analyzed by GFAAS (graphite furnace atomic absorption spectrophotometry) to determine the total dissolved Cu, Cd, Pb, Fe, Zn and Ni. Analytical results were checked using the international sea water standard (NASS-4, National Research Council of Canada, Division of Chemistry).

Trace metal speciation

Several analytical methods can be used to determine trace metal speciation. Trace metal complexation studies can be done using electrochemical techniques such as DPASV (differential pulse anodic stripping voltametry) and CSV (cathodic stripping voltametry) at different pH (Bruland, 1992; Boussemart *et al.*, 1992).

The solid phase extraction (SPE) techniques have also been widely used to study speciation of trace metals. In particular, C18 Sep-Pak cartridges (Millipore) were selected to evaluate the interaction of trace metals with dissolved organic matter (DOM). The method was first described by Mills and Quinn (1981) and was subsequently applied to estuarine (Mills *et al.*, 1982, 1987, 1989; Mills and Quinn, 1984) and oceanic waters (Hanson, 1981; Donat *et al.*, 1986). This technique is easy to use on board small research vessels and is expected to isolate the hydrophobic organic phase which represents a significant fraction of DOM. The C18 Sep-Pak extraction technique is strictly operational and only allows the determination of one fraction of the dissolved trace metals which are chelated by DOM. 300–500 ml of filtered water were passed through a previously conditioned Sep-Pak column with a peristaltic pump at a flow rate adjusted to 16 ml/min. Elution with 3 ml of 1:1 methanol–water solution (1/1) was also done on a clean bench and the eluates were frozen. The trace metals in the eluted phase were determined by GFAAS in the laboratory in a class 100 clean room.

Suspended matter and sediment

The concentration of trace metals in suspended matter and sediment samples was analyzed by GFAAS after a complete acid digestion with $HF + HNO_3 + HClO_4$ at 130–140°C in a closed teflon system. The concentration of aluminium was measured by colorimetry. All analytical results were checked using the international standard, MESS-1.

RESULTS AND DISCUSSION

Total trace metal concentrations

The concentration of total dissolved trace metals is given in Table 2.1 (a and b). The average concentrations of Cu, Cd, Pb, Zn, Fe and Ni in winter were 8.7, 0.1, 0.44, 31, 40 and 10 nM, respectively. Average spring concentrations of Cd and Pb (0.08 and 0.37 nM) were similar to those in winter. With other

Table 2.1   Concentration of dissolved trace metals in the Venice Lagoon, (a) winter survey; (b) spring survey; and (c) comparisons with concentrations in some rivers and oceans

	Station	Cu	Cd	*Element (nM)* Pb	Zn	Fe	Ni
(a)	1 (1 m)	6.3	0.30	0.33	47.9	12.8	7.3
	1 (1 m)	6.4	0.14	0.15	7.9	5.5	6.3
	1 (3 m)	6.5	0.14	0.28	8.7	6.3	6.4
	1 (6 m)	6.2	0.13	0.16	9.2	4.6	6.2
	1 (8 m)	5.9	0.10	0.14	7.2	5.6	6.3
	2 (1 m)	8.4	0.47	0.30	59.3	33.0	8.3
	2 (1 m)	5.6	0.12	0.29	10.7	10.4	6.8
	3 (1 m)	8.4	0.47	0.30	59.3	33.0	8.3
	3 (1 m)	5.6	0.07	0.27	16.2	16.2	7.9
	4 (1 m)	10.6	0.15	0.25	47.7	29.8	10.4
	4 (1 m)	11.7	0.17	0.60	49.6	89.0	13.1
	5 (1 m)	11.2	0.08	0.59	40.1	50.4	11.1
	6 (1 m)	11.4	0.08	0.48	44.3	40.1	11.3
	7 (1 m)	8.9	0.04	0.72	45.9	66.3	11.8
(b)	1 (1 m)	7.6	0.16	0.22	8.6	15.7	6.4
	2 (1 m)	7.3	0.11	0.34	9.6	16.3	8.4
	3 (1) 1 m	17.8	0.10	0.43	9.8	114.2	9.7
	3 (2) 1 m	21.1	0.08	0.40	12.4	89.4	10.3
	3 (3) 1 m	15.8	0.09	0.40	9.3	51.1	9.3
	3 (4) 1 m	10.6	0.02	0.27	4.2	48.9	9.6
	3b (1) 1 m	14.5	0.06	0.36	5.8	73.2	9.9
	3b (2) 1 m	18.7	0.06	0.26	4.7	63.6	10.6
	3b (3) 1 m	13.1	0.14	0.24	10.7	52.6	9.2
	4 (1 m)	17.2	0.08	0.30	21.2	64.9	11.0
	4b (1 m)	17.2	0.05	0.28	26.6	69.2	21.1
	5b (1 m)	19.2	0.06	0.42	29.1	146.9	25.9
	5 (1 m)	18.1	0.06	0.53	29.9	152.6	11.0
	6 (1 m)	13.0	0.04	0.42	32.9	84.0	10.5
	7 (1 m)	9.1	0.03	0.49	28.9	99.4	9.5
	0 (1 m)	10.2	0.15	0.31	19.2	12.7	6.7
(c)	*River*						
	Silone (I)	this study	0.04	0.72	45.9	66.3	11.8
	Silone (II)	this study	0.03	0.49	28.9	99.4	9.5
	Lena[1]	12.0	0.07	0.22	11.5	535	4.8
	Amazon[2]	23.8	0.09	–	0.5	724	5.1
	Changjiang[3,4,5]	26.7	0.03	0.27	0.9	10	2.6
	Mississippi[4]	22.7	0.12	–	2.9	30	22.7
	Orinoco[3]	18.9	0.04	–	2.0	–	3.4
	Krka[7]	1.8	0.04	0.08	–	24.1	2.5
	Rhône[8]	37.8	0.28	0.40	–	–	27.3
	World average[6]	23.6	0.09	0.15	9.2	716	8.5

Continued on next page

Continued

		Element (nM)				
		Cu	Cd	Pb	Zn	Fe	Ni

Sea							
Lido	this study	0.13	0.18	8.3	5.5	6.3
W. Mediterranean[5]	2.1	0.05	0.31	2.7	1.2	3.5
Laptev Sea[1]	6.9	0.15	0.26	2.4	7.6	5.4
Central Arctic[10]	4.7	0.24	–	2.9	–	–
East Pacific[11]	6.5	0.13	0.31	–	–	–
Adriatic[7]	4.4	0.12	0.12	–	7.2	7.4

1 = Martin *et al.*, 1993; 2 = Boyle *et al.*, 1982; 3 = Edmond *et al.*, 1985; 4 = Shiller and Boyle, 1987; 5 = Elbaz-Poulichet *et al.*, 1987; 6 = Martin *et al.*, 1991; 7 = Guan, 1991; 8 = Martin *et al.*, 1993; 9 = Moore, 1983; 10 = Elbaz-Poulichet *et al.*, 1991; 11 = Bruland *et al.*, 1992

elements, however, there were differences: average concentrations of Cu, Fe and Ni (13.7, 78 and 18 nM, respectively) were higher in spring than in winter, but that of Zn (20.6 nM) was about 30% lower. The slightly higher concentrations of Cu, Fe and Ni may be linked to the strong biological production of *Ulva* and/or phytoplankton during spring, and some pollution by Zn.

Table 2.1a gives the concentrations of dissolved trace metals at different depths for the Lido station (station 1, sea water) in winter. This profile indicates a homogeneous vertical mixing for both salinity (34–36) and trace metals. These concentrations are very close to the average values measured in different oceanic areas (Table 2.1c), with the exception of Zn. Donazzolo *et al.* (1981) reported very high Zn concentrations, ranging from 150 ppm to 1%, in a recent study of the sediments of the Venice Lagoon. These elevated concentrations may have their origin in industrial wastewater discharged directly into the Lagoon by a sphalerite ore processing plant that was operating until a few years ago. The Zn input might also originate from carbamate insecticides which are used on the surrounding mainland. However, all the stations in the study of Vitturi *et al.* (1987) were located in the northwestern part of the Venice Lagoon, far away from the present study area. Our values show that most of the Zn discharged into the Lagoon remains deposited in a restricted area. A limited amount may be circulated within the lagoon by tidal and wind-driven currents, as shown by the relatively important variations of the concentration of trace metals at the same station on different dates (stations 1–4 in winter) and at different times (station 3 in spring).

Table 2.1c shows that the concentrations of trace metals in the Silone river are comparable with those measured in large pristine rivers such as the Amazon (Boyle *et al.*, 1982; Edmond *et al.*, 1985), the Mississippi (Shiller and Boyle, 1987), the Changjiang (Edmond *et al.*, 1985; Shiller and Boyle, 1987; Elbaz-Poulichet *et al.*, 1987) and the world average (Martin *et al.*, 1992). The concentrations of Zn and Pb were slightly higher than world averages, which is probably due to contamination from human activities (see above).

Our observations show that the concentration of suspended matter in the Venice Lagoon is very low, ranging from 2.4 to 5.4 mg/l in winter and from 3.4 to 8.3 mg/l in spring. During storm events, the concentration of suspended matter increased suddenly, up to 60–80 mg/l (station 1 in winter and stations 2 and 3 in spring) owing to resuspension of surface sediments.

Tables 2.2a and 2.2b show the concentrations of trace metals in the suspended matter and sediments. Table 2.2c gives a comparison of the average concentrations of trace metals with those from some selected rivers. The averages of Cu, Pb, Zn and Fe are similar to, or even lower than, the world averages (Martin and Windom, 1991).

Concentrations of both the particulate and dissolved trace metals in the study area were comparable with those measured in pristine rivers and non-contaminated seawater. The short renewal time of Venice Lagoon water by Adriatic seawater is the most likely reason that enables the maintenance of the trace metal concentrations at a rather low level, even though the Lagoon is obviously subjected to different sources of contamination.

We express the partition coefficient K_d as:

K_d = Concentration of trace metals in suspended matter/concentration of dissolved trace metals.

The partition coefficient, K_d, between the particulate and dissolved phases ranged from 10,000 to 350,000 for Cd, 16,000 to 29,000 for Cu, 1,300,000 to 9,200,000 for Fe, 13,000 to 48,000 for Ni, 70,000 to 590,000 for Pb and 14,000 to 89,000 for Zn. These values show that most trace metals are associated with the particulate phase. However, the concentration of suspended matter in the Venice Lagoon is very low, being only a few milligrams per litre. Thus, it is very clear that for all trace metals except Fe, the dissolved phase is dominant in the Lagoon. About 90% of Ni and Zn, 80% of Cd and Cu, and 63% of Pb occurred in a dissolved phase in the lagoon. Only 10% of Fe is dissolved, and the particulate phase of Fe dominates its transfer in the lagoon.

It is worth noting that the concentrations of Al and Fe in suspended matter were very low compared with those of major world rivers. Metal concentrations depend upon the suspended matter composition, which can be roughly defined as a simple mixture of clay minerals, particulate organic matter, carbonate, Fe–Mn oxides and silicate. The average concentrations of Al in the suspended matter and sediment of the study area were 2.5 and 3.45%, respectively, much lower than the world average (9.4%). This can be explained by the high concentration of organic/inorganic carbon in suspended matter (10–14% of POC) and in sediment (4–5% of POC) (Cauwet, pers. comm.), corresponding to 25–35% organic matter in suspended matter and 10–12.5% in sediment. A high concentration of Ca was also observed, both in suspended matter and in sediment (4–8% corresponding to 20–40% $CaCO_3$).

Table 2.2 Concentration of trace metals in suspended matter (a) and in a sediment core sample (CT2A) taken from station 3 in May 1992 (b) and comparison with major rivers of the world (c)

Sample	Cd	Cu	Pb	Zn	Ni	Mn	Fe	Al
				Element (μg/g)				
(a) January								
1–1 m	–	17.1	6.0	33.5	23.1	525.4	11800	37600
1–3 m	–	10.0	2.5	16.4	14.2	369.3	9200	16500
1–6 m	–	14.8	3.9	23.4	16.6	389.3	8200	18800
1–8 m	–	10.8	3.1	18.9	16.1	447.3	8400	28000
3 m	–	37.2	11.7	108.4	40.0	1130.8	27100	
May								
0*	0.25	22.3	8.9	50.2	18.2	596.4	12800	15700
1	0.37	12.7	5.3	50.0	18.0	760.6	1100	24500
2	0.13	13.3	5.1	8.8	13.7	434.3	8400	24600
3 (1)	0.74	17.6	25.0	49.8	14.0	635.5	9700	17000
3 (3)	0.32	19.0	11.4	39.9	14.2	608.4	12800	19300
3 (4)	0.14	13.2	5.1	26.7	12.3	340.2	10000	19600
4	3.13	21.7	18.3	69.9	18.1	846.6	13800	23600
4b	1.00	29.3	51.8	105.0	23.9	1155.8	14800	25800
5b	0.73	34.6	51.2	104.6	19.8	1125.8	18200	26900
(b) CT2A 0–1 cm	0.26	19.1	7.7	58.6	24.5	859	18300	33700
CT2A 3–4 cm	0.28	18.1	5.7	43.3	20.8	500	12500	45600
CT2A 6–7 cm	0.21	15.3	5.9	51.8	19.1	449	14700	50700
CT2A 12–13 cm	0.24	17.6	6.0	57.1	21.4	431	13200	24000
(c) *Location*								
Lido (n = 5)	–	13.1	4.2	28.4	17.6	498	10100	25100
Silone (n = 7)	–	24.6	24.9	72.0	20.3	835	15200	26100
Station 0 (n = 1)	–	22.3	8.9	50.2	18.2	596	12800	15700
sediment (n = 4)	–	17.5	6.3	52.7	21.5	559	14700	38500
River								
Lena (10)	–	28	23	143	31	–	–	74000
Amazon (1,2,7)	–	66	42	426	105	1034	55500	115000
Changjiang (3,9)	–	43	29.5	120	40.3	–	52000	–
Huanghe (4,9)	–	30	34	70	124	886	37200	84000
Mississippi (5,6,7)	–	32	33	120	55	1261	44300	88000
Orinoco (3,7)	–	73	23	119	30	588	74000	113000
Rhône (3,7)	–	38–58	36–50	–	43–57	931	–	66000
World average (8)	–	100	35	250	90	1051	–	94000

1 = Irion, 1976; 2 = Gibbs, 1977; 3 = Elbaz-Poulichet, 1988; 4 = Huang and Zhang, 1990; 5 = Trefry *et al.*, 1986; 6 = Trefry and Presley, 1976; 7 = Martin and Meybeck, 1979; 9 = Huang *et al.*, 1992; 10 = Martin *et al.*, 1993; 11 = Emeis, 1985

The concentrations of Fe in both suspended matter and sediment were lower than the world average, but Mn oxides are important in the suspended matter of the Venice Lagoon. Concentrations of Mn in suspended matter varied from 369 to 1155 $\mu g/g$. Mn concentration was low at the marine station (Lido) but reached 1300 $\mu g/g$ at the Palude della Rosa station during the summer survey (Martin *et al.*, 1995) i.e. about four times the concentration at Lido and 2–3 times the world average (515 $\mu g/g$). This may be ascribed to flocculation of colloidal Mn oxides and/or to the massive growth of *Ulva* that leads to hyperoxygenation which, in turn, results in precipitation/co-precipitation of stable Mn oxides. The concentration of Mn in sediment core (station 3) varied from 431 to 550 $\mu g/g$, with the highest concentration in the surface layer. In the deeper layers of sediments, depletion of oxygen leads to reduction of Mn oxide, with the release of Mn^{2+} into the water.

Due to the enrichment in organic matter, carbonates and Mn oxides, the clay mineral content is relatively low. No relation between the concentration of Al and trace metals was observed. Conversely, a linear relation between POC and the concentration of trace metals (Figure 2.2) was found, suggesting that trace metal concentrations are predominantly controlled by organic matter rather than by clay minerals.

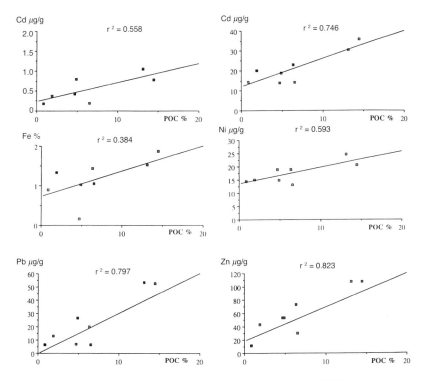

Figure 2.2 Concentration of particulate trace metals versus POC (%)

The mixing of freshwater and seawater

Particulate trace metals

Since the concentration of suspended matter was very low in winter, we were not able to collect enough samples to analyze particulate trace metals, especially in freshwaters; only some results at stations 1 and 3 were obtained (Table 2.2). Figure 2.3 shows the relationship between the concentration of trace metals in the suspended matter and salinity in spring. On the whole, concentrations of particulate trace metals and POC decrease from riverwater to seawater (Figure 2.4).

It is likely that concentrations of particulate trace metals are partly controlled by mixing of metal-poor marine suspended matter dominated by inorganic components (carbonate, silicate) with metal-rich riverine organic matter formed by *in situ* primary production of phytoplankton. During the mixing processes of riverine and marine particulates observed for most trace metals, the flocculation of inorganic (Fe–Mn oxides) and organic colloids as well as the resuspension of deposited sediments modify this distribution of trace metals.

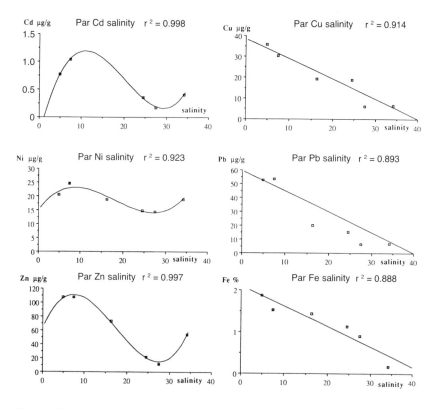

Figure 2.3   Concentration of particulate trace metals (Par) versus salinity

Particulate Cd, Ni, Mn and Zn show a similar distribution: the excess observed at station 4b (salinity 6.9 PSU) is probably related to flocculation of organic products of the decomposition of *Ulva* and/or to colloids transported by the Silone channel, as well as to precipitation of Mn oxides linked to hyperoxygenation. Indeed, one of the specific features of the Venice Lagoon is the large development of *Ulva* which may occupy vast areas, with standing stocks frequently exceeding 10 kg wet weight/m² (Sfriso, 1989). Their decomposition rates reach values as high as 250–1000 g wet wt/m²/day (Sfriso *et al.*, 1989). Table 2.3 shows the concentrations of trace metals in *Ulva* (dry) and the estimated budget of trace metals in *Ulva* (wet). It is obvious from this Table that *Ulva* represents an important reservoir of trace metals in the lagoon. The mobilization/removal processes between the particulate and dissolved phases mediated by decomposition of macroalgae and microalgae are still poorly known, but are likely to be important.

Dissolved trace metals

The relationships between the concentrations of dissolved trace metals and salinity in the surface water are shown in Figures 2.5a, b, c, d, e and f.

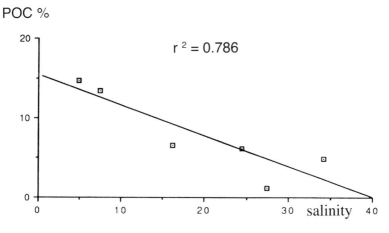

Figure 2.4 Particulate organic matter (POC) versus salinity (%)

Table 2.3 Trace metals in *Ulva*. Values based on standing crop of *Ulva* > 10 kg/m² (Sfriso, 1989); dry/wet weight = 5.4% (this study); *Budget = Conc. of TM (μg/g) × 0.054 × 10 kg/m³

	Ref	Fe	Mn	Cu		Ni	Pb		Cd	
μg/g dry weight	1	1546	54	9	16	–	3.5	15	0.06	5
μg/g dry weight	this study	540	90	6.14		2.6	0.17		0.13	
g/m² wet weight*	–	291.6	48.6	3.3		1.4	0.09		0.07	

1 = Romanin, 1989

A

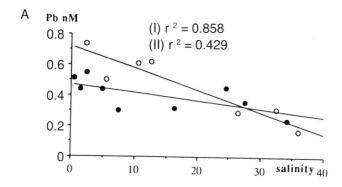

B

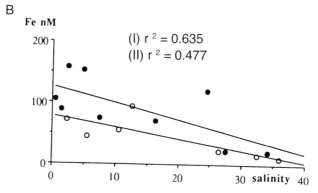

C

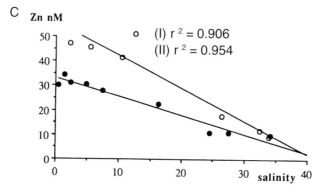

Continued on next page

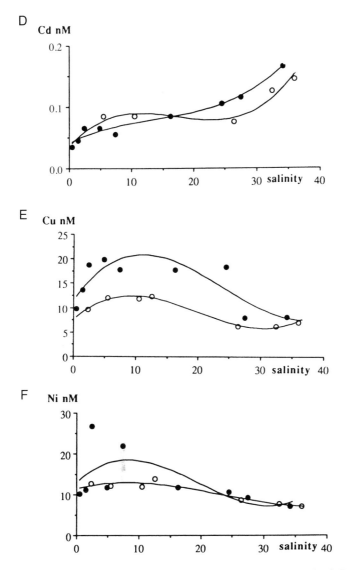

Figure 2.5 Concentration of dissolved element versus salinity. A, lead; B, iron; C, zinc; D, cadmium; E, copper; and F, nickel. o = Venice I and ● + Venice II

Iron, lead and zinc Despite some scatter, the concentration of Pb, Zn, and Fe decreases linearly with salinity, demonstrating conservative behaviour. This indicates that the mixing of riverwater with seawater occurs without significant particulate–dissolved exchange processes. As shown above, the concentrations of Zn and Pb during the winter survey were slightly higher than in spring. On the contrary, concentrations of Fe were higher during the spring survey. This may be due to slight local industrial contamination at the upstream station.

Cadmium   Contrary to the results obtained with Fe, Pb and Zn, a rather high concentration of Cd was observed in seawater as compared with the Silone channel. A similar observation has been made in the Lena delta where Cd concentration as a function of chlorinity followed a theoretical dilution line between riverwater and seawater (Martin *et al.*, 1993a). This observation disagrees with the classical picture described in many estuaries, where Cd is released from the suspended matter owing to the formation of Cd–chloro complexes and sulfate complexes (Elbaz-Poulichet *et al.*, 1987). In the Venice Lagoon the concentration increases regularly from 0.04 nM in winter and 0.03 nM in spring at the Silone river station to a rather high value (0.14 in winter and 0.16 nM in spring) in the seawater (Lido station). Such observations are also in agreement with results obtained in the Rhône estuary (Elbaz-Poulichet *et al.*, 1993) where the relatively short residence time of water in the mixing zone and the absence of turbidity maximum reduces the importance of solid– liquid interactions. This low reactivity in the Venice Lagoon may be further reduced by low solid discharge of the rivers flowing into the Lagoon. It must be added here that, for elements like Cd which are characterized by a higher concentration in seawater than in riverwater, the comparison of actual measurements with the theoretical dilution curve can hardly be used to predict any mobilization/desorption processes. A high concentration (0.17 nM) was observed at station 4 in winter; this is probably the result of contamination either during sampling or analysis.

Copper and nickel   For Cu and Ni, a significant increase was observed between freshwater and seawater at stations 3, 4 and 5. The concentrations in spring were relatively higher than in winter. This enrichment is associated with the intense production of *Ulva lactuca* in the Lagoon. Following the degradation of organic matter, high concentrations of Cu and Ni might be mobilized from suspended matter and/or sediments.

Trace metal chelation by the hydrophobic organic fraction

Donat *et al.* (1986) showed that the classical C-18 Sep-Pak technique suffers from a lack of selectivity and extracts one part of the DOM-associated copper that is complexed by non-polar, hydrophobic organic material. The percentages of C-18 Sep-Pak-isolated trace metals are shown in Table 2.4. The C18 Sep-Pak columns isolate only a small fraction of trace metals present in the Venice Lagoon except for Cu, as shown previously for estuarine and seawater environments (Donat *et al.*, 1986, Hanson, 1981). The trace metals are complexed in the following order: Cd, Pb < Fe < Zn < Ni < Cu. The hydrophobe

Table 2.4 Hydrophobic fraction of trace metals as a % of total trace metals in Venice Lagoon

	Ref	*Fe*	*Cu*	*Ni*	*Pb*	*Cd*	*Zn*
W. Mediterranean	1	0–2.5	28–40	0–0.25	ND	ND	0.2–4
Venice winter	this study 3	15	37–48	1 13	0–5	0–4	2 6
Venice spring	this study 2	9	31–52	0–28	0–5	0–4	2 24

1 = Martin *et al.*, 1993

fraction of Cd and Pb appears to be negligible while the Cu fraction ranges from 30 to 50%.

Figure 2.6 shows distribution of the hydrophobic fractions of Cu, Ni, Zn and Fe. Except for Fe, the values were higher in spring than in winter. A maximum for Ni, Zn, Fe and Cu was observed in the waters of intermediate salinity (stations 3, 4 and 5). The Cu maximum was very significant, especially in spring and may be attributed to the spring bloom of phytoplankton and/or macroalgae (*Ulva*).

Table 2.5 gives some comparative values of primary production in oceanic and coastal waters and the Venice Lagoon. It is surprising to note that primary productivity in the Venice Lagoon in spring is 900 times higher than average oceanic productivity (Heip, 1991) and 70 times higher than in the Gulf of Lyons (Minas and Minas, 1992). In the Venice Lagoon Bianchi (1992) measured an average primary production of 6.2 mg C/m^3/h in spring; this value was 10 times higher than in winter, but the percentages of the C18 Sep-Pak-isolated copper remained almost the same.

The C-18 Sep-Pak-isolated fraction of Ni and Zn, on the contrary, showed a substantial increase in the Venice Lagoon, especially in the area where the bloom of *Ulva* developed intensively in spring. Table 2.4 shows that during spring, at stations 3 and 4, the percentages of Sep-Pak-isolated Ni and Zn increase 10 and 6 times, as compared to the western Mediterranean (Martin *et al.*, 1993). It appears that Sep-Pak-isolated trace metals are not only simply related to primary productivity, but are also probably dependent upon the species of phytoplankton and macroalgae which release different ligands interacting differently with each trace metal.

Bada and Lee (1977) and Donat *et al.* (1986) suggest that DOM in the ocean consists of two pools: an old fraction which is resistant to decomposition, and a much more labile pool principally of recent biological origin. The more labile material is found in the productive surface water, where the organic matter released by recent biological activity is superimposed upon the pool of more resistant material. The fraction of labile DOM, consisting of complexing material released by organisms directly into the water, in association with a wide range of products of degradation and polymerization processes, appears to bind with some trace metals. As the composition of the organisms and products of degradation and polymerization are variable, the labile DOM becomes associated with different trace metals with different efficiencies. This

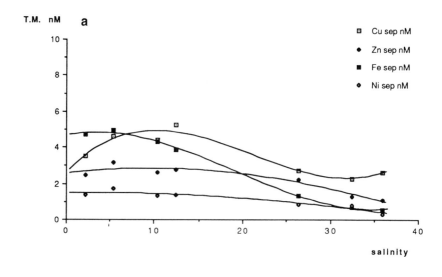

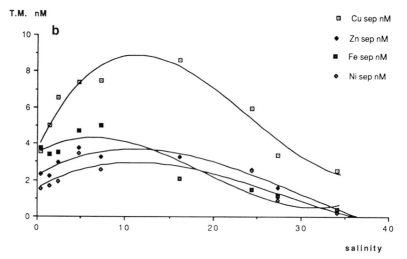

Figure 2.6 Hydrophobic fraction trace metals (TM) versus salinity (a: winter, b: spring)

may explain the variations of Sep-Pak-isolated trace metal fractions according to season and the geographical areas. However, the laboratory results of Moffet *et al.* (1990), obtained from a limited survey of four marine phytoplankton species, indicated that only the cyanobacterial genus *Synechococcus* produced a chelator able to form strong Cu complexes. *Synechococcus* is widespread in the open sea and can account for up to 50% of primary productivity in some regions (Glover *et al.*, 1986; Iturriaga and Marra, 1988).

Table 2.5 Primary production and hydrophobic fraction Cu (%) in different environments

Location	Primary production		Sep-Pak % Cu	
	g C/m³y	references	%	references
Ocean (surface)	0.6	1	9 44*	4.5
Gulf of Lions	7.3	2	40	6
Venice winter (mean)	53	3	42	this study
Venice spring (mean)	530	3	41	this study

1 = Heip, 1991; 2 = Minas and Minas, 1992; 3 = Bianchi, 1992; 4 = Hanson, 1981; 5 = Donat *et al.*, 1986; 7 = Elbaz-Poulichet *et al.*, 1990. * = Northwestern Atlantic

Bruland *et al.* (1991) suggested that this genus could have a major impact on Cu speciation in many oceanic regions. McKnight and Morel (1979) reported that, among freshwater phytoplankton taxa, only cyanobacteria would be able to produce strong Cu-binding organic ligands. In the case of the Venice Lagoon, high primary production during spring is characterized by massive proliferation of macroalgae (*Ulva*). Phytoplankton biomass showed only a slight increase during spring as compared to winter. The predominant phytoplankton species were pennate diatoms such as *Gyrosigma fasciola, Cylindrotheca closterium, Navicula* spp, *Amphiprora* spp and *Pleurosigma* sp. (Bianchi, 1992). The results of Donat *et al.* (1986) showed that the vertical distribution of Sep-Pak-isolated Cu in intermediate and deep water of the central North Pacific was similar to that observed in the North Atlantic but that a major discrepancy exists in oceanic surface water. This may result from differences in the species composition of phytoplankton; the relationship between phytoplankton and the organic complexed trace metals, therefore, needs further study.

The Sep-Pak columns, however, isolate only a small fraction of trace metal organic complexes present in the Venice Lagoon. Direct determination by DPASV (differential pulse anodic stripping voltametry) showed no detectable labile Cu until more than several nanomoles of Cu had been added. This observation indicates that the samples contained an excess of natural organic ligands which are able to complex the added Cu. Donat *et al.* (1986) and Bruland (1992) used the DPASV titration method to study complexation of trace metals by natural ligands in the open sea. Donat *et al.* (1986) reported that the major part of DOM responsible for Cu complexation, as determined by DPASV, is not isolated by Sep-Pak columns. Bruland (1992) reported that about 70% of dissolved Cd in surface waters was associated with strong complexes. The Sep-Pak-isolated fraction of Cd and Pb was not detected in the Western Mediterranean (Elbaz-Poulichet *et al.*, 1990; Martin *et al.*, 1993) but indicates some degree of complexation (0–5%) in the Venice Lagoon.

Dissolved trace metal inputs to the Venice Lagoon

Although it represents only a small percentage of the river discharge, the Silone channel has been considered as representative of the river input to the Lagoon. Hence, average trace metal concentrations at the upstream end of the Silone channel (station 7) in winter and spring surveys were multiplied by the total discharge of rivers flowing into the Lagoon to obtain a preliminary estimate of river input of trace metals (Table 2.6). The values for flux from the atmosphere (including dissolution of trace metals linked to atmospheric aerosols) were taken from Guieu (1991) and Guieu *et al.* (1991) and multiplied by the surface area of the Venice Lagoon to estimate inputs from the atmosphere.

This comparison indicates that river input of trace metals is small compared to atmospheric input. The predominance of atmospheric input of trace metals has been established for many coastal zones and the open ocean (Windom, 1981, 1986; Martin *et al.*, 1989; Dorten *et al.*, 1991). Our estimates show that about 90% of Fe, Zn, Cd and Pb found in the Lagoon waters probably come from the atmosphere. About 60% of Cu and 40% of Ni also originate from atmospheric input. However, it must be added that we did not take into account local atmospheric contamination problem which might increase atmospheric input. Therefore, the values taken from Guieu (1991) correspond to minimum values. The river inputs of some trace metals such as Fe, Zn, Cd and Pb were much lower than those from the atmosphere, so that slightly higher trace metal concentrations in the rivers flowing into the Lagoon, which have not been studied, would not significantly change the contribution of river input of trace metals.

As far as Cu is concerned, non-conservative behaviour was observed during the mixing of freshwater and seawater. At stations 3 and 4, where the *Ulva* develops intensively, the concentration of dissolved Cu was 2–3 times higher than in seawater or freshwater. This process provides an additional source of dissolved Cu to the Lagoon which was not taken into consideration in this budget estimate.

Table 2.6   Dissolved trace metal inputs in Venice Lagoon

	Cu		*Cd*		*Pb*		*Zn*		*Fe*		*Ni*	
	mol/y	*%*	*mol/y*	*%*	*mol/y*	*%*	*mol/y*	*%*	*mol/y*	*%*	*mol/y*	*%*
Rivers (1)	11	40	0.05	1	0.7	1	46	1	98	10	13	60
Atmospheric flux (2) (3)	16	60	5	99	10	99	660	99	900	90	8	40

(1) Q = 40 m³/sec (annual average river flux); (2) surface of lagoon = 549 km²; (3) flux: Guieu, 1991; River input = Concentration of trace metals at station 7 (average of winter and spring) × Q; Atmospheric flux = flux of reference (3) × surface of lagoon

CONCLUSION

Overall, the Venice Lagoon is a peculiar environment where the biological processes are very active. The freshwater input and solid discharge are limited and the Lagoon is shallow. This paper presents the first results on the distribution of dissolved (total concentration and hydrophobic fraction) and particulate (sediment core and suspended matter) trace metals in the Venice Lagoon.

One of the most obvious results of this preliminary study is the unexpected low level of contamination of the Venice Lagoon in the area located between Porto di Lido and Silone channel in good agreement with previous studies (Martin *et al.*, 1994). This is linked to the short residence time of water in the Lagoon which is quickly renewed by input of Adriatic seawater.

During the mixing of freshwater and seawater Pb, Zn, Ni, Fe and Cd behave conservatively; Cu was released in significant amounts from the particulate form to the dissolved form and non-conservative behaviour was observed.

As far as the complexation of trace metals is concerned, it is interesting to note that although gross production is several orders of magnitude higher than in the open seawater, the percentage that is complexed is not substantially different. This observation suggests that *Ulva* is probably not the main producer of complexing agents. Future work should consider these complex relationships in order to specify the exact role of trace metals in production and pollution of the Lagoon.

Finally, our rough estimates show that about 90% of Fe, Zn, Cd, Pb, 60% of Cu and 40% of Ni in the Venice Lagoon come from the atmosphere.

ACKNOWLEDGEMENTS

This work was carried out in the framework of the UNESCO project, Venice Lagoon Ecosystem, a part of the Venice Lagoon System Project coordinated and financed by the Italian Ministry for the Universities and for Scientific and Technological Research. We are grateful to Dr V. Chevtchenko for particulate Al and Mn analyses and to Dr G. Cauwet for POC analyses. We thank our colleagues at the Institut de Biogéochimie Marine and the Institute of Marine Biology (CNR-Venice) for their efficient help during fieldwork. We gratefully acknowledge the stimulating role of Prof. P. Lasserre and A. Marzollo in the implementation of this project.

REFERENCES

Anderson, D. M. and Morel, F. M. N. (1978). Copper sensitivity of *Gonyaulax tamarensis*. *Limnol. Oceanogr.*, **23**: 283–95

Bada, J. L. and Lee, C. (1977). Decomposition and alteration of organic compounds dissolved in seawater. *Mar. Chem.*, **5**: 523–53

Battiston, G. A., Degetto, S., Gerbasi, R., Sbrignadello, G. and Tositti, L. (1988). The use of ^{210}Pb and ^{137}Cs in the study of sediment pollution in the Venice lagoon. *The*

Science of the Total Environment, **77**: 15–23

Bianchi, F. (1992). A study of the processes of production and consumption and of the dynamics of substance in the waters of the Venice lagoon. *Project Venice lagoon ecosystem 929/ITA/41, UNESCO/IBM-CNR agreement SC 233.086.2*

Boussemart, M., Van den Berg, C. M. G. and Ghaddaf, M. (1992). The determination of the chromium speciation in seawater using catalytic cathodic stripping voltametry. *Anal. Chem. Acta,* **262**: 103–15

Boyle, E. A., Huested, S. S. and Grant, B. (1982). The chemical mass balance of the Amazon plume. II. copper, nickel and cadmium. *Deep Sea Res.,* **29**(11A): 1355–64

Bruland, K. W., Donat, J. R. and Hutchins, D. A. (1991). Interactive influences of bioactive trace metals on biological production in oceanic water. *Limnol. Oceanog.,* **36**(8): 1555–77

Bruland, K. W. (1992). Complexation of cadmium by natural organic ligands in the central North Pacific. *Limnol. Oceanog.,* **37**(5): 1008–17

Danielsson, L. G., Magnusson, B., Westerlund, S. and Zhang, K. (1982). Trace metal determinations in estuarine waters by electrothermal absorption spectrometry after extraction of dithiocarbamate complexes into freon. *Analytica Chimica Acta,* **144**: 183–8

Degobbis, D., Gilmartin, M. and Orio, A. A. (1986).The relation of nutrient regeneration in the sediments of the north Adriatic to eutrophication, with special reference to the lagoon of Venice. *The Science of the Total Environment,* **56**: 201–10

Donat, J. R., Statham, P. J. and Bruland, K. W. (1986). An evaluation of a C-18 solid phase extraction technique for isolating metal–organic complexes from central North Pacific ocean waters. *Mar. Chem.,* **18**: 85–99

Donazzolo, R., Merlin, O. H., Vittori, L. M., Orio, A. A., Pavoni, B., Perin, G. and Rabitti, S. (1981). Heavy metal contamination in surface sediments from the gulf of Venice, Italy. *Mar. Poll. Bull.,* **12**: 417–25

Dorten, W. S., Elbaz-Poulichet, F., Mart, L. R. and Martin, J. M. (1991). Reassessment of the river input of trace metals into the Mediterranean Sea. *Ambio,* **20**(1): 2–6

Edmond, J. M., Spivack, A., Grant, B. C., Chen, Z., Hu, M., Zeng, X. and Chen, S. (1985). Chemical dynamics of the Changjiang estuary. *Cont. Shelf Res.,* **4**: 17–36

Elbaz-Poulichet, F., Huang, W. W., Martin, J. M., Seyler, P., Zhong, X. M. and Zhu, J. X. (1990). Biogeochemistry behaviour of dissolved trace metals in the Changjiang. In Yu, G., Martin, J. M. and Zhou, J. (eds.), *Biogeochemical Study of the Changjiang Estuary.* China Ocean Press, Beijing, 293–311

Elbaz-Poulichet, F., Martin, J. M., Huang, W. W. and Zhu, J. X. (1987). Dissolved Cd behaviour in some selected French and Chinese estuaries. Consequences on Cd supply to the ocean. *Mar. Chem.,* **22**: 125–36

Elbaz-Poulichet, F. (1988). Apports fluviatiles et estuariens de plomb, cadmium et cuivre aux océans, comparaison avec l'apport atmosphérique, Thèse de doctorat d'etat. Université de Paris 6, pp. 288

Elbaz-Poulichet, F., Guan, D. M. and Martin, J. M. (1993). Riverine variability and estuarine conservativity of trace elements in the Rhône river delta (France). Submitted to *Est. Coast. Shelf Sci*

Emeis, K. (1985). Particulate suspended matter in major world rivers – II: Results on the rivers Indus, Waikato, Nile, St Lawrence, Yangtze, Parana, Orinoco, Caroni and Mackenzie. In Degens, E. T. and Herrera, R. (eds.), *Transport of Carbon and Minerals in Major World Rivers.* Mitt. Geol.-Palaont. Inst. Univ. Hamburg, *SCOPE/UNEP Sonderband,* **58**: 593–617

Foster, P. L. and Morel F. M. N. (1982). Reversal of cadmium toxicity in the diatom *Thalassiosira weissflogii*. *Limnol. Oceanogr.*, **2**: 745–52

Gibbs, R. J. (1977). Transport phase of transition metals in the Amazon and Yukon rivers. *Geol. Soc. Am. Bull.*, **88**: 829–43

Glover, H. E., Keller, M. D. and Guillard, R. R. L. (1986). Light quality and oceanic ultraphytoplankters. *Nature*, **319**: 142–3

Guan, D. M. (1991). Comportements biogéochimiques des metaux traces dissous dans deux estuaires mediterrannéens et en Mediterranée occidentale. Thèse doctorat de l'Université Paris 6, pp. 167

Guieu, C., Martin, J. M., Thomas, A. J. and Elbaz-Poulichet, F. (1991). Atmospheric versus river inputs of metals to the Gulf of Lyons. *Marine Pollution Bulletin*, **22**(4): 176–83

Guieu, C. (1991). Apports atmospheriques à la Mediterranée Nord-Occidentale. Thèse de doctorat de l'Université Paris 6, pp. 224

Hanson, A. K. (1981). The distribution and biogeochemical cycling of transition metal–organic complexes in the marine environment. Ph.D thesis, University of Rhode Island

Heip, C. (1991). Biogeochemical aspects of the biogeochemistry of estuaries. Delta Institute for Hydrobiological Research Yerseke, The Netherlands; Lecture notes of the Intensive Summer course on biogeochemical processes in estuaries, Melreux, Belgium. 19 August–6 September 1991, pp. 57

Huang, W. W. and Zhang, J. (1990). Effect of particle size on transition metal concentrations in the Changjiang and the Huanghe, China. *The Science of the Total Environment*, **94**: 187–207

Huang, W. W., Zhang, J. and Zhou, Z. H. (1992). Particulate element inventory of the Huanghe: A large, high-turbidity river. *Geochim. et Cosmochim. Acta*, **56**: 3669–80

Irion, G. (1976). Amazon: in mineralogische und geochimische untersuchungen an der pelitischen fraktion amazonischer uberboden und sedimente. *Biogeographica*, **7**: 7–25

Iturriaga, R. and Marra, J. (1988). Temporal and spatial variability of chroococcoid cyanobacteria *Synechococcus* spp. specific growth rates and their contribution to primary production in the Sargasso Sea. *Mar. Ecol. Prog. Ser.*, **44**: 175–81

Lancelot, C., Billen, G. and Barth, H. (eds), (1991). The dynamics of *Phaeocystis* blooms in nutrient-enriched coastal zones. *Water Poll. Res. Rept.*, **23**: 106 pp

Martin, J. M. and Meybeck, M. (1979). Elemental mass-balance of material carried by major world rivers. *Mar. Chem.*, **7**: 173–206

Martin, J. M., Guieu, C., Elbaz-Poulichet, F. and Loye-Pilot, M. D. (1989). River versus atmospheric input of material into the Mediterranean Sea: An overview. *Mar. Chem.*, **28**: 159–82

Martin, J.M. and Windom, H. (1991). Present and future role of ocean margins in regulating marine biogeochemical cycle of trace elements. In Mantoura, R. F. C., Martin, J. M. and Wollast, R. (eds), *Proc. Dalhem Conf. Marginal Seas Processes in Global Change*. J. Wiley & Sons, New York, 45–67

Martin, J. M., Huang, W. W. and Yoon, Y. Y. (1993a). Total concentration and chemical speciation of dissolved copper, nickel and cadmium in the western Mediterranean. In Martin, J. M. and Barth, H. (eds), EROS 2000 Fourth Workshop on the North-West Mediterranean sea. *Water Pollution Research Report*, **30**: 119–28

Martin, J. M., Guan, D. M., Ebaz-Poulichet, F., Thomas, A. J. and Gordeen V. V. (1993b). Preliminary assessment of the distributions of some trace elements (As,

Cd, Cu, Fe, Ni, Pb and Zn) in a pristine aquatic environment: the Lena River Estuary (Russia). *Mar. Chem.,* **43**: 185–99

Martin, J. M., Huang, W. W. and Yoon, Y. Y. (1994). Level and fate of trace metals in the lagoon of Venice (Italy). *Mar. Chem.,* **46**: 371–86

Martin, J. M., Dai, M. and Cauwet, G. (1995). Significance of colloids in the biogeo-chemical cycling of organic carbon and trace metals in a coastal environment – example of the Venice Lagoon (Italy). *Limnol. Oceanogr.,* **40**: 119–31

McKnight, D. M. and Morel, F. M. M. (1979). Release of weak and strong copper-complexing agents by algae. *Limnol. Oceanogr.,* **24**: 823–37

Mills, G. L. and Quinn, J. G. (1981). Isolation of dissolved organic matter and copper organic complexes from estuarine waters using reverse-phase liquid chromatography. *Mar. Chem.,* **10**: 93–103

Mills, G. L. and Quinn, J. G. (1984). Dissolved copper and copper organic complexes in the Nagarranset Bay Estuary. *Mar. Chem.,* **15**: 151–72

Mills, G. L., McFadden, E. and Quinn, J. G. (1987). Chromatographic studies of dissolved organic matter and copper organic complexes isolated from estuarine waters. *Mar. Chem.,* **20**: 313–23

Mills, G. L., Douglas, G. S. and Quinn, J. G. (1989). Dissolved organic isolated by C-18 reverse phase extraction in an anoxic basin located in the Pettaquamscutt river estuary. *Mar. Chem.,* **26**: 277–88

Mills, G. L., Hanson, A. K., Quinn, J. G., Lammela, W. R. and Chasteen, N. D. (1982). Chemical studies of copper organic complexes from estuarine waters using reverse-phase liquid chromatography. *Mar. Chem.,* **11**: 355–77

Minas, M. and Minas, H. J. (1992). Hydrological and chemical condition in the Gulf of Lyons and relationships to primary production encountered during Cybele. In Martin, J. M. and Barth, H. (eds), *EROS 2000, Water Pollution Report,* **28**: 127–38

Moffett, J. W., Zika, R. G. and Brand, L. E. (1990). Distribution and potential sources and sinks of copper chelators in the Sargasso Sea. *Deep-Sea Res.,* **37**: 27–36

Moore, R. M. (1981). Oceanographic distribution of zinc, cadmium, copper and aluminum in water of the central Arctic. *Geochim. et Cosmochim. Acta,* **45**: 2475–82

Moore, R. M. (1983). The relationship between distribution of dissolved cadmium, iron and aluminium and hydrography in the central Arctic ocean. In Wong, C. S., Boyle, E., Bruland, K. W., Burton, J. D. and Goldberg, E. D. (eds), *Trace metals in sea water.* Plenum Press, New York and London, 131–42

Qu, C. H. and Yan, R. E. (1990). Chemical composition and factors controlling suspended matter in three major Chinese rivers. *Sci. Total Environ.,* **97/98**: 335–46

Romanin, M. V. (1989). Composizione chimica di alghe marine raccolte nella laguna di Venezia e utilizzo, nella fertilizzazione del suolo. *Agricoltura delle Venezie,* **12**: 513–25

Sfriso, A., Pavoni, B., Marcomini, A. and Orio, A. A. (1988). Annual variations of nutrients in the lagoon of Venice. *Marine Pollution Bulletin,* **19**(2): 54–60

Sfriso, A., Pavoni, B. and Marcomini, A. (1989). Macroalgae and phytoplankton standing crops in the central Venice lagoon: primary production and nutrient balance. *The Science of the Total Environment,* **80**: 139–59

Shiller, A. M. and Boyle, E. A. (1987). Variability of dissolved trace metals in the Mississippi river. *Geochim. Cosmochim. Acta,* **51**: 3273–7

Sunda, W. G. and Guillard, R. R. L. (1976). The relationship between cupric ion activity and the toxicity of Copper to phytoplankton. *J. Mar. Res.,* **34**: 511–29

Tefry, J. H. and Presley, B. J. (1976). Heavy metal transport from the Mississippi river to the Gulf of Mexico. In Windom, H. L. and Duce, R. A. (eds), *Marine Pollution Transfer*. Lexington Books, Lexington, Mass, 39–76

Tefry, J. H., Nelsen, T. A., Trocine, R. P., Metz, S. and Vetter, T. W. (1986). Trace metal fluxes through the Mississippi river delta system. *Rap. Reun. Cons. Int. Explor. Mer.*, **186**: 277–88

Windom, H. L. (1981). Comparison of atmospheric and riverine transport of trace metals to the continental shelf environment. In Martin, J. M., Burton, J. D. and Eisma, D. (eds), *River Inputs to Ocean Systems*. Proceedings of a review workshop held at FAO. Rome. 26–30 March 1979, 360–9

Windom, H. L. (1986). The importance of atmospherically transported trace metals to coastal marine waters in relation to river transport. In *Estuarine Processes: an Application to the Tagus Estuary*, Proceedings of a UNESCO/CNA workshop held at Palacio Foz, Lisbon. Portugal. 13–16 December 1982, 193–202

CHAPTER 3

SIGNIFICANCE OF COLLOIDS IN THE BIOGEOCHEMICAL CYCLING OF ORGANIC CARBON

M. Dai, J.M. Martin and G. Cauwet

SUMMARY

Colloidal organic carbon and trace metals in the Venice Lagoon were separated by a cross-flow ultrafiltration device. A significant part of the trace metals which would previously have been considered as in the dissolved phase were actually found to be associated with colloidal material. This colloidal material has been shown to play an important role in biogeochemical cycling of organic carbon and trace metals in the Lagoon. The partitioning of trace elements between colloidal, macroparticulate and truly dissolved phases is also discussed, based on observations in the Venice Lagoon.

INTRODUCTION

Colloids are operationally defined as particles with a size range between 1 nm and 1 μm (Vold and Vold, 1983). Most of these colloid-size particles are thus not retained by commercially available filters with a pore size ranging from 0.22 to 0.7 μm. Therefore, in most aquatic chemistry studies, they are generally included in the so-called dissolved fraction.

Colloids are ubiquitous in aquatic environments and play an important role in transport and partitioning between dissolved and solid phases (Koike *et al.*, 1990; Honeyman and Santschi, 1991; Buffle and Van Leeuwen, 1992; Morel and Gschwend, 1987). In addition, studies on colloids would be useful in assessing bioavailable trace metal concentrations since colloids can only pass with difficulty through a cell membrane (Wells *et al.*, 1983; Rich and Morel, 1990), although uncertainties remain about the availability of truly dissolved organic trace metal complexes and the assimilation of colloidal iron when thermal or photochemical dissolution occurs.

The aims of this study were to assess:

(1) The distribution and significance of both colloidal organic carbon and trace metal complexes,
(2) The different behaviours of colloidal materials compared to real dissolved species, and
(3) The role of colloids in partitioning of metals between the dissolved and solid phases (K_d).

MATERIALS AND METHODS

Sampling and prefiltration

Samples were collected in July 1992 at seven stations along the salinity gradient from Porto di Lido to the Silone Channel (Figure 3.1). Surface samples were collected by hand in acid-cleaned 10 l polypropylene bottles from a small rubber boat. One subsurface sample (station 1, referred to as S1–2) was taken from a depth of 2 m with an air-driven (Teflon) pump.

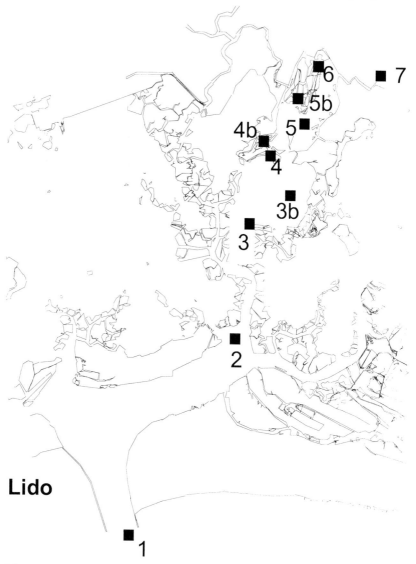

Figure 3.1 Map of sampling stations in the Venice Lagoon

About 15 l water samples were prefiltered immediately after sampling using acid-cleaned Nuclepore filters (0.4 μm, Ø 142 mm) and a Teflon filter holder, in a closed, pressurized system (pure N_2 gas, about 1 bar). In order to avoid the adsorption on, or release from, the Nuclepore filter, the first 1–2 l of the filtrate were discarded. Then, 50 ml of the filtrate were collected for dissolved organic carbon (DOC) analysis and preserved with $HgCl_2$. 500 ml of the filtrate were acidified to pH = 2 with concentrated HNO_3 (ultrapure) for trace metal analysis. The remaining prefiltered water (~10–12 l) was processed by cross-flow filtration so as to isolate colloids from the real dissolved phase. Another portion of the samples (~1 l) was filtered through precombusted Whatman GF/F (0.7 μm) filters with a Millipore glass kit and the filters were stored for later particulate organic carbon (POC) measurements.

Cross-flow ultrafiltration (CFF)

Our CFF system (Figure 3.2) is a modified Millipore Pellicon cassette system (Millipore Corp.) containing a polysulfone filter with a nominal molecular weight cutoff of 10^4 and a filter area of 0.462 m^2. The polysulfone filter has a pore size of about 3 nm which approximates to the lower size limit of colloids.

Chemical analyses

Organic carbon was determined in the total dissolved (DOC < 0.4 μm), truly dissolved (UOC < 3 nm), colloidal (COC 3 nm–0.4 μm) and particulate (POC > 0.7 μm) fractions. DOC, UOC and COC were measured with the high-temperature catalytic oxidation method (HTCO) using a Shimadzu TOC-5000 analyzer. POC was measured according to the method of Cauwet (1983).

Trace metals were measured in both prefiltrate and ultrafiltrate by graphite furnace atomic absorption spectrophotometry (GFAAS) after extraction (Danielsson *et al.*, 1982). The colloidal concentration was calculated as the difference between the total dissolved and true dissolved concentrations. Total trace metal concentrations in particulate matter (> 0.4 μm) were also determined by GFAAS after digesting the filter with a mixture of HNO_3, $HClO_4$ and HF.

RESULTS

Significance of the colloidal material

Concentrations of DOC, COC and UOC and their variations are shown in Table 3.1. COC, with an average concentration of 33.46 μM, represents from 10.4 to 26% of the total DOC, with a maximum percentage in the riverine water (station 7) indicating that most of the total DOC has a low nominal molecular weight (< 10^4) and corresponds to true dissolved organic carbon. The correlations between POC and the total dissolved fraction (POC =

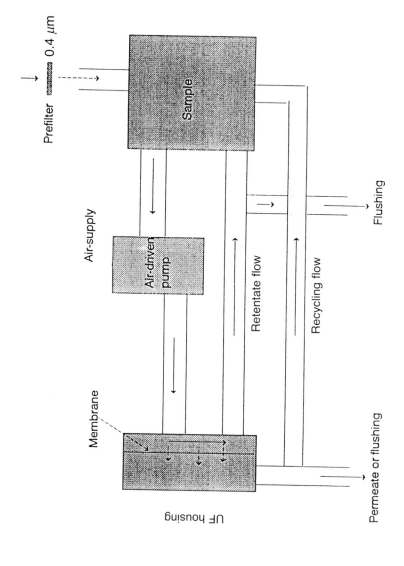

Figure 3.2 The cross-flow ultrafiltration (CFF) system

Table 3.1 Organic carbon fractions[#] in the Venice Lagoon in July 1992

Station	TSM (mg/g)	TOC (µM)	POC			DOC		UOC		COC		
			POC (µM)	POC/TOC (%)	POC** (%)	DOC (µM)	DOC/TOC (%)	UOC (µM)	UOC/TOC (%)	COC (µM)	COC/DOC (%)	COC/TOC (%)
Lido-S1	5.04	211.75	35.33	16.69	8.40	176.42	83.31	148.42	70.09	28.00	15.87	13.22
Lido-S1-1	3.96	187.75	17.33	9.23	5.30	170.42	90.77	140.50	74.83	29.92	17.56	15.93
Lido-S1-2*	4.86	188.50	13.33	7.07	3.40	175.17	92.93	140.08	74.31	35.08	20.03	18.61
Crevan-S2	10.11	226.75	43.42	19.15	5.20	183.33	80.85	152.33	67.18	31.00	16.91	13.67
Palude-S3	23.89	425.42	113.17	26.60	5.70	312.25	73.40	267.17	62.80	45.08	14.44	10.60
Silone-S4	11.06	277.67	43.33	15.61	4.70	234.33	84.39	210.00	75.63	24.33	10.38	8.76
Silone-S5	5.32	303.83	75.67	24.90	17.10	228.17	75.10	198.50	65.33	29.67	13.00	9.76
Silone-S6	6.26	253.50	61.58	24.29	11.80	191.92	75.71	144.17	56.87	47.75	24.88	18.84
Silone-S7	3.47	149.75	33.08	22.09	11.40	116.67	77.91	86.33	57.65	30.34	26.00	20.26

[#]UOC – truly dissolved organic carbon ($< 10^4$ Dalton, corresponding approximately to 3 nm)

DOC – total 'dissolved' organic carbon ($< 0.4 \, \mu m$)

COC – colloidal organic carbon (3 nm–0.4 µm)

POC – particulate organic carbon ($> 0.7 \, \mu m$)

TOC – total organic carbon

*Subsurface sample, all others are surface water samples

**Particulate organic carbon percentage in TSM

0.50 DOC − 44.83) and between DOC and TSM (DOC = 138.27 + 7.36 TSM) were positive and significant (r^2 = respectively 0.68 and 0.75) demonstrating a link between the suspended matter and the total DOC (including colloidal carbon). This relationship is even more obvious when all organic fractions and TSM are plotted against salinity (Figure 3.3). They all exhibit a maximum around station 3, where productivity was higher. Such a distribution of organic carbon is unlike many other estuaries where terrestrial sources of organic carbon dominate. The distribution of COC is somewhat more complicated, since it is also controlled by various physicochemical processes during the mixing of freshwater and salt water. POC ranged from 13.3 to 113 μM, accounting for 7.1–26.6% of the total organic carbon, i.e. a percentage similar to that of COC.

The importance of COC can be inferred from the fact that COC and POC represent an equivalent percentage of the total organic carbon. Since colloids have a much greater surface area than large particles, it is likely that COC plays as important a role as POC in the transport and biological cycling of chemical elements and compounds.

Despite its relatively low percent abundance (< 26% of DOC), the COC appears to play a major role in the transfer and transport of some trace metals (especially iron, lead, manganese and copper). As far as the trace metals are concerned, their colloidal fractions were rather significant (Table 3.2). On average, 33.5% of Cd, 46.3% of Cu, 87.4% of Fe, 18.0% of Ni, 58.1% of Pb and 54.0% of Mn in the so-called dissolved phase were actually associated with colloidal materials (Figure 3.4). Such a dominance of the colloidal fraction is also evident from the relationship between total dissolved and colloidal trace metal concentrations (Figure 3.5).

Colloids also have an important effect on the behaviour of trace metals during mixing of seawater and freshwater. For example, an apparently conservative distribution was observed for total dissolved Mn, while the truly dissolved Mn and colloidal Mn were actually non-conservative. Truly dissolved Mn showed a concave upward regression on salinity and colloidal Mn was found in excess relative to the theoretical dilution line (Figure 3.6). This discrepancy probably results from a dynamic equilibrium between the colloidal and truly dissolved phases, i.e. a transformation from the truly dissolved phase into the colloidal phase. This transformation results from both adsorption (or complexation) with colloids and uptake by colloid-size phytoplankton species which are included in the colloidal fraction. At the same time, the formation of fresh colloids via precipitation of Mn oxides is also possible. The *in situ* biological production of colloids is probably more important than input from freshwater. Thus, flocculation of riverine colloids is less important than in other estuaries such as the Rhône river estuary (Dai *et al.*, 1995) and Ob/ Yenisey river estuaries (Dai and Martin, 1995) where colloid flocculation evidently occurs. As a result, the total dissolved Mn shows apparent conservative behaviour in the Venice Lagoon.

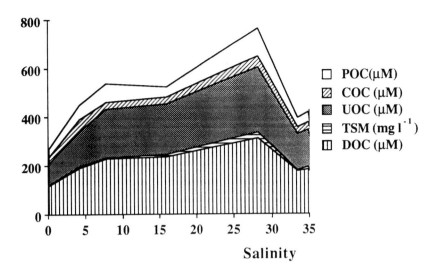

Figure 3.3 Organic fractions and TSM against salinity

Table 3.2 Trace metal concentrations in the truly dissolved (C_u, nM), total dissolved (C_p, nM), colloidal (C_c, nM) and macroparticulate fractions (PP, nM, per water volume, and PP,μg g^{-1}, per macroparticulate mass) in the Venice Lagoon. (The unit of macroparticulate Fe concentration per particulate mass is %)

Station	C_u	C_p	C_c	C_c/C_p (%)	C_c/C_t (%)	PP (nM)	PP/C_t (%)	PP (μg g^{-1})
	Cd							
Lido-S1	0.075	0.110	0.035	31.82	28.24	0.014	11.25	0.85
Lido-S1-1	0.079	0.080	0.001	1.25	–	–	–	–
Lido-S1-2	0.083	0.126	0.043	34.13	–	–	–	–
Crevan-S2	0.061	0.077	0.016	20.78	15.00	0.030	27.81	0.90
Palude-S3	0.007	0.010	0.003	30.00	9.46	0.022	68.46	0.36
Silone-S4	0.025	0.039	0.014	35.90	20.48	0.029	42.95	0.39
Silone-S5	0.007	0.019	0.012	63.16	–	–	–	–
Silone-S6	0.014	0.024	0.010	41.67	–	–	–	–
Silone-S7	0.016	0.028	0.012	42.86	16.61	0.044	61.25	1.24
	Cu							
Lido-S1	4.50	11.05	6.55	59.24	48.66	2.41	17.90	83.12
Lido-S1-1	3.76	5.64	1.88	33.37	–	–	–	–
Lido-S1-2	4.52	10.52	6.00	57.05	–	–	–	–
Crevan-S2	4.56	5.74	1.18	20.52	14.51	2.39	29.40	40.99
Palude-S3	3.15	5.69	2.54	44.64	33.03	2.00	26.01	18.74
Silone-S4	4.22	8.54	4.32	50.59	10.71	31.80	78.83	239.98
Silone-S5	6.07	13.08	7.00	53.55	–	–	–	–

Continued on next page

Table 3.2 *Continued*

Station	C_u	C_p	C_c	C_c/C_p (%)	C_c/C_t (%)	PP (nM)	PP/C_t (%)	PP ($\mu g\ g^{-1}$)
Silone-S6	7.07	15.41	8.34	54.11	–	–	–	–
Silone-S7	8.85	15.82	6.96	44.04	32.68	5.48	25.73	87.02
Fe								
Lido-S1	1.46	11.94	10.48	87.74	3.92	255.33	95.53	0.78
Lido-S1-1	2.48	7.69	5.21	67.75	–	–	–	–
Lido-S1-2	2.50	22.95	20.45	89.12	–	–	–	–
Crevan-S2	2.63	9.63	7.00	72.72	0.76	908.50	98.95	1.37
Palude-S3	4.04	299.27	295.23	98.65	12.63	2037.36	87.19	1.68
Silone-S4	3.25	1124.92	1121.67	99.71	32.18	2361.09	67.73	1.57
Silone-S5	5.60	544.24	538.64	98.97	–	–	–	–
Silone-S6	5.76	58.81	53.06	90.21	–	–	–	–
Silone-S7	11.37	62.29	50.92	81.74	4.20	1149.27	94.86	1.61
Ni								
Lido-S1	7.10	8.62	1.51	17.55	15.15	1.35	0.14	43.08
Lido-S1-1	6.50	6.72	0.22	3.23	–	–	–	–
Lido-S1-2	8.13	11.89	3.76	31.63	–	–	–	–
Crevan-S2	6.59	6.66	0.08	1.13	0.98	0.96	0.13	15.22
Palude-S3	12.62	14.73	2.11	14.33	12.47	2.19	0.13	18.94
Silone-S4	11.02	14.57	3.55	24.36	19.78	3.38	0.19	23.60
Silone-S5	11.46	17.25	5.79	33.55	–	–	–	–
Silone-S6	12.62	17.41	4.79	27.52	–	–	–	–
Silone-S7	19.17	20.89	1.72	8.24	7.33	2.57	0.11	37.73
Pb								
Lido-S1	0.033	0.050	0.017	34.00	13.08	0.08	61.54	8.76
Lido-S1-1	0.049	0.060	0.011	18.33	–	–	–	–
Lido-S1-2	0.045	0.115	0.070	60.87	–	–	–	–
Crevan-S2	0.062	0.064	0.002	3.13	0.98	0.14	68.63	7.90
Palude-S3	0.021	0.368	0.347	94.29	45.18	0.40	52.08	12.08
Silone-S4	0.100	1.078	0.978	90.72	57.26	0.63	36.88	15.50
Silone-S5	0.024	0.441	0.417	94.52	–	–	–	–
Silone-S6	0.026	0.120	0.094	78.33	–	–	–	–
Silone-S7	0.156	0.306	0.150	49.02	28.52	0.22	41.82	11.61
Mn								
Lido-S1	5.93	21.58	15.65	72.51	29.71	31.10	59.04	928.62
Lido-S1-1	4.95	13.59	8.63	63.55	–	–	–	–
Lido-S1-2	6.61	11.12	4.50	40.52	–	–	–	–
Crevan-S2	9.02	13.69	4.68	34.14	6.47	58.65	81.08	870.81
Palude-S3	6.39	22.55	16.16	71.67	4.71	320.79	93.43	2599.31
Silone-S4	11.68	33.96	22.28	65.60	10.56	176.98	83.90	1154.71
Silone-S5	8.00	34.42	26.42	76.75	–	–	–	–
Silone-S6	27.79	44.26	16.47	37.21	–	–	–	–
Silone-S7	24.26	31.95	7.69	24.06	8.06	63.44	66.51	871.33

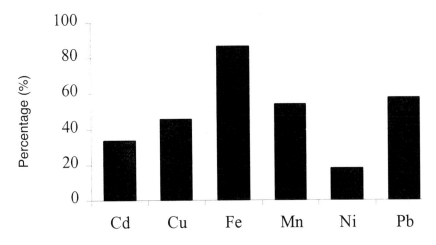

Figure 3.4 Average percentages of colloidal trace metals in the total 'dissolved' fraction

Importance of colloids in partitioning of trace metals between the dissolved and particulate phases

The distribution coefficient represents the distribution of one given element between the dissolved and solid phases.

$$K_d = Me_p/Me_d \qquad (3.1)$$

where Me_p is the particulate metal concentration (nmol/kg) and Me_d is the dissolved metal concentration (nM). In many aquatic environments, however, the experimental observations do not fit with physicochemical theory of the sorption mechanism. Among others, a particle concentration effect has been reported (O'Connor and Connolly, 1980; Morel and Gschwend, 1987).

When the colloidal fraction is considered, an apparent distribution coefficient (K_d') may be defined as:

$$Kd' = Me_p/(Me_c + Me_d) \qquad (3.2)$$

where Me_c is the metal concentration in the colloidal phase (nmol/l).

Defining K_c as the coefficient for distribution of metals between the colloidal and the truly dissolved phases, and K_d as the coefficient between the macro-particle phase and the truly dissolved phase:

$$K_c = Me_c/(Me_d\ M_c), \qquad K_d = Me_p/Me_d$$

Then

$$K_d' = K_d/(1 + Me_c/Me_d) = K_d/(1 + K_c\ M_c) \qquad (3.3)$$

where M_c represents the mass of colloids (mg/l).

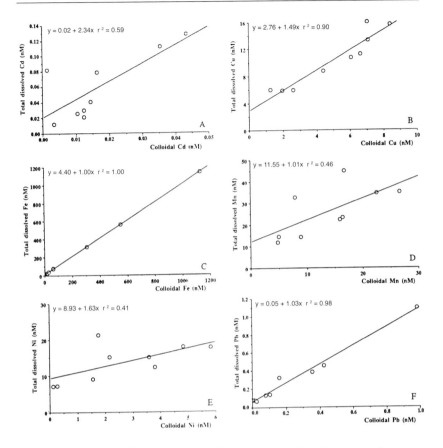

Figure 3.5 Relationship between total dissolved and colloidal trace metal concentrations

The influence of colloids on K_d can then be illustrated by the relation between K_d and the TSM (total suspended matter) as a function of M_c/TSM ratios (r) (Figure 3.7). The curves show that the maximum particle concentration effect decreases with r, i.e. the 'particle concentration effect' becomes more significant when M_c becomes important relative to TSM. Thus, the conventionally determined distribution coefficient, namely the apparent distribution coefficient (K_d') may significantly decrease as a function of TSM in most natural aquatic systems with the occurrence of colloids. This confirms the assumption of Morel and Gschwend (1987).

CONCLUSIONS

Our data (summer period) indicate that the Venice Lagoon ecosystem (at least the northern part covered in this study) does not appear to be significantly polluted with respect to the trace metals that we studied, as already observed

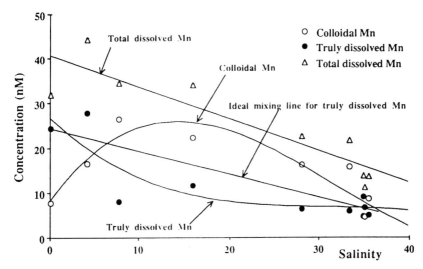

Figure 3.6 The concentration of total dissolved, truly dissolved and colloidal manganese as of salinity

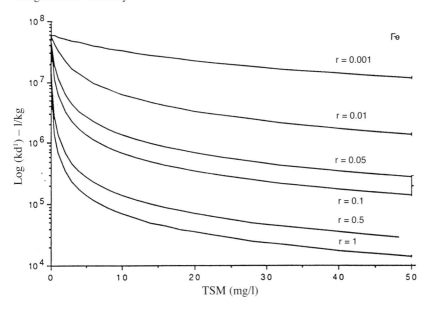

Figure 3.7 Relation between K_d and the TSM (total suspended matter) as a function of Mc/TSM ratios (r)

in winter and spring (see Martin *et al.*, Chapter 2). This is especially valid when the truly dissolved concentration is considered. Conversely, the total DOC concentrations in the Lagoon, particularly in Palude della Rosa, are markedly higher than in other Mediterranean coastal regions due to the high primary productivity. In turn, this high primary productivity largely determines the distribution of organic carbon, leading to spatial distribution patterns with maxima at intermediate salinities (Palude della Rosa) rather than at the river-end member as in other estuaries. This suggests that organic matter is mainly autochthonous in the study area.

The colloidal fraction usually exhibits a behaviour different from the real dissolved fraction, even for organic carbon, although it differs from one element to the other. Information on the involvement of colloids in the biogeochemical processes may greatly enhance understanding of their distribution in the aquatic environment. Colloidal organic carbon may act as the major ligand for formation of organic–metal complexes in spite of its relatively low mass (< 26% of the total DOC). A simple three-phase partitioning model confirmed that a particle concentration effect may be an important artifact in traditional computation of K_d normally used to model the fate of trace metals in coastal and estuarine environments, where colloids and particles are abundant.

REFERENCES

Buffle, J. and Van Leeuwen, H. P. (1992). Environmental Particles. Vol. 1, *IUPAC Environmental Analytical Chemistry Monograph Series*. Lewis Publ., CRC Press Inc., Boca Raton, Florida, pp. 554

Cauwet, G. (1983). Distribution du carbone organique dissous et particulaire en Méditerranée occidentale. *Rapp. Comm. Int. Mer Médit.*, **28**: 101–5

Dai, M., Martin, J. M. and Cauwet, G. (1995). The significant role of colloids on the transport and transformation of organic carbon and trace metals in the Rhone delta, France. *Mar. Chem.*, **51**: 159–75

Dai, M. H. and Martin, J. M. (1995). First data on the trace metal level and behavior in two major Arctic river/estuarine systems (Ob and Yenisey) and in the adjacent Kara Sea. *Earth Planetary Sci. Lett.*, **131**: 127–41

Danielsson, L. G., Magnusson, B., Westerlund, S. and Zhang, K. (1982). Trace metal determination in estuarine waters by electrothermal AAS after extraction of dithiocarbamate complexes into freon. *Anal. Chim. Acta*, **144**: 183–8

Honeyman, J. T. and Santschi, P. H. (1991). Coupling adsorption and particle aggregation: laboratory studies of 'colloidal pumping' using ^{59}Fe-labelled hematite. *Environ. Sci. Technol.*, **25**: 1739–47

Koike, I., Hara, S., Terauchi, K. and Kogure, K. (1990). Role of submicrometre particles in the ocean. *Nature*, **345**: 242–4

Morel, F. M. M. and Gschwend, P. M. (1987). The role of colloids in the partitioning of solutes in natural waters. In Stumm, W. (ed.), *Aquatic Surface Chemistry*. John Wiley & Sons Inc., New York, 405–22

O'Connor, D. J. and Connolly, J. P. (1980). The effect of concentration of adsorbing solids on the partition coefficient. *Water Res.*, **14**: 1517–26

Rich, H. W. and Morel, F. M. M. (1990). Availability of well-defined colloids to the marine diatom *Talassiosira weissflogii*. *Limnol. Oceanogr.*, **35**: 652–62

Vold, R. D. and Vold, M. J. (1983). *Colloid and Interface Chemistry*. Addison-Wesley, Reading, Massachusetts, pp. 694

Wells, M. L., Zorkin, N. G. and Lewis, A. G. (1983). The role of colloid chemistry in providing a source of iron to phytoplankton. *J. Mar. Res.*, **41**: 731–46

CHAPTER 4

CHARACTERIZATION OF PARTICLES AND BIOLOGICAL ACTIVITY IN THE VENICE LAGOON BY FLOW CYTOMETRY

M. Lepesteur and J.M. Martin

SUMMARY

Particles in the Venice Lagoon were characterized by flow cytometry between January and April 1992. The measurement of particle autofluorescence and induced fluorescence was used to differentiate organic/inorganic, living/non-living, active/inactive algae and free-living/particle-bound bacteria.

The Venice Lagoon was characterized by high organic production. Regardless of the season, organic particles corresponding to algal cells and detritus were dominant. The consequence of this might be primary production in the coastal area of the Adriatic Sea and eutrophication in the Lagoon, particularly in the Palude della Rosa.

During the low productive period, most of the cells were particle-bound. Bacteria were particularly immobilized on algal cells, allowing the recycling of organic matter. In spring, the release of high amounts of dissolved and colloidal organics favoured the free-living, active bacteria. Bound, active bacteria were detected in the Palude on the benthic alga *Ulva* and their activity led to nutrient release and phytoplankton blooms.

INTRODUCTION

The input of domestic sewage, agricultural drainage and various wastes from the industrial area of Mestre has led to a change in the composition and abundance of macroalgae species in the Venice Lagoon (Sfriso *et al.*, 1988). One of the consequences of the high primary production is an increase of biological oxygen demand (BOD) in both sediments and waters due to increased sedimentation of organic matter.

When nutrients are high in the Venice Lagoon, spermatophyte species are replaced by nitrophile species, particularly the Ulvaceae (Sfriso *et al.*, 1988). The decomposition of the macroalgae, at rates between 250 and 1000 g wet weight m^{-2} day^{-1} during peak decay (Sfriso *et al.*, 1989), releases enough nutrients to trigger phytoplankton blooms and enhanced organic matter flux to the sediments, leading eventually to anoxic conditions. The measurement of biological activity, particularly bacterial respiratory activity, is of interest for an understanding of these anoxic conditions.

Characterization of particle fractions with differentiation of organic/inorganic, living/non-living, active/inactive algae and free-living or particle-bound bacteria was made by flow cytometry and discussed to deepen our understanding of seasonal variations in bacterial activity with respect to the origin of particles of a lagoon environment. The bacterial activity, which leads to anoxic events and eutrophication of the Lagoon, may be dependent on season and nature of the particles. Thus, to restore the Lagoon and improve its health, it becomes more and more important to determine the origin of the particles and their influence on bacterial activity.

MATERIAL AND METHODS

The traditional analytical methods used to characterize suspended particles present some limitations. The use of the microscope is time-consuming and often does not provide quantitative estimations. Particle size analysis with a Coulter counter does not provide any insight into the composition of different fractions.

The cytometer system (ACR type) from Bruker is a versatile, computer-controlled flow cytometer that can quantify and qualify fluorescent signals. Total automation adds speed and convenience to extremely sensitive fluorescent analyses. Flow cytometry allows rapid measurements of the optical properties of a large number of particles in a short time (Chisholm *et al.*, 1988; Premazzi *et al.*, 1989). Excitation and emission filters can be chosen to obtain a combination of wavelengths that optimizes the fluorescence signal for each application. Light scatter and natural or induced fluorescence are detected simultaneously with photomultiplier tubes, which convert the light pulses into equivalent electrical pulses (Burkill, 1987). The simultaneous detection of fluorescence intensities permits differentiation of algae and bacteria, either living or non-living, and of organic from inorganic particles, as a function of particle size (Steen, 1986, 1990), giving some new insight into the biological and chemical characterization of individual particles (Chisholm *et al.*, 1988; Yentsch and Horan, 1989).

Sampling

Between January and April 1992, water samples were collected from seven different stations using the Italian research vessel *Orata*. The stations were chosen as a function of the salinity gradient, from the Lido entrance to the Silone Channel (Figure 4.1). After collection, the samples were held in sterilized dark glass bottles until they were brought back to the laboratory for further processing, which was usually within a few hours. The sampling period, January–April 1992, permits a comparison of winter and spring situations. Samplings during summer and autumn were not done. The collections in spring corresponded only to the beginning of the degradation of benthic species and did not include any anoxic events. Moreover, to estimate the day-to-day variations, a diurnal study was carried out in the Palude della Rosa.

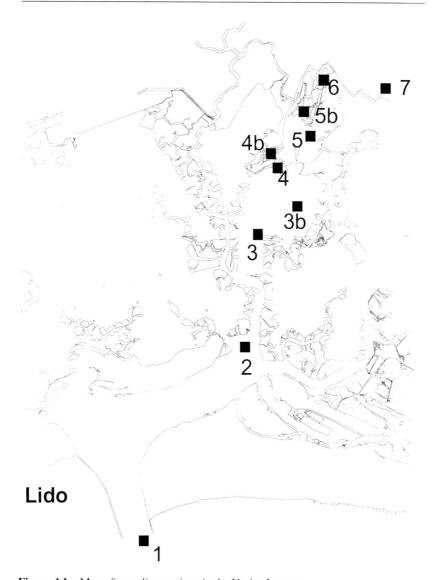

Figure 4.1 Map of sampling stations in the Venice Lagoon

Characterization of particles

Phytoplanktonic particles were characterized by the direct measurement of phycobili-proteins (PHY) and chlorophyll (CHL) autofluorescence (Cucci *et al.*, 1989; Premazzi *et al.*, 1989).

Organic particles were determined by staining with fluorescein isothiocyanate (FITC), a specific protein stain.

Detrital particles were stained with propidium iodide (PI) and living particles fluoresced with fluorescein diacetate (FDA). PI inserts into double-stranded nucleic acids and is excluded from living cells by intact membranes. FDA is lipophilic, thus it is able to cross the plasma membrane and remains non-fluorescent until it is cleaved and converted to fluorescein by intracellular esterases. The resulting fluorescein molecule is fluorescent and hydrophilic; it cannot therefore easily cross back through the cell membrane and thus is retained by the cell if the membrane is intact (Olson *et al.*, 1991). But when the same cell can fluoresce with PI and FDA, it was considered to be degenerating. Chorophyll, phycobiliproteins or phaeopigments could all be observed in the same cell.

The total bacterioplankton was estimated by staining with DAPI (diamidinophenylindole) (Robertson and Button, 1989; Monfort and Baleux, 1992).

Active bacteria were characterized by the CTC (5-cyano-2,3,di-tolyl-terazolium dichlorure) method (Zimmermann *et al.*, 1978; Stellmach and Severin, 1987; Rodriguez *et al.*, 1992). CTC acts as an electron acceptor. It is reduced when the electron transport system (ETS) in the cell is active and forms fluorescent formazan crystals. This method can also be considered as specific for respiratory activity measurement of heterotrophic bacteria (Dutton *et al.*, 1983; Rodriguez *et al.*, 1992).

Microscopic observations and counting with Malassez cells were carried out concurrently.

In order to separate bacteria from the particles or one type of particles from others, samples were subsequently deflocculated with tetrasodium pyrophosphate (TSPP) incubation followed by sonification with ultrasound (US) (Velji and Albright, 1986) and injected for a second time into the cytometer.

After incubation with different fluorochromes, the subsamples were prescreened through a 48 μm mesh, which corresponds to the largest particle size detectable with our cytometer. All samples were analyzed with the Bruker ACR 1000, using different excitation wavelengths, with light from a high pressure mercury arc of 100 W. Sheath fluid was filtered through 0.1 μm pore size filter. 100 μl of each filtered subsample were injected into the cytometer. Calibration was carried out using 1.96 μm fluorescent beads (Polysciences, Warrington, PA, USA).

Results were stored in list mode. This mode allowed the sample acquisition to be simulated in the laboratory, hence the data could be computed later. Each of the measured parameters was recorded on a three-decade logarithmic scale onto 256 channels for monoparametric, and 64×64 channels for biparametric, histograms. Particles were divided into two groups (< 3 μm and > 3 μm), according to their position on the histogram with respect to that of the standard beads.

The fluorescent components were isolated by passing through various specific emission light filters (Table 4.1).

Table 4.1 Optimal excitation and emission wavelengths of different fluorochromes (induced or natural fluorescence) and cellular characteristics analysed by flow cytometry

Fluorochromes	Excitation wavelength (nm)	Emission wavelength (nm)	Characteristics
FITC	488	LP 520	Protein staining
FDA	488	LP 520	Intracellular esterase activity
PI	488	LP 630	Membrane integrity
CHL	488	LP 630	Chl autofluorescence
PHY	488	BP 590	Phycobiliprotein autofluorescence
DAPI	390	BP 450	Bacterial DNA
CTC	415	LP 520	Bacterial respiratory activity

Enzymatic activity

Enzymes were detected using the API ZYM galleries. This semiquantitative method permits the systematic study of 19 enzymatic reactions from complex non-purified samples.

Oxygen measurements

Dissolved oxygen was measured using an oxymeter (Orbisphere Model 2609 Oxygen Indicator) with *in situ* probes.

Particulate organic carbon (POC) and suspended particulate matter (SPM) measurements

SPM was determined by filtration through dried and pre-weighed GF/F Whatman glass fibre filters (Ø 47 mm, 0.7 μm pore size) followed by re-weighing. POC was analysed by combustion of the filters and measurements as CO_2 by non-dispersive infrared (NDIR) detection (LECO CS 125 carbon analyser) (Cauwet, 1983). POC content was taken as representative of the percentage of organic matter contained in the suspended matter.

RESULTS AND DISCUSSION

Characterization of particles

Organic particles

Organic particles were dominant (~ 70% of total particles) in winter. At Crevan (station 2) 60% of the particles were organic. The maximum of organic particles (85%) was observed at the Palude entrance (station 3) (Figure 4.2A).

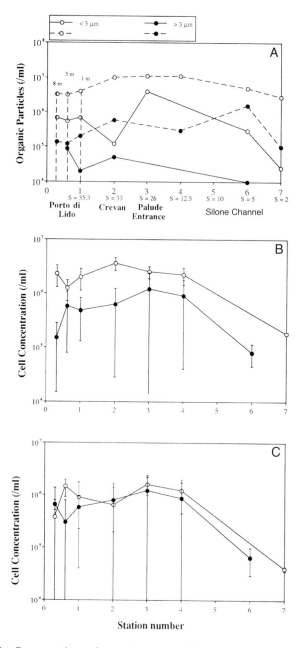

Figure 4.2 Concentrations of organic particles (A), living cells (B) and non-living cells (C) along the salinity gradient (January, 1992). The continuous lines represent the sample before deflocculation treatment (particles and aggregates), the dotted lines, the samples after deflocculation treatment. The error bars on living and non-living cells correspond to degenerate cells (cells detected as living as well as non-living)

After that, the percentage of organic particles remained constant (80–90%) along the Silone Channel. In terms of concentration, a slow decrease of organic particles was observed from the Lido mouth (station 1) to the Silone Channel, with a maximum at the Palude entrance.

It must be pointed out that the intense production of mucopolysaccharide exudates by algae and macrophytes in spring led to great difficulties in the characterization of particles by flow cytometry. In fact, with several samples, the cytometer capillaries were blocked by coagulation of colloidal organic matter or by flocculation of samples, thus preventing any further analysis. However, all the samples that could be analysed showed the same degree of dominance by organic particles. Moreover, a good correlation between POC and SPM was observed in the Lagoon (station 3b) during the diurnal study in the Palude, except at 0400 h when there was a resuspension of inorganic particles induced by stormy weather (Figure 4.3). Very high values of POC (around 10%) were measured, indicating a high fraction of organic particles. A decrease in the percentage of POC was observed when SPM increased. This can be ascribed to input of typical terrigenous particles from the Silone and the phenomenon of resuspension.

Living/non-living particles

In winter, organic particles at Porto di Lido (station 1) corresponded to living algal cells. Microscopic observation at Porto di Lido (8 m depth) showed an abundance of detritus and aggregates larger than 50 μm in size. This can explain the low percentage of detritus (20–30%) detected with the cytometer in the sample from this depth (Figure 4.2B).

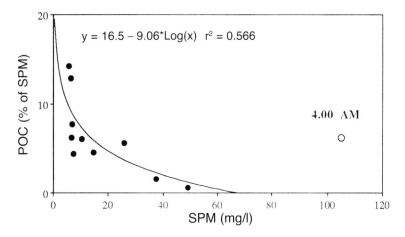

Figure 4.3 Correlation between suspended particulate matter (SPM) and particulate organic carbon (POC) (% of the SPM) in spring in the Venice Lagoon. O corresponds to the sample collected at 0400 h during stormy weather

At Crevan (station 2), detritus corresponded to ~ 60% of the organic particles and ~ 30% of all particles.

The concentrations of living and non-living cells increased along the Silone
Channel with a minimum at 2 PSU salinity (station 7), indicating that organic
particles were not coated particles but cells and particle-bound cells. The low
percentage of organic particles can be correlated with the increase of detritus
and terrigenous particles of > 50 µm size, observed microscopically.

Algae

In winter, at Porto di Lido (station 1), a relative maximum concentration of
algae was observed at 1 m depth (Figure 4.4), corresponding to a maximum
concentration of particles < 3 µm compared to those > 3 µm (Figure 4.2C).
But the sample from 3 m depth showed the maximum percentage of algae
corresponding to the totality of organic particles.

At Crevan (station 2), 70% of the organic particles were algal cells but these
cells seemed to be degenerate (decrease after sonification). Thus, small living
cells and dead particles were very abundant.

At 5 PSU salinity (station 6), a minimal concentration of algae occurred
even though algae accounted for the major part of the organic pool.

In spring, phytoplankton were apparently not very abundant because the
major part consisted of *Skeletonema*, which had a size of > 50 µm and hence
could not be detected with the cytometer. Thus, the cyanobacterial fraction
constituted by cells showing phycobiliprotein fluorescence and ultraphyto-
plankton (2–5 µm) seemed to be the most important. But station 7 showed
abundant particle-bound cells with chlorophyll fluorescence (Figure 4.4). As
in winter, the maximum of organic particles corresponded to the minimum of
living algal cells (58%) and to a high concentration of total bacteria. The
Palude della Rosa (stations 3 and 3b) was supposed to be the most eutro-
phicated area and hence more attention was paid to it. In spring, the apparent
lack of phytoplankton was also observed in the Palude, where the macroalga
Ulva dominated during the diurnal study. The increase of cells with CHL
fluorescence compared with those with PHY fluorescence during the diurnal
study in the Palude suggests that their doubling time was shorter. Moreover,
the two-fold increase of both fractions during the night indicates that
phytoplankton had nyctoperiodic multiplication or migration.

Biological activity

Demers *et al*. (1989), using a flow cytometer to examine all CHL-containing
particles, observed that phytoplankton from surface waters had low fluorescence per volume compared with cells from the subsurface chlorophyll
maximum. They interpreted this as due to an effect of nutrient-limitation at the
surface. At the Porto di Lido (station 1) in winter, phytoplankton from the

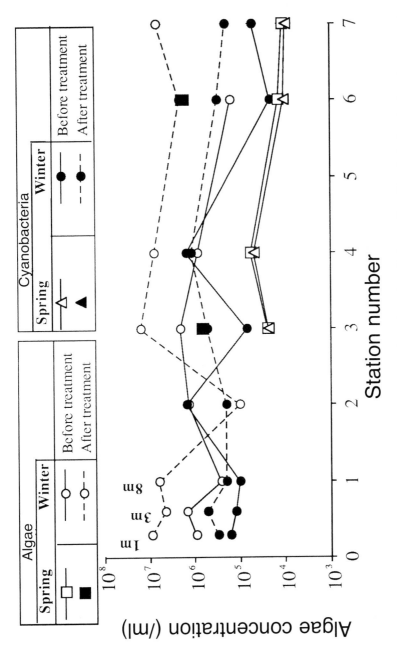

Figure 4.4 Variations in algae and cyanobacteria concentrations along the salinity gradient. Results before treatment correspond to cells + algal aggregates and after treatment to total cells (free-living and particle-bound cells)

surface appeared to have higher fluorescence per volume than those from 3 and 8 m depths. Data on nutrient concentrations and flux measurements at the three inlets would be necessary to reach conclusions concerning the input of nutrients from the Lagoon to the Adriatic Sea. In the same way, concentrations of PHY fluorescent cells (10^4–10^7 ml^{-1}) were higher than those of *Synechococcus* obtained from surface seawater (10^3–10^5 ml^{-1}) (Olson *et al.*, 1991), even in the Adriatic Sea samples (10^5–10^7 ml^{-1}).

At station 7 in winter, the small algal cells showing dim chlorophyll fluorescence are probably prochlorophytes. This occurrence of particle-bound prochlorophytes has not been reported before. Heterotrophic bacteria dominate in regions with high organic matter concentrations but in the case of station 7 in winter, the substrates were mainly inorganic, inducing a dominance of autotrophic bacteria i.e. cyanobacteria.

The spring measurements were carried out at the beginning of the period of the decomposition of macroalgae. Therefore, the correlations between the POC data and cell concentrations measured by flow cytometry (Figure 4.5) showed a decrease of algal cells (CHL fluorescence) and an increase in active bacteria when POC increased, suggesting further degradation of algal matter during spring.

In the Palude (station 3b) in spring, the macroalgae (*Ulva*), lifted to the surface by oxygen bubbles, formed wide 'green blankets' which, in a few days, turned white because of strong light (Sfriso *et al.*, 1989). Experiments on deflocculation of bacteria were carried out in the Palude using 'fresh' green and 'white' *Ulva*. The fraction of the *Ulva*-bound bacteria appeared to be over

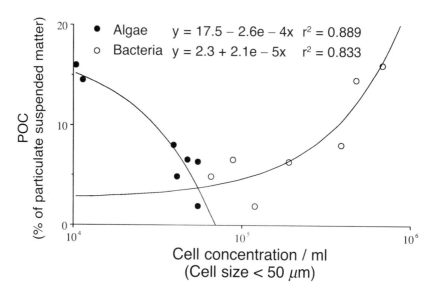

Figure 4.5 Correlation between the total (free-living and particle-bound) active cell concentrations and percentages of particulate organic matter (April 1992)

100 times higher than other bacterial fractions. The variations of CTC fluorescence intensity, corresponding to bacterial respiratory activity, were negligible
along the stations, except in the *Ulva*-bound samples which exhibited high
intensities. Both facts implicate important bacterial activity and a high rate of
bacterial degradation linked to *Ulva*. During the diurnal study, an increase of
active bacterial concentration was observed at night, with a minimum at first
light. The bacterial activity was maximum in the Silone Channel at station 4b,
near the Palude entrance. *Ulva*-bound bacteria showed high and diversified
enzymatic activity. The enzymatic activity and diversity decreased during the
night. This result corresponded to those observed with flow cytometry
(bacteria less active during the night) but might be attributed to tidal variations
and input of terigenous particles. The enzymatic activities and diversities in
the Palude were higher than at Porto di Lido or upstream at station 7.

In winter, the enzymes from bacteria in Porto di Lido samples showed weak
diversity and low activity except for valine arylamidase at 3 m and 8 m depths.
No activity involving sugars can be detected with this method. The diversity
and intensity of enzyme activity increased at Crevan, particularly for those
activities involving amino acids, although Crevan (station 2) was considered
as a non-contaminated station for hydrocarbons and heavy metals (Fossato
and Lowe, 1992). These results indicate organic, rather than inorganic, contamination. A low value for concentration of bacteria appeared before treatment
with TSPP and US. Under the microscope, we observed cyanobacteria and
bacteria bound to detritus. After treatment, the total concentration of bacteria
increased, indicating dominance of particle-bound bacterial fraction.

Bacterial enzyme diversity varied along the salinity gradient. At 2 PSU
(station 7), enzyme activity and diversity were very weak. Before treatment,
concentrations of active bacteria (corresponding to active free-living bacteria)
remained approximately constant along the Silone Channel except at stations
2 and 7 (Figure 4.6). The concentration increased after deflocculation
indicating that the active bacteria were particle-bound. The activities involving sugars were maximum at 5 PSU (station 6), and correlated with a maximum
of aggregation and flocculation and a maximum of detritus and degenerate
cells. This may be an indicator of the 'glueing' role of polysaccharides in
coagulation of particles in estuarine systems.

Seasonal variations

Seasonal variations were difficult to estimate because of the lack of sampling
in summer and autumn. The period of collection in spring corresponded to the
beginning of decomposition of macroalgae in the Palude but we did not detect
any anoxic events. During the period of low production, the organic fraction
was very significant and corresponded totally to algal cells, except for station
7. The percentage of detritus exceeded 50% of the organic particles, but the
greatest part of this was degenerate cells. The particle-bound bacteria were

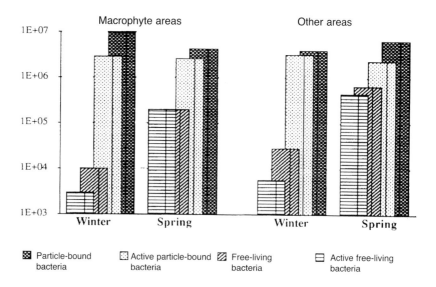

Figure 4.6   Seasonal variations in bacterial distribution

dominant and the particles were principally of algal origin. Thus, the biodegradation of algae by attached bacteria and a rapid recycling of organic matter is expected in such an environment. The eutrophication, increased sedimentation of organic matter and increased nutrient release in the water column, results in uncontrolled growth of benthic algae when light conditions become more favourable.

Regardless of season (winter or spring), organic particles dominated (~ 70% of total particles). The apparent mean percentage of algae was less important in spring than in winter due to the detection limit of the cytometer which did not allow analysis of *Skeletonema* which were greater than 50 μm in size. The cyanobacterial fraction was the most important algal fraction detected and increased during the spring. The Silone Channel showed winter and spring prochorophyte production.

The percentage of active bacteria increased during spring (> 50%). The main fraction was bound to particles or to *Ulva* in the Palude whereas the free-living fraction was dominant in winter in the same location. The ratio of active/total bacteria increased along the Silone Channel from downstream to upstream and all the bacterial populations appeared to be active at station 7. The enzymatic activities showed important variations with light during spring but mean activity in the Palude was higher in spring than winter. Elsewhere, bacteria were principally particle-bound in winter and free-living in spring. In the Palude, biodegradation of macroalgae by *Ulva*-bound bacteria led to nutrient release and phytoplankton blooms.

The phytoplankton production was difficult to estimate in spring due to the detection limit of the cytometer. However, two patterns could be observed for most of the algae, as well as for the bacteria. In winter a large fraction of the cells were particle-bound, while the free-living fraction was dominant in spring. Moreover, almost all the free-living bacteria were active in spring. These patterns can be explained by the release of high amounts of dissolved and colloidal organic matter in spring. In fact, the bacteria were not bound to particles in spring in the Palude but to the benthic alga *Ulva*. The active, bound, bacterial fraction was thus detected linked to *Ulva* in the Palude.

CONCLUSIONS

The main characteristic of the Venice Lagoon is its high organic production caused by anthropogenic inputs. In spring, when light conditions are favourable, phytoplankton blooms occur in the whole Lagoon and benthic algal production occurs in the Palude. The bacterial activity which leads to nutrient release may contribute, as a function of the hydrodynamic conditions, either to increase primary production in the coastal area of the Adriatic Sea or to eutrophication in the Lagoon.

Flow cytometry analysis showed that, regardless of season, detritus accounted for a large amount of the organic particles. But the distribution of the fractions of bacteria differed between seasons. In spring, the particle-bound bacterial fraction was found on the benthic alga *Ulva*. In winter, most of the bacteria were bound to organic, algal, and other particles, and enabled recycling of organic matter. Thus, this study essentially shows an alternation of benthic/planktonic systems according to season. Unfortunately, this study did not continue through summer and autumn, hence no conclusions on the influence of these systems on anoxic events could be drawn.

REFERENCES

Burkill, P. H. (1987). Analytical flow cytometry and its application to marine microbial ecology. In Sleigh, M. A. (ed.). *Microbes in the Sea*, Ellis Horwood Ltd, Chichester, 139–66

Cauwet, G. (1983). Distribution du carbone organique dissous et particulaire en Méditerranée occidentale. *Rapp. Comm. Int. Mer. Médit.*, **28**: 101–5

Chisholm, S. W., Olson, R. J. and Yentsch, C. M. (1988). Flow cytometry in oceanography: Status and prospects. *Eos*, **3**: 570–2

Cucci, T. L., Shumway, S. E., Brown, W. S. and Newell, C. R. (1989). Using phytoplankton and flow cytometry to analyse grazing by marine organisms. *Cytometry*, **10**: 659–69

Demers, S., Davis, K. and Cucci, T. L. (1989). A flow cytometric approach to assessing the environmental and physiological status of phytoplankton. *Cytometry*, **10**: 644–52

Dutton, R. J., Bitton, G. and Koopman, B. (1983). Malachite green-INT (MINT) method for determining active bacteria in sewage. *Appl. Environ. Microbiol.*, **45**(6): 1263–7

Fossato, V. and Lowe, D. M. (1992). Population response to salinity and pollution gradients. *UNESCO–MURST Project Venice Lagoon Ecosystem*

Monfort, P. and Baleux, B. (1992). Comparison of flow cytometry and epifluorescence microscopy for counting bacteria in aquatic ecosystems. *Cytometry*, **13**: 188–92

Olson, R. J., Zettler, E. R., Chisholm, S. W. and Dusenburry, J. A. (1991). Advances in oceanography through flow cytometry. In Demers, S. (ed.). *Particles analysis in oceanography*. NATO ASI series, Springer-verlag, Berlin, Heidelberg, Vol G 27, 351–99

Premazzi, G., Buonaccorsi, G. and Zilio, P. (1989). Flow cytometry for algal studies. *Water Res.*, **23**(4): 431–42

Robertson, B. R. and Button, D. K. (1989). Characterizing aquatic bacteria according to population, cell size and apparent DNA content by flow cytometry. *Cytometry*, **10**: 70–6

Rodriguez, G. G., Phipps, D., Ishigura, K. and Ridgway, H. F. (1992). Use of a fluorescent redox probe for direct visualization of actively respiring bacteria. *Appl. Environ. Microbiol.*, **58**(6): 1801–8

Sfriso, A., Pavoni, B. and Marcomini, A. (1989). Macroalgae and phytoplankton standing crops in the central Venice Lagoon: primary production and nutrient balance. *The Science of Total Environ.*, **80**: 139–59

Sfriso, A., Pavoni, B., Marcomini, B. and Orio, A. A. (1988). Annual variations of nutrients in the Lagoon of Venice. *Mar. Poll. Bull.*, **19**(2): 54–60

Steen, H. B. (1986). Simultaneous separate detection of low angle and large angle light scattering in an arc lamp-based flow cytometer. *Cytometry*, **7**: 445–9

Steen, H. B. (1990). Light scattering measurement in an arc lamp-based flow cyto-meter. *Cytometry*, **11**: 223–30

Stellmach, J. and Severin, E. (1987). A fluorescent redox dye: influence of several substrates and electron carriers on the tetrazolium salt formazan reaction of Ehrlich ascites tumor cells. *Histochem. J.*, **19**: 21–26

Velji, M. I. and Albright, L. J. (1986). The dispersion of adhered marine bacteria by pyrophosphate and ultrasound prior to direct counting. *GERBAM.IFREMER, Actes de colloques, 3, CNRS, Brest, 1–5 octobre 1984*, 249–59

Yentsch, C. M. and Horan, P. K. (1989). Cytometry in the aquatic sciences. *Cytometry*, **10**: 497–9

Zimmermann, R., Iturriaga, R. and Becker-Birck, J. (1978). Simultaneous determination of the total number of aquatic bacteria and the number thereof involved in respiration. *Appl. Environ. Microbiol.*, **36**: 926–35

CHAPTER 5

SURFACE REACTIVITY OF SUSPENDED MATTER AND SEDIMENTS

J.M. Garnier, J.M. Martin and C. Turetta

SUMMARY

The potential ability of sediments and suspended matter of the Venice Lagoon to associate dissolved pollutants was analyzed through three operational surface parameters: the specific surface area (SSA), the heat of immersion (HI) in water and polar or non-polar organic compounds, and the cation exchange capacity using ammonium as the exchangeable cation (ASI).

The particulate surface reactivity appears to be controlled by hydrophilic characteristic of surface functional groups which are related to the amount and composition of particulate organic matter.

INTRODUCTION

It has long been shown that interactions at solid–liquid interfaces control the fate of organic xenobiotics and heavy metals in the aquatic environment (see Stumm, 1987). Both for living and non-living particles, the interactions depend on particulate surface reactivity, characterization of which is one of the key factors in understanding and prediction of processes such as sorption or desorption of chemical species.

Characterization of particles can be made on fresh, natural water samples (see Chapter 4) or later in the laboratory on freeze-dried samples. Several parameters are used to assess particulate surface properties: the specific surface area (SSA), the heat of immersion (HI) in water and in several organic compounds, and the ammonium saturation index (ASI). These parameters have been selected in order to specify three main characteristics of particulate reactivity. The specific surface is a basic parameter directly linked to the number of reactive sites involved in solid–liquid interactions. The heat of immersion allows (i) assessment of the hydrophilic or hydrophobic nature of the particulate surface, and (ii) specification of the nature of the particulate functional groups. The ammonium saturation index provides an index of the number of cations (associated to solid surface ligands) easily exchangeable by other cations during competitive sorption processes.

Biological and physicochemical variations occurring in estuarine systems can modify bulk particulate surface properties and may thus affect their sorptive behaviours. In the case of the Venice Lagoon, due to high primary production, the biology must be particularly complex.

MATERIALS AND METHODS

Two types of particulates in the Venice Lagoon were collected during two surveys. Sediments were sampled during spring 1992 (campaign code 2) in the Paluda della Rosa (station 3; Figure 5.1) with a box corer. The core was sliced at each centimetre and each sample was centrifuged to separate sediments and interstitial waters. Suspended matter (SM) samples were

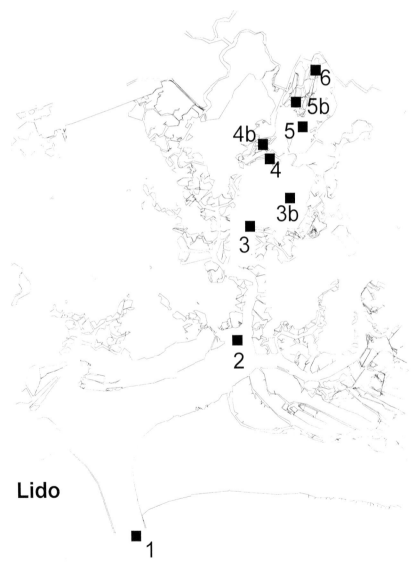

Figure 5.1   Map of the study area of the Venice Lagoon

collected during summer 1992 (campaign code 3) along the salinity gradient between Porto di Lido (station 1) and the upstream of the Silone Channel (Figure 5.1). Sample 1–2 corresponds to a sample collected at 2 m depth near station 1. SM were collected by pumping large-volume water samples (800 to 1000 l). Suspended matter was preconcentrated by tangential flow filtration at 0.2 μm and later centrifuged at 2000 g using a continuous flow separator to recover a solid–water mixture containing about 50% water. Prior to analysis, all samples were freeze-dried and homogenized.

Details of the analytical procedures have been reported in Garnier *et al.* (1991, 1993). Briefly, specific surface area (SSA) is measured either by a multipoint analyzer (Carlo Erba, Sorptomatic) according to the BET method (SSA_{BET}) using N_2 (Gregg and Sing, 1982) or by the amount of ethylene glycol mono-ethyl sorbed at the solid surface (SSA_{EGME}) (Heilman *et al.*, 1965). Usually, the differences of SSA obtained by these two methods relate to the amount and nature of the particulate organic matter. Measurements of heat of immersion (HI) of the particulates were done using either a polar compound such as water, ether, MIBK (methyl-isobutylketone), octanol or a non-polar molecule (heptane). Each of these molecular probes is expected to specify the characteristics of particulate surface functional groups. The heat of immersion was measured using a microcalorimeter (Setaram). For cation exchange capacity, we applied the most common procedure (Peech *et al.*, 1962) which uses ammonium ions as exchangeable cations and thus is termed ammonium saturation index (ASI), to refer clearly to the operational nature of this para-meter (Sposito, 1984). Conceptually, any of the cation exchange capacity methods will reflect the potential ability of particulate to associate charged pollutants.

RESULTS

Results are given in Tables 5.1 and 5.2. CT, VE3, and Rh refer respectively to Venice Lagoon sediments (campaign code 2), Venice Lagoon suspended matter (campaign code 3) and the suspended matter of the Rhône estuary (the Rhône river and its plume; Garnier *et al.*, 1991). Labels for sediments (CT) refer to the layer (in cm) sampled in the core and, for suspended matter (VE3 and Rh), refer to the sampling location (see Figure 5.1).

SSA_{BET}, HI_w and HI_{hept} measured at each centimetre (Table 5.1) of the core were very similar. Therefore, only four sediment samples (at 1, 4, 7 and 13 cm; see labels in Table 5.1), in which the results were slightly different, were taken for complete analytical measurements (Table 5.2). The data available on suspended matter were too few to generalize. The results, are therefore, discussed to include selected data obtained from a similar salinity gradient and range of particulate organic carbon (the Rhône estuarine system).

The specific surface given by the SSA_{BET} method was low (6 to 10 m^2 g^{-1}). However, SSA_{EGME} values were four to six-fold higher than SSA_{BET} values. On

Table 5.1 Specific surface area (SSA), heat of immersion (HI) in water (w) and in heptane (Hept) of sediment collected in the Venice Lagoon

Sample	SSA_{BET} $m^2 g^{-1}$	HI_w $J g^{-1}$	HI_{hept} $J g^{-1}$
CT.1	7.1	14.9	1.5
CT.2	6.7	14.3	1.8
CT.3	6.5	13.4	1.0
CT.4	8.3	12.8	1.0
CT.5	8.2	14.3	1.1
CT.6	7.8	13.1	1.8
CT.7	6.3	12.1	1.2
CT.8	6.5	11.9	0.8
CT.9	7.8	13.3	0.9
CT.10	7.2	11.9	1.0
CT.11	7.5	12.1	0.9
CT.12	7.3	11.4	1.2
CT.13	7.9	11.9	0.8

the whole, the surface area for sediments was more homogeneous and slightly lower than for suspended matter, indicating a larger potential number of reactive sites for the latter. As observed elsewhere (Garnier *et al.*, 1993), there was no relationship between the results of surface estimates obtained with these two methods.

Another index of change in specific surface area can be obtained from the heat of immersion in heptane (HI_{hept}), a non-polar molecule. HI_{hept} must be proportional to the surface since interactions between heptane, as with N_2 and the solid surface, only involves ubiquitous dispersive forces. The comparison between SSA_{BET} and HI_{hept} (Figure 5.2) did not show any relationship for the Venice samples but a positive trend was observed for the Rhône estuary SM. It has been shown that SSA_{BET}, SSA_{EGME} and HI_{hept} are well correlated for clean mineral phases (Garnier *et al.*, 1995). For natural samples, the discrepancy between SSA_{BET} and SSA_{EGME} may be due to the masking effect of organic matter. Unfortunately, the comparison between the two SSAs and the POC does not show any significant relationship. On the whole, the discrepancy between all these data sets underscores the problem of determination of true reactive surface in the case of natural samples. Under these conditions, working from the ratio between surface parameters rather than working with absolute values appears to be a more suitable approach.

The hydrophilic characteristic of the particulate can be estimated using the ratio between the heat of immersion in water (HI_w) and in an hydrophobic organic compound such as octanol (HI_{oct}). It has been shown that this ratio is close to one for clean solid phases constituting natural suspended matter (Garnier *et al.*, 1995). The same ratio is about 20 for a peat, used as representative of the organic end member, and lower than one in the case of

Table 5.2 Water salinity, specific surface area (SSA), heat of immersion (HI), ammonium saturation index (ASI) and particulate organic matter (POC) of sediment and suspended matter collected in the Venice Lagoon and in the Rhône estuary

Sample	Salinity $g\,l^{-1}$	SSA_{BET} $m^2\,g^{-1}$	SSA_{EGME} $m^2\,g^{-1}$	ASI meq $100\,g^{-1}$	HI_w $J\,g^{-1}$	HI_{hept} $J\,g^{-1}$	HI_{MIBK} $J\,g^{-1}$	HI_{ether} $J\,g^{-1}$	HI_{oct} $J\,g^{-1}$	POC $g\,g^{-1}$
CT.1		7.1	34.1	15.6	14.9	1.5	7.2	6.4	3.5	3.8
CT.4		8.3	34.4	13.0	12.8	1.0	9.3	3.1	3.4	5.2
CT.7		6.3	35.0	12.0	12.1	1.2	9.2	4.6	3.9	4.1
CT.13		7.9	32.4	13.3	11.9	0.8	7.7	3.4	2.9	4.5
VE St 1	35.1	9.2	nd	54.1	48.4	1.0	15.2	2.7	3.5	11.9
VE St 1–2	33.4	6.3	23.2	10.8	22.3	1.4	6.2	3.2	3.2	5.6
VE St 3	28.1	6.0	39.7	23.5	35.3	1.1	11.0	2.8	3.5	11.1
VE St 4	16.0	9.1	33.7	10.5	20.9	1.2	6.2	5.7	3.0	4.2
VE St 7	0.1	9.8	59.4	51.5	43.1	0.6	8.0	6.6	4.2	12.2
Rh 17	0.01	6.9	22.1	10.4	10.3	1.8	10.8	5.2	nd	1.4
Rh 01	0.01	15.8	65.1	28.8	18.4	2.1	12.0	11.5	1.9	1.1
Rh 02	0.01	12.6	69.0	29.3	21.2	3.2	9.9	7.4	5.8	4.5
Rh 03	29.4	9.7	64.2	32.5	16.4	1.7	5.3	3.4	2.7	5.9
Rh 04	19.7	9.1	49.0	46.5	20.8	1.4	4.1	2.4	2.1	6.7
Rh 05	37.2	10.0	62.6	28.7	16.8	1.6	10.0	7.4	4.7	2.7
Rh 07	11.2	10.7	83.6	42.8	27.4	2.0	6.7	2.4	2.6	8.0
Rh 08	0.01	14.1	79.9	28.1	26.8	2.7	8.8	8.2	5.0	5.6
Rh 09	0.01	6.8	nd	35.9	38.4	1.4	6.7	2.7	5.1	11.7

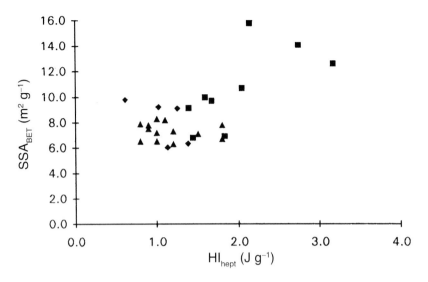

Figure 5.2 Relationship between the specific surface area (SSA) and the heat of immersion in heptane (HI_{hept}) for the Venice Lagoon sediments (triangle) and suspended matter (diamond) and the Rhône estuary suspended matter (square)

hydrophobic solids such as coal and SEP-PAK ($Si(CH_3)_2C_{18}H_{37}$, Millipore). Obviously for organically rich suspended matter, the HI_w/HI_{oct} ratio for the Venice and the Rhône particulates is higher than one. This ratio varies significantly (about four-fold) and appears to be controlled by the amount of organic matter (Figure 5.3). The increase of the hydrophilic property implies a greater ability of such particles to associate water-shielded compounds (hydrated cations, hydroxyl-bearing molecules) than hydrophobic species, such as non-polar molecules.

Some characteristics of functional groups of the particulate surface can be determined from the characteristics of the probe molecules used as a marker. For example, MIBK and ether have similar chemical properties but differ by their dipole moment (2.9 and 1.3 Debye, respectively). Therefore, the HI_{mibk}/HI_{eth} ratio reflects the polar contribution and consequently the potential ability of the particles to participate in polar interactions with other organic and inorganic polar compounds. Due to low heat pretreatment, the particle polar characteristics mainly result from the remaining hydroxyl particulate surface groups. However, the polar characteristics of the particles are also related to ASI, which represents the amount of exchangeable cations.

The relationship between HI_{mibk}/HI_{eth} ratio and ASI is shown in Figure 5.4. For SM (except for two samples) two different positive relationships, for the Venice Lagoon and Rhône estuary data, were obtained. The two slopes were rather similar. This suggests a control of both the polar and the exchangeable

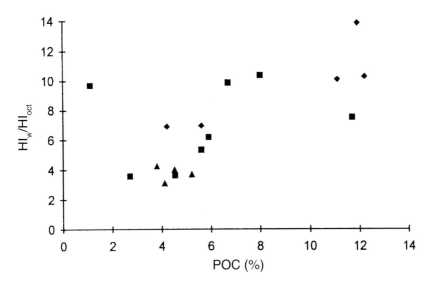

Figure 5.3 Relationship between the ratio of heat of immersion in water and in octanol ratio (HI_w/HI_{oct}) and the particulate organic carbon (POC) for the Venice Lagoon sediments (triangle) and suspended matter (diamond) and the Rhône estuary suspended matter (square)

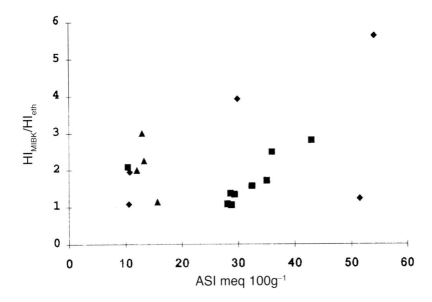

Figure 5.4 Relationship between the ratio of heat of immersion in MIBK and in ether (HI_{mibk}/HI_{eth}) and the ammonium saturation index (ASI) for the Venice Lagoon sediments (triangle) and suspended matter (diamond) and the Rhône estuary suspended matter (square)

properties by a common component of SM but of a different nature according to the area studied. But the feature of sediments shows that the potential polar ability does not depend on the same component of SM as that implicated in the exchangeable property.

CONCLUSIONS

From the large range in surface characteristics for SM and/or sediments, the following conclusions can be drawn. (1) Apart from biological effects, electrostatic (polar) interactions including effects such as H-bonding between hydroxyl groups and the dispersive (non-polar) interactions vary according to the nature of the particles (nature of the mineral phase as well as the organic phase) and the ecosystem under consideration. (2) The composition and distribution of particulate organic matter appear to be the major determining parameters. (3) According to the mode(s) of interaction involved for sorption of a given pollutant, the amount and distribution of sorbed chemical species will be variable and related to the above characteristics.

REFERENCES

Garnier, J. M., Martin, J. M., Mouchel, J. M. and Thomas, A. J. (1991). Surface reactivity of the Rhône suspended matter and relation with trace element sorption. *Mar. Chem.*, **36**: 267–89

Garnier, J. M., Martin, J. M., Mouchel, J. M. and Thomas, A. J. (1993). Surface properties characterization of suspended matter in the Ebro Delta (Spain); with an application of trace metal sorption. *Estuarine Coastal and Shelf Science*, **36**: 315–32

Garnier, J. M., Martin, J. M., Mouchel, J. M. and Chen, M. (1995). Calorimetric enthalpy of adhesion between water, organic solvents and isolated lithogenous particles. *Colloids and Surface A: Physicochemical and engineering aspects*, **97**: 203–15

Gregg, S. J. and Sing, K. S. W. (1982). *Adsorption, Surface Area and Porosity*, 2nd edn. Academic Press, London, pp. 303

Heilman, H. D., Carter, D. L. and Gonzalez, C. L. (1965). The ethylene glycol monoethyl ether (EGME) technique for determining soil-surface area. *Soil Science*, **100**: 409–13

Peech, M., Cowan, R. L. and Baker, J. H. (1962). A critical study of the $BaCl_2$–triethanolamine and the ammonium acetate methods for determining the exchangeable hydrogen of soils. *Proc. Soil. Sci. Soc.*, 37–40

Sposito, G. (1984). *The Surface Chemistry of Soils*. Oxford University Press, New York, pp. 234

Stumm, W. (1987). *Aquatic Surface Chemistry*. Wiley Interscience, New York, pp. 520

CHAPTER 6

ORGANIC MICROPOLLUTANTS AND TRACE METALS IN WATER AND SUSPENDED PARTICULATE MATTER

V.U. Fossato, G. Campesan, L. Craboledda, F. Dolci and G. Stocco

SUMMARY

Concentrations of hydrocarbons, chlorinated hydrocarbons and trace metals dissolved in water and sorbed on suspended particulate matter were measured along a salinity gradient in the northern Venice Lagoon, in the River Sile and in the Adriatic Sea using time-integrating water samplers.

Contribution of the chemical pollutants to the dissolved and particulate forms varied as functions of the trace element and the organic compound, and the quantity and nature of the particulates. Because of paucity of similar data, temporal and spatial comparisons were generally difficult but when such comparisons were possible, they showed that pollutant levels in the study area were relatively low.

INTRODUCTION

Adherence to a combination of controls on discharge of pollutants and the quality of the receiving waters has often been proposed as the best pollution control strategy for the Venice Lagoon and other sensitive coastal areas in Italy. This requires routine monitoring of water quality within the Lagoon. However, the data available on concentrations of hydrocarbons, chlorinated hydrocarbons and trace metals in the waters of Venice Lagoon are sparse compared to those on biota and sediments. This is due to three major problems in the determination of micropollutants in water. To begin with, their concentrations are so low that analysis is expensive, laborious and prone to a high degree of contamination. Second, temporal and spatial variations in their concentrations are highly amplified by the intermittent flow of effluents and hydrological factors such as tides and currents, thus necessitating a large number of analyses before averages can be obtained. Finally, such analyses in the water column are less useful in predicting the impact of micropollutants on biota compared to direct measurements on the biota themselves (Phillips, 1980). However, time-integrated water sampling equipment overcomes some of these problems (Green *et al.*, 1986). Thus, in the framework of the Venetian Lagoon Ecosystem Project, sampling for, and analysis of, aliphatic, poly-aromatic and chlorinated hydrocarbons and trace metals in water and suspended particulate matter (SPM) were included to test this technique and evaluate the levels in, and partitioning of, these pollutants between water and the suspended phase. The latter is important for prediction of the fate of a pollutant, since its dissolved and particulate forms are processed through different pathways (dilution, dispersion, sedimentation, bioaccumulation) in the environment.

The sampling sites (Figure 6.1) are representative of a wide range of environmental conditions. Station 1 (Platform), located eight miles offshore in the Gulf of Venice, has no direct source of pollution other than that from shipping and the movement of small boats, and hence is a relatively clean site. Stations 2 (Lio Grande), 3 (Crevan) and 4 (Torcello) are located along a salinity gradient in the northern basin of the Lagoon and station 0 (Businello), is in the Sile River.

MATERIAL AND METHODS

The samples were collected during five field trips between April 1991 and April 1993. In each of these, 6–13 water and SPM samples were collected using *in situ* water samplers (SeaStar, Sydney BC, Canada) equipped with filtration assembly and extraction columns specific for organic and inorganic pollutants, i.e. glass fibre filters (1 μm pore size) with Amberlite XAD-2 columns for the former and polycarbonate filters (0.4 μm pore size) with 8-hydroxyquinoline-bonded columns for the latter. Samplers were deployed and kept pumping at pre-set flow rates (150 cm^3/min for organics and 100 cm^3/min for trace metals) for 5 h each during ebb and flood tides. Water temperature and salinity were recorded during each deployment. On retrieval, the sample volume was recorded, the columns were disconnected, sealed and stored at room temperature, and the filters were removed from the apparatus and stored frozen. All analyses were carried out within three months, as recommended in the SeaStar manual.

Hydrocarbons and chlorinated hydrocarbons adsorbed on the resin columns were recovered by serial elution with methanol (200 cm^3) and dichloro-methane (200 cm^3), as suggested in the SeaStar manual. The dichloromethane fraction was concentrated by rotary evaporation, combined with the methanol fraction, with 400 cm^3 of pre-extracted seawater added, back-extracted three times with 50 cm^3 of *n*-hexane and dried over anhydrous sodium sulfate.

Filters were Soxhlet-extracted for 8 h with a mixture of 60% acetone and 40% *n*-hexane. The extracts were partitioned against pesticide-free seawater and the aqueous phase was extracted twice with 50 cm^3 of *n*-hexane (Fossato *et al.*, 1988).

The hexane extracts were concentrated to about 1 cm^3 and fractionated by chromatography on an alumina/silica gel column. When necessary, sulfur compounds were removed by shaking with activated copper powder and the organic compounds that interfere in gas chromatographic (GC) analyses were removed by treatment with concentrated sulfuric acid (Murphy, 1972).

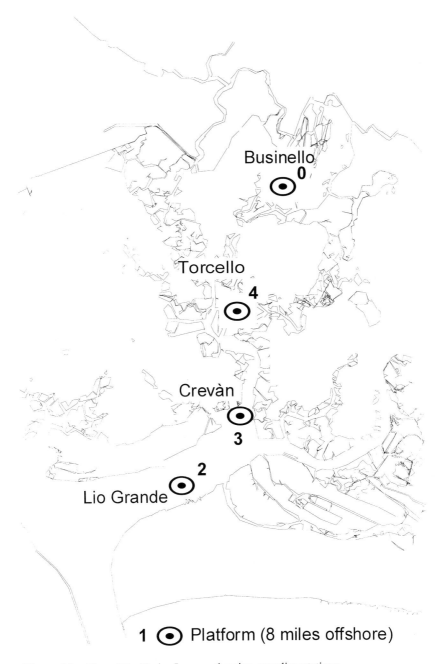

Figure 6.1 Map of the Venice Lagoon showing sampling stations

Individual fractions were analysed by capillary gas chromatography with flame ionization and electron capture detectors. Parameters measured by GC included total aliphatic hydrocarbons, unresolved complex mixture (UCM) and *n*-alkanes from *n*-C_{15} to C_{30}, chlorinated pesticides (αHCH, γHCH, pp'DDT, pp'DDD, pp'DDE) and polychlorinated biphenyls [individual CB No. 52-101-110-(118+149)-138-153-180, and total PCBs as Aroclor 1254].

Selected polyaromatic hydrocarbons (PAHs) were quantified by reverse-phase high performance liquid chromatography (HPLC) with programmed fluorescence detection, using a mixture of 12 external standards: fluorene (Fl), phenanthrene (Phe), fluoranthene (Ft), pyrene (Py), benzo(a)anthracene (BaA), chrysene (Chy), benzo(b)fluoranthene (BbF), benzo(k)fluoranthene (BkF), benzo(a)pyrene (BaP), dibenzo(a,h)anthracene (DBA), benzo(ghi)perylene (BPe) and indeno(1, 2, 3-cd)pyrene (InP). The detection limits were 20 ng/g for *n*-alkanes, 0.1 ng/g for PCBs and 1.0 to 5.0 ng/g for PAHs, depending on the compound.

Heavy metals adsorbed on 8-hydroxyquinoline-bonded columns were eluted three times with 100 cm^3 × mixtures of 1 M HCl and 0. 1 M HNO_3, as suggested in the SeaStar manual. Measurements of cadmium, iron, manganese, nickel, lead and zinc were carried out by graphite furnace atomic absorption spectrometry (GFAAS), using the standard addition method. Blanks were prepared with acid eluates from cleaned and conditioned columns. The detection limits were: Cd 2 ng/dm^3, Mn 0.5 $\mu g/dm^3$, Zn 0.5 $\mu g/dm^3$, Pb 10 ng/dm^3, Fe 0.10 $\mu g/dm^3$, Ni 0.20 $\mu g/dm^3$.

The polycarbonate filters were digested at 120°C in a Teflon cylinder for 4 h with 5 cm^3 of concentrated HNO_3 and the resultant solution made up to 50 cm^3 with Q-water. Analyses of Cu, Fe, Zn, Mn, Hg, Cd, Ni, Co, Cr and Pb were done in the same way as in the case of the organisms (Fossato *et al.*, Chapter 25). Blanks prepared by digesting clean polycarbonate filters were analysed with each set of samples. The detection limits were: Cd 0.1 ng/dm^3, Cu 0.10 $\mu g/dm^3$, Mn 0.05 $\mu g/dm^3$, Cr 0.03 $\mu g/dm^3$, Zn 0.03 $\mu g/dm^3$, Pb 0.10 $\mu g/dm^3$, Fe 1 $\mu g/dm^3$, Ni 0.10 $\mu g/dm^3$, Co 4 ng/dm^3, Hg 1 ng/dm^3.

The SPM recovered on separate filters was dried to constant wt at 65°C and its particulate organic carbon (POC) content was measured by wet oxidation (Strickland and Parsons, 1972).

RESULTS AND DISCUSSION

On the basis of hydrological characteristics, three groups of samples can be distinguished: freshwater from the River Sile (station 0) with SPM concentrations between 2.3 and 17.1 mg/dm^3, lagoon samples (stations 2, 3, 4) with 24.4 to 36.5 PSU salinity and SPM in the range 7.4–51.5 mg/dm^3, and offshore samples (station 1) with salinity between 35.0 and 37.8 PSU and SPM < 2. 2 mg/dm^3.

Tables 6.1–6 summarise the average concentrations of organic micropollutants and trace metals in water and SPM. The concentrations, expressed both on volume and dry weight basis, permit easy estimation of the sorption ability of the particulate matter. Figures 6.2 and 6.3 show the relative abundance of organic pollutants and trace metals in particulate and dissolved fractions in the northern Lagoon. Partitioning between dissolved and particulate fractions varied as a function of a metal or organic compound, amount of SPM and its nature. These variations might also reflect the problems associated with separating dissolved and particulate fractions by filtration. The membrane and glass fibre filters can adsorb dissolved pollutants and retain particles smaller than the nominal pore size, leading to an underestimation of dissolved, and overestimation of particulate, concentrations (Palmork and Villeneuve, 1980).

The following sections describe the levels and behaviour of each class of compounds in the dissolved and particulate fractions.

Hydrocarbons

An unresolved complex mixture (UCM) constituted 90–97% of the aliphatic fraction. The resolved part was mainly composed of *n*-paraffins and some biogenic hydrocarbons, such as pristane, phytane and C_{21} *n*-alkanes. A high percentage of unresolved components and an abundance of *n*-alkanes with odd/even carbon number ratio $\cong 1.0$, taken together with the dominance of few unsubstituted PAH in the aromatic fraction, point to either a continuous exposure to low levels of petroleum or a combustion product of fossil fuels, most probably resulting from extensive boat traffic in the Lagoon.

Hydrocarbon concentrations of the whole (dissolved plus particulate) sample varied between 2.0 and 14.5 μg/dm^3, of which 63–71% were adsorbed on SPM. Figure 6.4 shows, as an example, the distribution of *n*-paraffins dissolved in the water and sorbed on SPM in the northern Venice Lagoon in May 1991.

Concentrations of dissolved and adsorbed PAHs were low (range: < 0.02 to 1.22 ng/dm^3) and often below detection limits. Phenanthrene, pyrene and fluoranthene were the only PAHs that gave consistently measurable peaks in all chromatograms of the dissolved phase.

Others, such as benzo(a)anthracene, chrysene, benzo(b)fluoranthene, benzo(k)fluoranthene and benzo(a)pyrene were present in quantifiable amounts in the suspended phase, which accounted for 55 to 96% of total PAHs of the whole sample. Figure 6.5 shows, as an example, the distribution of PAH in water and SPM in the northern Lagoon in December 1991. To the best of our knowledge, there are no other data on PAH concentrations in water and SPM, either from the Venice Lagoon or the Adriatic Sea.

Chlorinated hydrocarbons

Concentrations of DDT and its metabolites DDD and DDE in water and SPM were generally at around or below the detection limits of the method used

Table 6.1 Concentrations (mean ± SD, in ng/dm³) of aliphatic, polyaromatic and chlorinated hydrocarbons in the water

	n	nC15-30	UCM	PAH	HCH	DDT	PCB Ar. 1254	PCB 8C
Northern Lagoon	32	78 ± 76	1510 ± 922	0.93 ± 0.59	0.44 ± 0.38	0.07 ± 0.07	1.03 ± 0.44	0.56 ± 0.25
River Sile	10	62 ± 42	1896 ± 781	0.15 ± 0.11	0.15 ± 0.08	0.12 ± 0.05	1.27 ± 1.02	0.93 ± 0.76
Gulf of Venice	8	25 ± 11	1150 ± 261	0.63 ± 0.48	0.21 ± 0.12	0.03 ± 0.05	0.62 ± 0.21	0.35 ± 0.11

Table 6.2 Concentrations (mean ± SD, in ng) of aliphatic, polyaromatic and chlorinated hydrocarbons in SPM of 1 dm³ of water

	n	nC15-30	UCM	PAH	HCH	DDT	PCB Ar. 1254	PCB 8C
Northern Lagoon	32	213 ± 175	4220 ± 2050	4.58 ± 3.84	0.08 ± 0.12	0.07 ± 0.11	1.43 ± 0.78	0.92 ± 0.55
River Sile	10	372 ± 231	4858 ± 1986	1.46 ± 0.75	0.03 ± 0.04	0.07 ± 0.05	2.01 ± 1.09	1.48 ± 0.59
Gulf of Venice	8	70 ± 40	1918 ± 662	0.78 ± 0.32	<0.01	<0.05	0.83 ± 0.43	0.50 ± 0.32

Table 6.3 Concentrations (mean ± SD, in ng/g dry weight) of aliphatic, polyaromatic and chlorinated hydrocarbons in SPM

	n	nC15-30	UCM03	PAH	HCH	DDT	PCB Ar. 1254	PCB 8C
Northern Lagoon	32	6.8 ± 4.5	167 ± 134	121 ± 146	3.36 ± 5.61	3.30 ± 5.85	53 ± 36	30 ± 24
River Sile	10	65.0 ± 46.3	2037 ± 2300	562 ± 366	5.17 ± 6.83	17.01 ± 18.08	287 ± 126	193 ± 57
Gulf of Venice	8	73.9 ± 42.2	2020 ± 697	647 ± 633	< 1.0	< 1.0	696 ± 854	419 ± 643

Table 6.4 Concentrations (mean ± SD in $\mu g/dm^3$, except Cd and Pb – ng/dm^3) of trace metals in the water

	n	Cd	Mn	Zn	Pb	Fe	Ni
Northern Lagoon	32	24 ± 12	6.15 ± 4.30	1.31 ± 0.34	124 ± 96	3.43 ± 6.31	0.49 ± 0.20
River Sile	10	12 ± 7	5.73 ± 1.96	2.93 ± 0.62	135 ± 85	6.92 ± 5.44	1.22 ± 0.50
Gulf of Venice	8	34 ± 11	1.60 ± 0.55	0.76 ± 0.17	<10	<0.10	0.83 ± 0.22

Table 6.5 Concentrations (mean ± SD) of trace metals in SPM of 1 dm^3 water

	n	SPM mg/dm^3	Cd ng/dm^3	Cu $\mu g/dm^3$	Mn $\mu g/dm^3$	Cr $\mu g/dm^3$	Zn $\mu g/dm^3$	Pb $\mu g/dm^3$	Fe $\mu g/dm^3$	Ni $\mu g/dm^3$	Co ng/dm^3	Hg ng/dm^3
Northern Lagoon	32	26.8 ± 11.3	7.74 ± 3.98	0.69 ± 0.29	8.86 ± 4.87	0.81 ± 0.45	1.47 ± 0.76	0.53 ± 0.35	339 ± 154	0.68 ± 0.36	164 ± 102	18.48 ± 12.2
River Sile	10	7.5 ± 4.2	8.58 ± 7.32	0.52 ± 0.38	1.63 ± 0.85	0.41 ± 0.17	0.9 ± 0.41	0.77 ± 0.63	78 ± 37	0.32 ± 0.37	45 ± 45	2.36 ± 1.34
Gulf of Venice	8	1.3 ± 0.6	0.38 ± 0.33	<0.1	0.17 ± 0.14	0.04 ± 0.02	0.04 ± 0.02	0.12 ± 0.03	4 ± 3	<0.1	<4	<1.0

Table 6.6 Concentrations (mean ± SD in $\mu g/g$ except for SPM – mg/m^3 – and Fe – mg/g)

	n	SPM mg/dm^3	Cd	Cu	Mn	Cr	Zn	Pb	Fe	Ni	Co	Hg
Northern Lagoon	32	26.8 ± 11.3	0.30 ± 0.14	26.9 ± 8.9	336 ± 133	31.9 ± 17.3	57.6 ± 24.0	20.0 ± 9.6	13.0 ± 3.4	26.1 ± 10.0	6.2 ± 2.9	0.68 ± 0.31
River Sile	10	7.5 ± 4.2	1.48 ± 1.47	77.3 ± 47.5	317 ± 252	72.8 ± 57.2	172.7 ± 122.6	113.4 ± 80.1	15.1 ± 11.6	52.9 ± 70.4	6.8 ± 6.3	0.52 ± 0.34
Gulf of Venice	8	1.3 ± 0.6	0.39 ± 0.37	<77.0	194 ± 241	44.0 ± 39.9	47.4 ± 37.9	99.6 ± 94.6	4.1 ± 4.6	<77.0	<0.30	<0.08

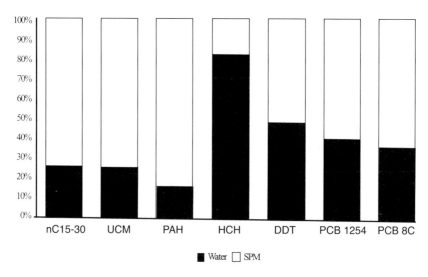

Figure 6.2 Percentage contribution of dissolved and particulate forms to the total concentrations of organic micropollutants in the northern Lagoon

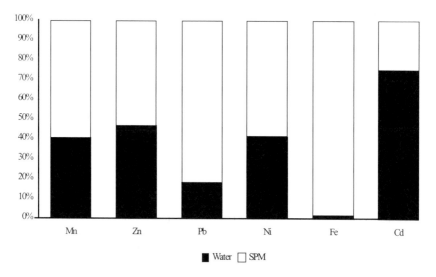

Figure 6.3 Percentage contribution of dissolved and particulate forms to the total concentrations of trace metals in the northern Lagoon

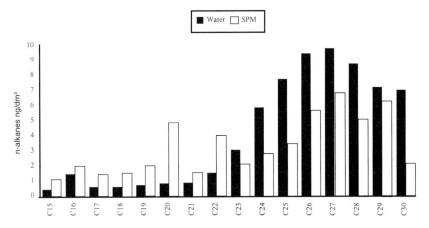

Figure 6.4   Distribution of *n*-alkanes (mean, *n* = 12) dissolved in the water and sorbed on the SPM in the northern Venice Lagoon, May 1991

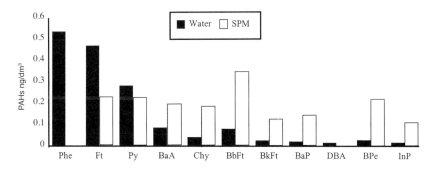

Figure 6.5   Distribution of polyaromatic hydrocarbons dissolved in the water and sorbed on the SPM in the northern Venice Lagoon, December 1991

(0.05 ng/dm³); only DDE was detected in 40% of the Lagoon samples with an equal distribution between water and suspended matter. Σ HCH (sum of α and γ isomers) was quantifiable in all River Sile samples, in 66% of Lagoon samples and in only one out of eight Adriatic samples. These were in the range of 0.01 to 1.40 ng/dm³ of the whole sample, with a predominance of γ over α isomers. As expected on the basis of relative solubility, most of the HCH occurred in the dissolved state even at those stations where particle loading was at its highest.

In spite of the restrictions on the industrial application of polychlorinated biphenyls (DPR No.82/915), they still remain a major class of pollutants in the Venice Lagoon. In fact, some CB congeners with 4 to 7 chlorine atoms substituted were present in quantifiable amounts in all the water and SPM samples analysed. Figure 6.6 shows the relative abundance of PCBs in water and SPM of eight individual congeners from the northern Lagoon in December 1991.

Total PCB concentrations of the whole sample ranged from 1.8 to 4.8 ng/dm^3 in the northern Lagoon, from 0.8 to 6.8 ng/dm^3 in the River Sile and from 0.6 to 2.4 ng/dm^3 in the Gulf of Venice. Their distribution thus appears to be relatively uniform in the study area, although slightly lower values were found in the open Adriatic Sea. PCBs exhibit a strong tendency for adsorption onto particulates (44 to 78% of PCBs of the whole sample) because of their hydrophobic nature (Duinker, 1986).

Published data on the concentrations of organochlorine compounds from the Venice Lagoon waters are scarce but have been available since 1970 from the Mediterranean Sea. Most of these data are older by more than a decade, but refinements to analytical methods in recent years may have been responsible for the observed decrease in organochlorine concentrations in marine waters (UNEP/FAO/WHO/IAEA, 1990). It is worth noting that Villeneuve *et al.* (1980) reported an average PCB concentration of 0.7 ng/dm^3 (range: 0.1–2.5 ng/dm^3) from the open waters of the Mediterranean Sea.

Trace metals

Cd, Cu, Mn, Cr, Zn, Pb, Fe, Ni, Co and Hg are minor components of seawater and exist partly in solution and partly associated with SPM. The proportions in each fraction vary with the element and with the abundance and nature of the particulates. Cadmium was present mainly as a solute while Pb and Fe

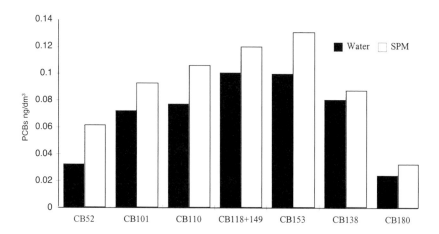

Figure 6.6 Distribution of chlorobiphenyls (average from 8 measurements) dissolved in the water and sorbed on the SPM in the northern Lagoon, December 1991

were present predominantly in the particulate fraction at all stations. Mn, Zn and Ni occurred mostly in the dissolved fraction in fresh (station 0) and marine (station 1) waters but were fairly equally distributed between dissolved and particulate fractions in the Lagoon waters (stations 2, 3, 4) where particle loading was high. Figure 6.7 shows the distribution of dissolved and SPM-sorbed trace metals in the northern Lagoon in December 1991.

Martin *et al.* (1994) studied the distribution of trace metals in the same area, but using different sampling and analytical procedures. A comparison of their data with the results in the present study for some stations of the northern Lagoon (Table 6.7) shows a good agreement, both for dissolved and particulate metals, with the exception of Pb, the concentrations of which were generally higher in our analyses.

Levels of dissolved elements were generally low (at μg/l level) and comparable to the 'unpolluted' environments of the Adriatic Sea (Scarponi *et al.*, 1989; Elbaz-Poulichet *et al.*, 1991), with the exception of Zn and Ni in freshwaters. Some of them correlated significantly ($p < 0.05$) with each other, either positively (Ni/Mn: $r = 0.66$; Fe/Pb: $r = 0.49$) or negatively (Pb/Mn: $r = -0.52$; Pb/Ni: $r = -0.75$), indicating that they are subjected to similar/different physicochemical processes. When metal concentrations in the particulate fractions were expressed on a volume basis (Table 6.5), all the elements correlated positively with the abundance of SPM and most of them with each other.

Several investigators (Sloot *et al.*, 1981; Duinker, 1986; Fossato *et al.*, 1991) have reported increased sorption properties for organic micropollutants and trace metals on particles of small size, large specific surface area and high organic matter content. This was verified with the present data by calculating correlation coefficients in linear and exponential regressions between SPM content (mg/dm^3) and its pollutant (ng/g dry weight of SPM) and POC contents (% dry weight of SPM). The results (Tables 6.8 and 6.9) show that the organic micropollutant contents were generally well correlated among

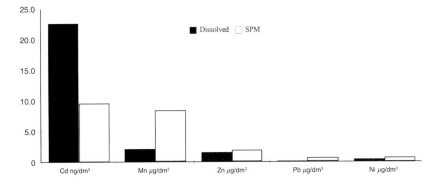

Figure 6.7 Distribution of trace metals (average from 8 measurements) dissolved in the water and sorbed on the SPM in the northern Lagoon, December 1991

Table 6.7 Comparison of trace metal concentrations (mean ± SD) reported in the present study with those of Martin *et al.* (1994) for some northern Lagoon stations

	Cd	Zn	Pb	Fe	Ni	Cu	Mn
Dissolved	ng/dm³	µg/dm³	µg/dm³	µg/dm³	µg/dm³		
Martin *et al.*	17 ± 14	1.10 ± 1.16	59 ± 17	2.02 ± 1.76	0.47 ± 0.09	—	—
Present study	24 ± 12	1.31 ± 0.34	124 ± 96	3.43 ± 6.31	0.49 ± 0.20	—	—
In SPM	µg/g	µg/g	µg/g	mg/g	µg/g	µg/g	µg/g
Martin *et al.*	0.34 ± 0.22	35.5 ± 27.0	7.9 ± 6.5	10.7 ± 6.3	18.2 ± 7.8	16.6 ± 7.5	564 ± 227
Present study	0.30 ± 0.14	57.6 ± 24.0	20.0 ± 9.6	13.0 ± 3.4	26.1 ± 10.0	26.9 ± 8.9	336 ± 133

Table 6.8 Correlation coefficients among the parameters in exponential ($y = aX^b$) relationships

	SPM	POC	nC_{15-30}	UCM	PAH 12C	HCH	DDT	PCB 1254	PCB 8C
SPM	1.00								
POC	-0.70	1.00							
nC_{15-30}	-0.85	0.67	1.00						
UCM	-0.85	0.71	0.91	1.00					
PAH 12C	-0.63	0.54	0.55	0.76	1.00				
HCH	0.10	-0.04	-0.02	-0.04	-0.25	1.00			
DDT	-0.28	0.39	0.40	0.47	0.40	0.24	1.00		
PCB1254	-0.88	0.58	0.80	0.82	0.71	-0.04	0.21	1.00	
PCB 8C	-0.87	0.54	0.78	0.85	0.72	-0.16	0.38	0.99	1.00

Abbreviations: SPM = suspended particulate matter; POC = particulate organic carbon; nC_{15-30} = *n*-alkanes with 15 to 30 carbon atoms; UCM = unresolved complex mixture of aliphatic hydrocarbons; PAH 12C = polyaromatic hydrocarbons, sum of 12 individual compounds; HCH = hexachlorocyclohexane, sum of (and isomers); DDT = sum of pp'DDT, pp'DDD, pp'DDE; PCB1254 = total polychlorinated biphenyls expressed as Aroclor 1254; PCB 8C = polychlorinated biphenyls, sum of 8 individual congeners

Table 6.9 Linear regression correlation coefficients between concentrations of trace metals in SPM

	SPM	POC	Cd	Cu	Mn	Cr	Zn	Pb	Fe	Ni	Co	Hg
SPM	1											
POC	0.094	1										
Cd	-0.071	-0.299	1									
Cu	-0.451**	0.034	0.162	1								
Mn	-0.130	-0.220	-0.164	0.623**	1							
Cr	-0.179	-0.380*	0.605**	0.217	0.085	1						
Zn	-0.300	0.073	0.008	0.316	0.267	0.109	1					
Pb	-0.119	-0.141	0.540**	0.168	-0.002	0.645**	-0.051	1				
Fe	-0.405*	-0.014	0.319	0.611**	0.443*	0.403*	0.468**	0.575**	1			
Ni	-0.185	-0.108	-0.069	-0.012	0.165	0.071	0.467**	0.223	0.607**	1		
Co	-0.049	-0.101	-0.038	0.029	0.158	0.221	0.601**	0.148	0.455***	0.654**	1	
Hg	-0.005	-0.191	0.125	-0.012	0.051	0.258	0.060	0.526**	0.512**	0.632**	0.452**	1

$*p \leq 0.05$; $**p \leq 0.01$

themselves and with those of SPM and POC. The only exceptions were with HCH, because it was mainly present in the dissolved phase, and with DDT, because of its very low concentrations. Trace metals in general correlated poorly with SPM and POC, but 21 out of the 66 couples showed significant correlations ($p < 0.05$) indicating that their changes are governed by the same processes.

On the whole, particle loading is an important factor to be taken into consideration in evaluating the levels and distribution of persistent chemical pollutants in coastal marine areas where variable amounts of inorganic and organic particulate material are present in the water column.

CONCLUSIONS

This study shows that reliable measurements of organic micropollutants and trace metals in water and suspended particulate matter can be obtained using time-integrating water samplers. This technique reduces the need for repetitive water sampling and analyses to evaluate the average concentrations of chemical pollutants that might prove to be expensive.

Hydrocarbons, chlorinated hydrocarbons and trace metals were present partly in solution and partly associated with SPM; the proportions in each fraction varied with element and compound, and the concentration of SPM and its nature.

Temporal and spatial comparisons were generally difficult, given the paucity of data obtained by comparable methods. However, when such comparisons were possible, they showed that the pollutant levels in the study area were relatively low.

Future investigations must increase the areal coverage in the Venice Lagoon and include detailed sampling of the major streams and drains flowing into the Lagoon.

REFERENCES

Duinker, J. C. (1986). The role of small, low density particles on the partition of selected PCB congeners between water and suspended matter (North Sea Area). *Neth. J. Sea Res.*, **20**(2/3): 229–38

Elbaz-Poulichet, F., Guan, D. M. and Martin, J. M. (1991) Trace metal behaviour in a highly stratified Mediterranean estuary: the Krka (Yugoslavia). *Mar. Chem.*, **32**: 211–24

Fossato, V. U., Van Vleet, E. S. and Dolci, F. (1988). Chlorinated hydrocarbons in sediments of Venetian canals. *Archo. Oceanogr. Limnol.*, **21**: 151–61

Fossato, V. U., Craboledda, L. and Dolci, F. (1991) Polychlorinated biphenyls in water, suspended particulate matter and zooplankton from the open Adriatic Sea. In

Gabrielides, G. P. (ed.), *Proceedings of the FAO/UNEP/IAEA/Consultation Meeting on the Accumulation and Transformation of Chemical Contaminants by Biotic and Abiotic Processes in the Marine Environment (LaSpezia, Italy 24–28 September 1990).* MAP Technical Reports Series No. 59 UNEP, Athens, 199–207

Green, D. R., Stull, J. K. and Heesen, T. C. (1986) Determination of chlorinated hydrocarbons in coastal waters using a moored *in situ* sampler and transplanted live mussels. *Mar. Pollut. Bull.,* **17**: 324–9

Martin, J. M., Huang, W. W. and Yoon, Y. Y. (1994) Level and fate of trace metals in the Lagoon of Venice (Italy). *Mar. Chem.,* **46**: 371–86

Murphy, P. G. (1972) Sulphuric acid for the cleanup of animal tissues for analysis of acid-stable chlorinated hydrocarbon residues. *J. Assoc. Off. Agricult. Chem.,* **55**: 1360–2

Palmork, K. H. and Villenueve, J. P. (1980) Outline of the method to be used for the determination of chlorinated hydrocarbons in sea water. In *Workshop on the intercalibration of sampling procedures of the IOC/WMO/UNEP pilot project on monitoring background levels of selected pollutants in open-ocean waters (Bermuda, 11–26 January 1980).* IOC Workshop Report No 25: 25–55

Phillips, D. J. H. (1980) *Quantitative Biological Indicators: their Use to Monitor Trace Metal and Organochlorine Pollution.* Applied Science Publishers, London, pp. 488

Scarponi, G., Capodaglio, G. and Cescon, P. (1989) Determinazione di Zn, Pb e Cu nell'acqua sub-superficiale del Mare Adriatico mediante voltammetria di ridissoluzione anodica. *Bull. Oceanol. Teor. Appl.,* (Special Issue), 233–40

Sloot, H. A. and Van Der Duinker, J. C. (1981) Isolation of different suspended matter fractions and their trace metal contents. *Environ. Technol. Letters,* **2**: 511–20

Strickland, J. D. H. and Parsons, T. R. (1972) *A Practical Handbook of Seawater Analysis,* Fisheries Research Board of Canada, Bulletin, 167–311

UNEP/FAO/WHO/IAEA (1990) *Assessment of the state of pollution of the Mediterranean Sea by organohalogen compounds.* MAP Technical Reports Series No. 39, UNEP, Athens, pp. 224

Villeneuve, J. P., Elder, D. L. and Fukai, R. (1980) Distribution of polychlorinated biphenyls in seawater and sediments from the open Mediterranean Sea. *V^es Journées Etud. Pollutions, CIESM, Cagliari,* 251–6

CHAPTER 7

BIOLOGICAL VARIABILITY IN THE VENICE LAGOON

F. Bianchi, F. Acri, L. Alberighi, M. Bastianini, A. Boldrin,
B. Cavalloni, F. Cioce, A. Comaschi, S. Rabitti, G. Socal
and M.M. Turchetto

SUMMARY

Short-term variations of temperature, salinity, pH, dissolved oxygen, nutrients, particulate material, phytoplankton and zooplankton were studied in the Palude della Rosa area of the northern part of the Venice Lagoon. Variations on hourly time scales were driven mainly by tidal dynamics in late autumn and by biological processes in spring and summer. Environmental conditions in the area were influenced by both macro- and microalgal blooms, respectively in April and July. The massive proliferation of Ulvales in spring caused a depletion in nitrate as well as interference with phytoplankton growth. In summer, a bloom of the diatoms *Cylindrotheca closterium* and *Cyclotella* sp. caused an increase in particulate organic material and a decrease in dissolved nutrients (except ammonium). Comparisons were made with previous data sets collected by IBM/CNR in adjacent areas since 1975.

The results showed, between 1975 and 1992, increases in average winter nitrate concentration and summer phytoplankton cell counts, chlorophyll, POC and PTN, a decrease in phosphate concentrations in the northern basin and a shift in zooplankton species composition. Comparisons of tidal and annual changes showed that the variations of total suspended matter and zooplankton are of the same order of magnitude.

INTRODUCTION

In brackish water ecosystems, such as estuaries, conservative and non-conservative properties show wide spatiotemporal variations (Postma, 1969; UNESCO, 1981; Morris *et al.*, 1982) and processes in this type of environment are regulated mainly by tidal action which influences the distribution of temperature, salinity, particulate matter and nutrients, as well as light penetration. The spatial variability becomes more pronounced when freshwater inputs are significant, since there is now a salinity gradient from the inner areas to the open sea along which dissolved and suspended matter diminish seawards (Lapointe and Clarke, 1992; Powell *et al.*, 1989; Cole and Cloern, 1987). As coastal lagoons generally behave like estuaries tides can be expected to be the major factors controlling the distribution of properties in lagoons.

Tidal dynamics and spatial heterogeneity may also be associated with other physical processes, such as wind-induced waves (Demers *et al.*, 1987). This synergy produces marked changes in biological associations, such as those of planktonic organisms, on time scales of tidal, diurnal or seasonal cycles. Hence, if the processes occurring in a lagoon ecosystem are to be correctly described, and the variations in space and time are to be quantified, then field studies must include a sampling frequency designed to match the periodicity of the event under investigation (Lara-Lara *et al.*, 1980; Morales Zamorano *et al.*, 1991; Cloern *et al.*, 1989).

The present study had three objectives. The first was to obtain a better understanding of the way in which tidal dynamics control short-term distribution of dissolved and particulate matter and plankton communities in the Venice Lagoon. The second was a comparison of these results with data collected by the Istituto di Biologia del Mare in the northern basin (Palude di Cona and neighbouring channels and rivers) of the Lagoon since 1975. The last was to verify whether the short-term variations can be studied by comparing hourly variations with the annual variations, described sometimes to be of the same order of magnitude (Cadee, 1982).

MATERIAL AND METHODS

The Venice Lagoon is a brackish-water environment that has been considerably altered by human activities over the last few centuries. At present, water exchange between the sea and the Lagoon takes place through three openings and three sub-basins with areas where current speeds are close to zero can be identified (Avanzi *et al.*, 1979). After diversion of the main rivers, fluvial inputs since the seventeenth century have only been from small rivers and a few channels, all characterised by low discharge rates (Cavazzoni, 1973).

As outlined in Carrer *et al.* (Chapter 16) some freshwater inputs (River Dese and Silone Channel, with a maximum discharge of 90 and 50 m^3/s, respectively – Cavazzoni, 1973) still exist in the northern part of the Lagoon. This area is also subject to considerable runoff from agricultural wastes (Zingales *et al.*, 1980). In this northern sub-basin, characterised morphologically by shallow waters (marshes or 'paludi', mean depth 0.5 m) and channels as deep as 10 m, we studied the hydrochemical and biological variations in a well-defined area, the Palude della Rosa (area = 3.5 km^2, water volume approximately 2×10^6 m^3 at zero datum). Here, most of the water (about 60% of the volume exchanged) flows through the Torcello Channel while limited amounts of freshwater (about 10%) enter through the northern opening. The remaining fraction of the water volume moves through the small lateral channels. Therefore, assuming the Torcello Channel to be the main exchange point, mass balance of major parameters may be approximately estimated.

From the biological point of view, this shallow area has been characterised by increasing proliferation of macroalgae (mainly Ulvales) over the last ten years, that occurs in spring and covers the whole bottom, with a biomass up to 10 kg of wet weight/m² (Sfriso, 1994). The bloom is of short duration, culminates in hypertrophic conditions in early summer and virtually disappears shortly thereafter (Sfriso *et al.*, 1992; Solazzi *et al.*, 1991). An increase in phytoplankton biomass generally occurs after the macroalgal bloom. The high biomasses of macro- and microalgae result in large nycthemeral variation in dissolved oxygen concentrations.

The study was carried out in three seasons: late autumn (November 26–December 1, 1991), spring (April 26–May 1, 1992) and summer (July 26–31, 1992). Discrete samples, at a depth of 0.5 m, were obtained simultaneously from the centre of the Palude della Rosa (station 1) and the entrance (Torcello Channel, station 2) over a period of 72 hours (Figure 7.1). To follow the variations of the hydrological features in more detail, continuous measurements by means of a multiparametric profiler (CTD Idronaut OCEAN 7) were performed from the R/V *U. D'Ancona*, moored in the Torcello channel. The recorded measurements were temperature, salinity, dissolved oxygen, pH, *in-situ* fluorescence and transmittance, with an acquisition frequency of 1 min and for a four-day period.

Current speed and direction were recorded with a Savonious current meter Marine Adviser, moored at 1 m depth in the Torcello channel. Meteorological data were recorded with a self-recording data logger, located at the IBM (9 km from the study site).

Transparency was measured with a Secchi disk, temperature with reversing thermometers, and salinity with a Guildline laboratory salinometer. Dissolved oxygen concentrations were analysed by the Winkler titration method (Strickland and Parsons, 1972), and the saturation percentage computed following Weiss (1970); pH was measured utilizing a Beckman Research pH meter. Samples for dissolved nutrients, total suspended matter (TSM), particulate organic carbon (POC), and particulate total nitrogen (PTN) were filtered onto Whatman GF/C filters (porosity 1 μm) and stored at $-30°C$. Nutrient analyses were performed with a System-Alliance autoanalyzer (Hansen and Grasshof, 1983; Strickland and Parsons, 1972). TSM was determined gravimetrically after Strickland and Parsons (1972); the organic fraction was estimated after incineration at 480°C. POC and PTN were analysed by means of a Perkin-Elmer CHN (Hedges and Stern, 1984). Particle size spectra were determined immediately after sampling, with a multichannel Coulter counter particle analyzer: the observed particle size ranged from 2 to 100 μm (Sheldon *et al.*, 1972). Chlorophyll *a* and phaeopigment fluorometric analyses were performed following Holm-Hansen *et al.* (1965) with a Perkin-Elmer luminescence spectrometer; standardization was made against pure chlorophyll *a* (Sigma Chemical Co.).

Photosynthetic rate measurements were carried out using the radioactive carbon uptake method (Steemann-Nielsen, 1952; Saggiomo *et al.*, 1990): samples were transferred into dark and light bottles (125 ml), inoculated with

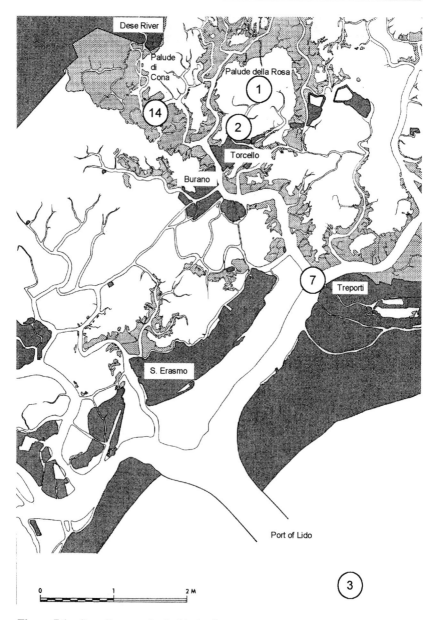

Figure 7.1 Sampling area in the Venice Lagoon

1 ml of $NaH^{14}CO_3$ (4 μCi activity) and placed in incubators with continuous flowing water, for 4 h; then filtered on Whatman GF/C and the sample activity detected with a Packard liquid scintillator. Potential primary production was estimated from the empirical function of Cole and Cloern (1987), based on

chlorophyll *a*, TSM and surface irradiance measurements. When not available, radiometric irradiance was converted to PAR (400–700 nm) applying a correction factor of 0.47 (Vollenweider, 1974).

Samples for phytoplankton cell counts and species composition were fixed in formalin; counting was carried out on subsamples obtained after sedimentation in 5 to 25 ml chambers, according to Uthermöhl (1958), using a Zeiss inverted microscope. Zooplankton were collected after 5 min horizontal trawlings with a Clarke-Bumpus sampler (200 *μ*m net), then fixed in formalin; subsamples, obtained by means of a Folsom Plankton Splitter (McEven *et al.*, 1954), were counted and classified under a stereomicroscope.

Some parameters were not sampled continuously due to meteorological and technical problems. Because of their non-normal distribution, the biological data (particle number, chlorophyll *a*, phaeopigments, phyto- and zooplankton abundance) were transformed to $\log_{10}$ (n + 1) (Sokal and Rohlf, 1969). The fluxes of dissolved nutrients and suspended matter inside and outside the Palude della Rosa through the Torcello Channel were estimated from the algebraic sums of ebb and flood values integrated over the whole sampling period (72 h).

RESULTS

The three seasons are characterised by different biological situations: low activity in late autumn, extensive bloom of Ulvales in spring, and phytoplankton bloom in summer. Table 7.1 presents the range and the average with standard deviation of each parameter for the whole data set (Palude della Rosa and Torcello Channel).

Continuous measurements

Tidal and meteorological conditions

The tidal range was from 70 cm at a neap tide to 90 cm at spring tide during all field surveys; strong winds due to a storm increased the tide level in April by about 20 cm. Current speeds showed a maximum of 50–60 cm/s in the Torcello Channel during both ebb and flood, with discharges of up to 100 m³/s.

In November 1991, air temperature ranged between 2 and 12 °C. High air pressure and calm weather caused a thick fog, especially at night. These are typical autumn weather conditions. Photosynthetically active radiation (PAR) measured in autumn was low and averaged 8570 *μ*E/m²/day. The spring survey was characterised by rapid meteorological variations ending in a storm with strong winds from the NNE lasting about 20 hours. The air temperature consequently decreased from 16 to 10 °C. The average PAR at this time was 22,387 *μ*E/m²/day. In July 1992, typical summer conditions – high pressure, air temperatures above 31°C, and a moderate SSE wind during the day – were observed. The average PAR was 47,246 *μ*E/m²/day.

Table 7.1 Number of observations, minimum, maximum, average and standard deviation of indicated parameters, calculated for the whole data set

Variable	Unit	November 1991					April 1992					July 1992				
		n	min	max	avg	std	n	min	max	avg	std	n	min	max	avg	std
Temperature	°C	62	7.3	9.9	8.7	0.6	68	13.3	23.8	17.9	2.3	71	25.1	30.4	28.2	1.4
Salinity	PSU	62	16.6	29.8	23.2	2.9	68	19.6	30.3	24.7	2.2	72	23.8	33.4	28.3	2.4
Dissolved oxygen	cm^3/dm^3	62	6.1	13.5	8.6	1.9	68	4.1	13.7	6.2	2.2	72	0.2	11.5	4.1	2.5
Relative oxygen	%	62	88.2	188.7	122.0	26.2	68	69.7	257.7	108.4	42.2	71	5.0	248.3	89.6	54.7
pH		62	8.05	8.78	8.31	0.17	68	8.05	9.24	8.59	0.26	71	7.51	8.58	8.06	0.20
N-NH$_3$	μM	62	0.5	13.5	6.6	2.5	68	1.3	9.6	4.3	1.8	72	0.5	22.7	6.8	3.9
N-NO$_2$	μM	62	1.0	4.0	2.1	0.6	68	0.1	0.7	0.3	0.2	72	0.1	3.3	0.4	0.4
N-NO$_3$	μM	62	39.1	145.8	79.0	23.6	68	0.3	9.4	3.4	2.6	72	0.0	9.3	3.0	2.2
DIN	μM	62	46.4	160.1	87.8	24.9	68	1.8	19.6	8.0	4.3	72	0.6	28.0	10.2	5.5
Si-Si(OH)$_4$	μM	62	20.3	63.4	36.3	8.8	68	19.6	63.3	38.5	9.0	72	0.4	26.2	9.9	6.1
P-PO$_4$	μM	62	0.05	1.10	0.35	0.24	68	0.05	1.33	0.32	0.21	72	0.03	0.34	0.09	0.05
N/P	mol	62	122.4	1505.4	345.9	225.2\	68	4.9	289.4	32.9	36.2	72	6.7	479.3	132.5	83.9
TSM	mg/dm^3	62	2.6	22.4	8.0	5.0	68	3.7	125.7	19.4	18.5	72	5.6	101.9	29.4	19.7
TSM Organic fraction	%	62	13.0	46.4	25.7	6.4	68	12.2	57.9	25.5	6.3	72	11.3	64.9	34.2	12.8
Particle number	n/cm^3	59	117532	177355	143769	13523	68	738933	3874806	1380774	582850	72	1315987	3100588	2300689	511307
POC	μg/dm^3	62	131.6	817.1	284.3	138.3	68	231.3	1746.8	615.6	353.7	70	469.7	4109.7	1794.7	941.7
PTN	μg/dm^3	62	18.4	106.9	36.8	17.9	68	33.6	258.8	97.4	49.5	70	67.8	755.1	305.4	168.8
C/N	mol	62	7.4	13.0	9.1	0.9	68	5.4	8.9	7.2	0.8	70	6.0	8.2	7.0	0.6
Chlorophyll *a*	μg/dm^3	62	0.1	1.8	0.5	0.3	66	0.7	5.2	1.9	1.0	72	3.7	31.4	16.6	7.8
Phaeopigments	μg/dm^3	62	1.2	3.7	2.0	0.6	66	1.3	17.9	5.4	3.6	72	4.3	25.8	13.1	5.5
Total phytoplankton	cells/cm^3	30	233	948	447	144	34	86	2040	687	536	36	1292	95794	33146	25874
Diatoms	cells/cm^3	30	128	414	213	56	34	43	1929	573	513	36	1020	92468	30824	24709
Dinoflagellates	cells/cm^3	30	0	11	1	3	34	0	11	1	3	36	0	443	49	97
Nanoflagellates	cells/cm^3	30	0	820	232	171	34	11	621	112	108	36	89	7207	2248	1909
Total zooplankton	ind/m^3	32	29	4646	992	1178	34	6	5480	892	1215	35	1522	88026	19072	19021
Copepods	ind/m^3	32	29	4450	954	1136	34	4	4089	627	910	35	1439	87665	18615	19059

Multiparametric probe measurements

A comparison of the continuous measurements at the two sampling sites was made by using two Idronaut CTD probes in November. The first of these was moored in the centre of Palude della Rosa and transmitted data via radio to the R/V *U. D'Ancona*. The second was connected to a microcomputer on the ship moored in the Torcello Channel. As no differences between these two sets of data were detected, only one probe was used in the following season (Torcello Channel). The data from the continuous measurements correlated well ($p < 0.01$; Figure 7.2) with those from discrete samples obtained with Niskin samplers.

Discrete samplings

There were no statistically significant differences between the data from the two stations, confirming that the Torcello Channel is representative of the major hydrochemical and biological features of the Palude. Hence, the results of discrete samplings from both Palude and Channel are reported and discussed together in the following sections.

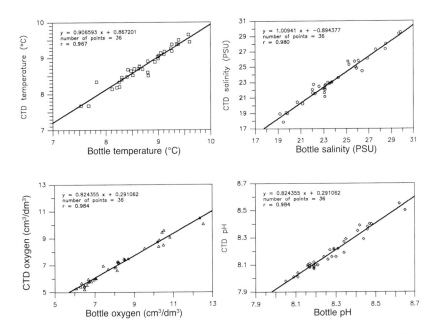

Figure 7.2 Continuous measurements from CTD probes vs. discrete sampling from Niskin samplers: common data related to temperature, salinity, dissolved oxygen and pH (November 1991, station 2)

Hydrology

Water temperature followed seasonal trends, with a range of 7–10 °C in late autumn, 13–24 °C in spring, and 25–30 °C in summer. Salinity varied from 16.6 to 33.4 PSU, mainly as a function of the state of the tides. Salinity values in November were low due to increased advection of freshwater from the rivers (rainfall from 1–27 November was 108.6 mm; Ufficio Idrografico e Mareografico). pH values showed a narrow range in November, attained a maximum of 9.24 in late April, and decreased to a low value of 7.51 in July. Correlations between pH and dissolved oxygen were highly significant for the whole data set ($r = 0.91$ in November, 0.76 in April, and 0.82 in July; $p < 0.01$).

Dissolved oxygen concentrations had a narrow range of variations in November (Table 7.1) when they were related more to physical than biological processes, as shown by the negative correlations with tide and salinity ($r = -0.81$, $p < 0.01$ and $r = -0.28$, $p < 0.05$, respectively). In April and July, they showed large variations and an obvious nycthermeral cycle, both due to the high biological activity (Figure 7.3). The significant negative correlation with tide found in April ($r = -0.41$, $p < 0.01$) was due to advection of oxygen-rich waters from the centre of the Palude through the Torcello Channel during ebb tide. Peaks of oxygen saturation occurred on 28 April, coincident with high PAR. In July, oxygen maxima resulting from primary production were detected in the afternoon (18:00) whereas hypoxic conditions, as a consequence of respiration and high temperature, prevailed in the early morning (06:00) hours.

Dissolved nutrients

Average dissolved inorganic nitrogen (DIN, as sum of ammonium, nitrite and nitrate) concentrations were 87.8, 8 and 10 μM respectively in November, April and July. The late autumn maximum of 160 μM measured in November was due to increased freshwater influx, as previously observed in an adjacent area (Bianchi *et al.*, 1987a and b). Low biological uptake in November led to a conservative distribution of DIN, as evident from its significant relationship ($r = -0.85$, $p < 0.01$; Figure 7.4) with salinity. In spring, the waters flowing out of the Palude were depleted of DIN as a result of uptake by macroalgae and the correlation between DIN and salinity at this time was positive ($r = 0.72$; $p < 0.01$). No correlation between DIN and salinity was found in July. Among the three DIN compounds, nitrate showed the greatest depletion; from 90% of DIN concentrations in November to 31% in spring and 29% in summer (Figure 7.5).

Phosphate concentrations were generally low (Table 7.1), rarely exceeded 1 μM, and had a conservative relation with salinity in late autumn ($r = -0.67$, $p < 0.01$) and summer ($r = -0.60$, $p < 0.01$), but not in spring. Changes in its concentrations did not bear any relation to tides, probably because of their

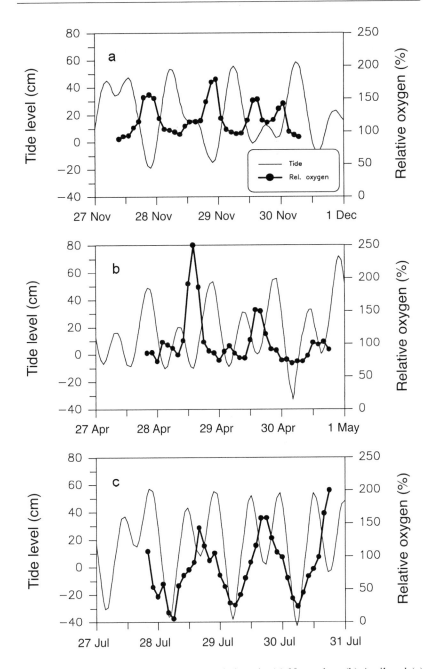

Figure 7.3 Tide and relative oxygen variations in (a) November, (b) April and (c) July; discrete samples for 72 hours at station 2

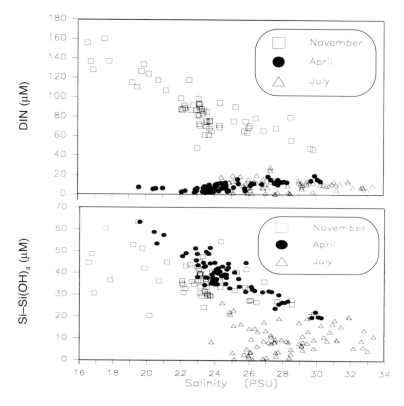

Figure 7.4 Scatter plots of dissolved inorganic nitrogen (DIN) and silicates vs. salinity in 1991–92 (all data)

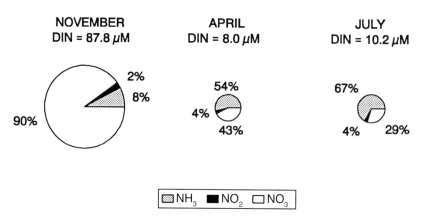

Figure 7.5 Proportional pie-charts showing average concentrations of ammonium, nitrite and nitrate in 1991–92 (all data)

interaction with sediments (Degobbis *et al.*, 1986). The N/P ratios varied widely, with extremely high values in November (Table 7.1). Correlations between silicate and salinity were significant in November ($r = -0.56$, $p < 0.01$) and April ($r = -0.89$, $p < 0.01$) showing that rivers are their main source. Silicate concentrations in summer were lower than in spring and late autumn due to uptake by the diatom bloom (Figure 7.4).

The budgets (Table 7.2) show an efflux of nutrient through the Torcello Channel, except in the case of DIN in spring when it (mainly as nitrate) was trapped in the Palude by the *Ulva* bloom.

Seston

In November, total suspended matter (TSM) values were low (av. 8 mg/dm^3), showed low variations (Figure 7.6a) and correlated with the current speed of the Torcello Channel. In April, the highest values were recorded after the storm (maximum = 126 mg/dm^3), coincident with a maximum wind speed (Figure 7.6b). The ratio of inorganic to organic matter of the SPM did not change after the storm, showing that benthic organisms were also brought into resuspension, along with the inorganics, by water turbulence. TSM flux through the channel increased from 1 to 7 kg/s an hour after the storm. In July, TSM changes showed peaks coinciding with maxima in current speeds ($r = 0.42$, $p < 0.01$) and an increase, up to 34%, in the organic fraction content. SPM budgets showed a net transfer from the Palude through the Torcello Channel over the whole sampling period (19 kg in November, 59 kg in April and 49 kg in July – Table 7.2).

Particle size spectra did not reveal the prevalence of any particular size class in November, whereas in spring and summer a mode around 7–10 μm, indicating the presence of both detrital particles and phytoplankton cells, was observed. Average particle number and volume doubled from spring (1.4×10^6/cm^3 and 3.5 mm^3/dm^3) to summer (2.3×10^6/cm^3 and 6.5 mm^3/dm^3).

The particulate organic carbon (POC) content was less than 300 μg/dm^3 in late autumn, higher than 600 μg/dm^3 in spring, and was about 1800 μg/dm^3 in

Table 7.2 Dissolved nutrients and total suspended matter budget in Palude della Rosa over whole sampling period (72 h) (+ = input; – = discharge)

Variable	Unit	November	April	July
DIN	mmol	−7898	+8275	−3227
P-PO$_4$	mmol	−85	−241	−57
Si-Si(OH)$_4$	mmol	−7202	−36051	−4870
TSM	Kg	−19	−59	−49

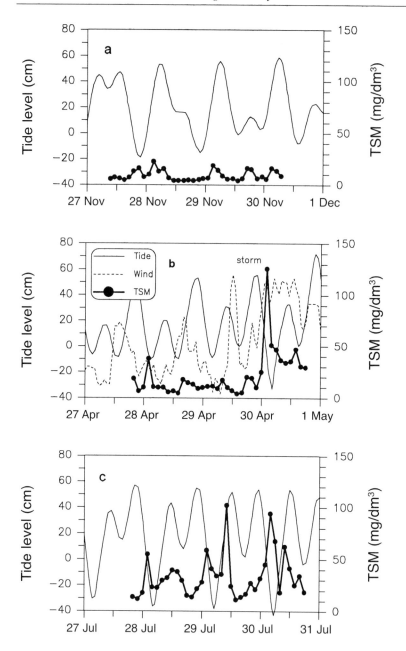

Figure 7.6 Variations in tide and total suspended matter (TSM) in (a) November, (b) April and (c) July at station 2

summer. Corresponding particulate total nitrogen (PTN) concentrations were 37, 97 and 305 $\mu g/dm^3$. Molar C/N ratios decreased from November (av. 9.1) to April (av. 7.2) and July (av. 7.0) indicating the importance of inorganic detritus in particulate matter in winter while the biological component became more evident in spring and summer.

Chlorophyll a, phytoplankton and primary production

Chl *a* concentrations increased by about 30 times from late autumn to summer (averages from 0.5 to 16.6 $\mu g/dm^3$ – Table 7.1). Chl *a*/phaeopigment ratios were 0.3 in November, 0.4 in April and 1.3 in July. Chl *a* concentrations in summer correlated inversely with salinity and tide height ($r = -0.76$ and -0.64; $p < 0.01$) showing that the species responsible for the phytoplankton bloom found favourable growth conditions in the brackish environment (Figure 7.7). Average POC/chl *a* ratios decreased progressively in the 3 surveys from November to July (665, 362 and 114, respectively). The high ratio in late autumn was due to the predominance of a detrital component in POC, while in summer the phytoplankton carbon fraction, calculated from cell plasma volume (Edler, 1979), was more pronounced.

Phytoplankton cell counts were $< 10^6$ cells/dm^3 in late autumn and spring and $> 10^7$ cells/dm^3 in summer, with a generally inverse trend with tide height (Figure 7.8). The April storm caused an increase in both chl *a* (2.5 times) and phytoplankton counts (3 times), because of resuspension of benthic diatoms in the water column.

Nitzschia, Navicula, Amphora, Pleurosigma and *Gyrosigma* were the representative phytoplankton species in the three seasons. In November, nanoflagellates accounted for up to 50% of the total population and *Cryptomonas* sp., among them, attained a concentration of 0.4×10^6 cells/dm^3. In April, due to seawater intrusion, some centric diatoms, such as *Chaetoceros* spp. and *Skeletonema costatum*, appeared more frequently. In summer, a bloom of *Cylindrotheca closterium*, with cell densities up to 66×10^6 cells/dm^3 (52% of total), occurred. *Cyclotella* sp., a centric diatom typical of brackish waters, also had a noticeable abundance (up to 23×10^6 cells/dm^3). Both these species belong to the 7–10 μm size class.

Phytoplankton production rates ranged from 0.2 to 145.4 mg C/m^3/h with seasonal averages of 0.5, 5.8 and 83.9 mg C/m^3/h, respectively, in late autumn, spring and summer (Table 7.3).These rates are comparable with those measured more than thirty years ago (Vatova, 1960) and with the more recent observations in the central basin (Battaglia *et al.*, 1983; Degobbis *et al.*, 1986). The photosynthesis index, defined as carbon assimilation per unit chl *a* per hour (Marra and Heinemann, 1987), was about 1 mg C/mg Chl *a*/h in late autumn, confirming the low level of phytoplankton growth, but rose to 3.8 and 3.4 mg C/mg Chl *a*/h in spring and summer (Table 7.3).

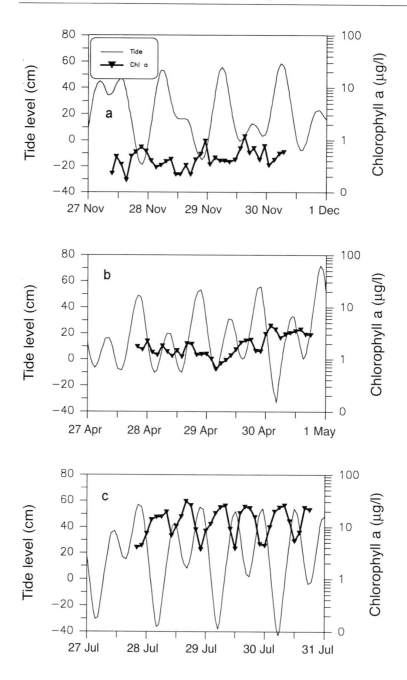

Figure 7.7   Tide and chlorophyll *a* variations at station 2

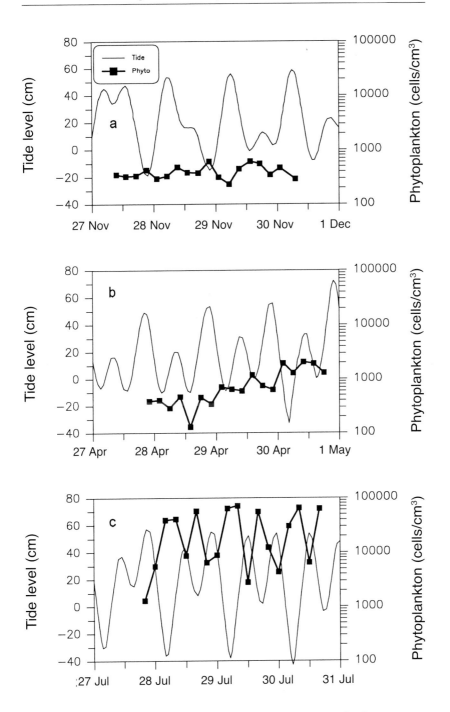

Figure 7.8 Tide and variations in phytoplankton abundance at station 2

Table 7.3 Total photosynthetically active radiation (PAR), chlorophyll a, primary production and assimilation number at station 2)

Date	TOT PAR μEin/m^{2}day	CHL a mg/m^3	PROD mgC/m^{3}h	P:B mgC/mgChl/h
27.11.91	11597	0.2	0.2	1.0
28.11.91	7733	0.4	0.7	1.7
29.11.91	6380	0.4	0.4	1.1
average	8570	0.3	0.5	1.3
28.04.92	43160	1.1	6.1	5.4
29.04.92	16345	1.2	4.9	4.0
30.04.92	7656	3.2	6.4	2.0
average	22387	1.9	5.8	3.8
28.07.92	46230	19.4	38.3	2.0
29.07.92	48303	25.1	145.4	5.5
30.07.92	47205	25.8	68.2	2.6
average	47246	23.4	83.9	3.4

Zooplankton

Zooplankton numbers ranged from a minimum of 6 (April) to a maximum of 88026 individuals/m^3 (July). Species composition was very monotonous in November. Copepods were the most abundant and, among them, *Acartia clausii* was dominant at flood tide and the epibenthic species, represented by harpacticoid copepods (mainly *Tisbe*), at ebb tide. In April, *A. clausii* and harpacticoid copepods were still abundant, along with amphipods and larvae of various phyla. Positive correlations between zooplankton and tide ($r = 0.52$; $p < 0.01$) found in this survey, as in earlier studies (Comaschi-Scaramuzza, 1987; Socal *et al.*, 1987; Comashic *et al.*, 1995), demonstrates the coastal origin of this population. In July, the correlation of total zooplankton with tide was inverse ($r = -0.51$; $p < 0.01$; Figure 7.9) and this was due to the very high abundance of *Acartia tonsa* (up to 87,318 individuals/m^3, 99% of the total), never previously observed in the lagoon (Cioce *et al.*, 1979; Lombardo and Comaschi Scaramuzza, 1977; Comaschi *et al.*, 1994). According to many authors (Paffenhofer and Stearns, 1988; Reeve and Walter, 1977; Roman, 1984), this species prefers inner areas characterised by an abundance of high phytoplankton and organic detritus (correlation coefficients of *A. tonsa* vs chl a and *A. tonsa* vs POC were 0.35 and 0.39, respectively; $p < 0.01$).

Comparison with previous data set

In order to provide a general picture of variability in the northern basin of the Lagoon in the following sections, the results collected in recent years by the Istituto di Biologia del Mare are reported and discussed. The data relate to two

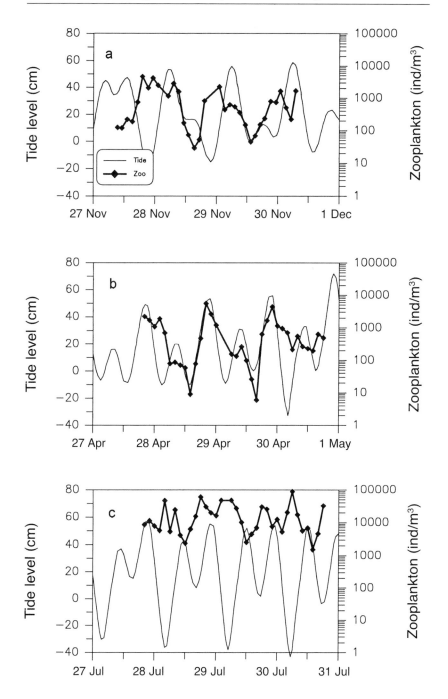

Figure 7.9 Tide and variations in zooplankton abundance at station 2

annual cycles in 1975–76 and 1977–78 at three stations along the salinity gradient from the Lido Lagoon inlet to the freshwater zone, and several short-term studies between 1982 and 1985 in the adjacent Palude di Cona (Figure 7.1).

Annual cycles

Samples in the annual cycles were collected at fortnightly intervals, at high and low tides, along the salinity gradient from the open sea to the inner Lagoon (stations 3, 7 and 14 – Figure 7.1; Comaschi Scaramuzza and Lombardo, 1977; Lombardo and Comaschi Scaramuzza, 1977; Cioce *et al.*, 1979; Rossi, 1979; Socal, 1981; Comaschi Scaramuzza, 1987). The results, based on good correlations of the whole set of environmental parameters with salinity, confirmed that tides regulate all of them.

Land runoff was the main source of DIN and silicate. Mean DIN values increased from the sea towards inner areas. Percentages of ammonium and nitrate in DIN were more or less similar but the former was slightly higher in the coastal station whereas the latter was highest at station 14. Concentrations of ammonium showed peaks in summer and those of nitrate, in winter (Figure 7.10a). Silicate concentrations showed peaks in summer and winter (Figure 7.10b) and correlated inversely with diatom abundance. The higher phosphate concentrations at station 14 (av. 1.5 μM in 1975–76 and 2.2 μM in 1977–78) owe their origin to freshwater inputs. Phosphate changes showed summer peaks (Figure 7.10c) as described by Nixon (1982) for many lagoon systems.

Seston distribution was regulated by freshwater runoff and resuspension of sediments in shallow waters: in fact, the average TSM content increased from station 3 to station 14. Seston concentrations showed positive correlations with nutrients and biological parameters. POC changes reflected those of seston and organic particulates, with higher values in the inner Lagoon and a seasonal distribution characterised by a summer maximum and a late autumn minimum (Figure 7.10d).

Phytoplankton cell counts did not show significant differences between the open sea and the inner Lagoon but the taxonomic composition did: prevalence of diatoms decreased from 54% in the open sea to 45% (at station 14) in the Lagoon whereas that of nanoflagellates increased from 40 to 48%. Seasonally, the maximum in phytoplankton abundance was in summer and the minimum, in winter (Figure 7.10e). Zooplankton numbers decreased along the salinity gradient to low values at the inner station, demonstrating the absence of typical lagoon associations. Peaks in their abundance were evident in early summer (Figure 7.10f).

Palude di Cona data set (1982–85)

The research carried out from 1982–85 focused on short-term variations (tidal and nycthemeral) in different seasons at the nearby Palude di Cona area

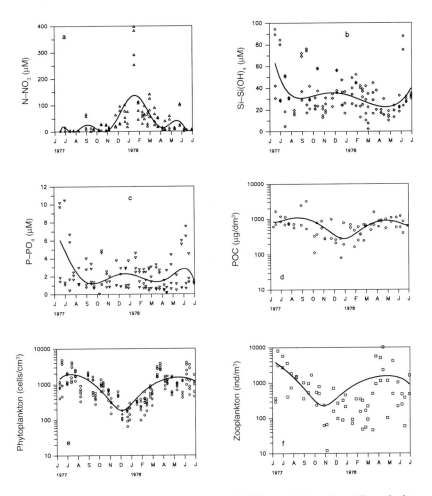

Figure 7.10   Seasonal changes in (a) nitrate, (b) silicate, (c) phosphate, (d) particulate organic carbon, (e) phytoplankton and (f) zooplankton abundance in 1977–78 at station 14. Best-fit curves are shown

(Figure 7.1). Five surveys, each with a sampling frequency of 2 h over a period of 12 or 24 h, were carried out in four separate areas (river, high-dynamic channel, low-dynamic channel, marsh) representing various lagoon environments (Barillari *et al.*, 1985; Bianchi *et al.*, 1987a and b; Boldrin *et al.*, 1987; Comaschi *et al.*, 1995) during September 1982, May and October, 1983, July 1984 and March 1985.

River discharge was generally high in autumn. Dissolved oxygen concentrations followed a marked nycthemeral cycle in summer and anoxic conditions prevailed in the low-dynamics channels. DIN compounds had a conservative distribution in autumn but their concentrations were reduced by autotrophic uptake in summer. Phosphate concentrations correlated inversely with salinity in both autumn and summer.

In July 1984 a massive *Cryptomonas* sp. bloom (> 10^8 cells/dm³, chl a > 500 μg/dm³, C/N ratio 5.8) was observed in the freshwater areas (Socal, unpublished). Consequently, DIN decreased markedly whereas silicate and phosphate did not. A bloom of the dinoflagellate *Peridinium foliaceum* was also observed (Socal, unpublished) in low-dynamic areas, with counts exceeding 10^7 cells/dm³ (chl a > 150 μg/dm³). No evidence of phytoplankton bloom was found in the surrounding marshes and channels in summer.

Zooplankton was dominated by copepods but their abundance was lower than in later years. *Acartia margalefii* had a higher abundance in summer and autumn. A bloom of the cyclopod *Metacyclops* sp., typical of brackish waters, was observed in 1984–85.

DISCUSSION

Present study (Palude della Rosa)

Our observation show that tide plays a significant role in hydrobiological variations but other factors, linked to seasonal variations, may superimpose on this trend.

(1) In late autumn, when the biomass of auto- and heterotrophs was low, tidal dynamics was the main controlling factor and the hydrological parameters showed significant relationships with salinity. Distribution of dissolved nutrients was strikingly conservative (Loder and Reichard, 1981). Inorganic matter was the main component of SPM.

(2) Depletion of dissolved nutrients, especially nitrate, in spring was due to the growth of macroalgae, while silicate concentrations did not vary much. The marsh areas act as traps for nitrogen compounds (Odum, 1959). Macroalgal growth seriously interferred with the development of phytoplankton (Sfriso *et al.*, 1989; Sfriso and Pavoni, 1994) which did not attain the biomass levels observed in other areas of the Lagoon (Voltolina, 1973; Alberighi *et al.*, 1992) in spring.

(3) Progressive decrease in the abundance of *Ulva* in summer was followed by a pennate diatom bloom. As a consequence, the organic fraction of particulate matter increased and the concentrations of all dissolved nutrients, except ammonium, decreased.

The trend of phosphate changes in the three surveys appears to be complex. In particular, its concentrations were much lower than expected in a lagoon environment in summer (Nixon, 1982) which led to very high N/P ratios. In spite of this, phosphorus did not appear to limit algal growth, probably because it is rapidly remineralised in sediments. As a consequence, the trophic dynamics of the system are mostly driven by nitrogen from land runoff, as also confirmed by the very high N/P ratio.

Seston concentration was linked not only to riverine runoff but also to resuspension processes that take place in shallow marshy areas, with a consequent transport along channels, as reported by several authors (Demers *et al.*, 1987; Carrick *et al.*, 1993). Sudden events, like the April storm, can mobilise large amounts of sediment: TSM removal from the Palude della Rosa through the Torcello Channel in a few hours was ten-times higher than at other times. This also resulted in an increase in the abundance of benthic diatoms, POC and chlorophyll *a*. C/N ratios were higher in late autumn when biological activity was low and the detrital fraction prevailed, but decreased through spring to a minimum in summer when phytoplankton blooms occurred (Roman, 1978, 1980; Redfield *et al.*, 1963).

The phytoplankton population was characterised by the presence of benthic diatoms, always abundant in shallow waters. Roman and Tenore (1978) identified, in resuspension, a mechanism to make algal food available for zooplankton swarms. Such an availability of food in summer probably supported the development of *Acartia tonsa*, typical of estuarine ecosystems characterised by high abundance of particulate organic matter (Roman, 1978) and small diatoms (Sanders and Kuenzler, 1979; Andreoli and Tolomio, 1988). *Acartia tonsa* occupies the habitat of the congeneric species *A. margalefii* that was quantitatively the most important species in the inner Lagoon in summer and autumn but has tended to disappear over the last few years, not only from the Venice Lagoon but also from other estuaries in Italy (Ferrari and Belmonte, pers. comm.).

Pluriannual trend

To highlight possible long-term trends, all the data from the period 1975–1992 were grouped by salinity and the data set in the salinity range 18–34 PSU was taken to represent Lagoon conditions. This was then divided to give two strongly different seasonal situations: winter (temperature $< 12\,°C$) and summer (temperature $> 23\,°C$). A comparison of the mean values of the parameters that best describe the trophic features led to the following observations:

(1)   Winter DIN concentrations increased, and summer DIN concentrations decreased over the years (Figure 7.11). This may be related to changes in both the quantity and quality of fertilisers used in agriculture (Zingales *et al.*, 1980). The summer decrease was due to an uptake by autotrophs, as evident from the growth of macroalgae (Sfriso *et al.*, 1992) and phytoplankton.

(2)   Both winter and summer phosphate concentrations in 1991–92 were significantly lower than in 1982–85 (Figure 7.11). This trend, confirmed

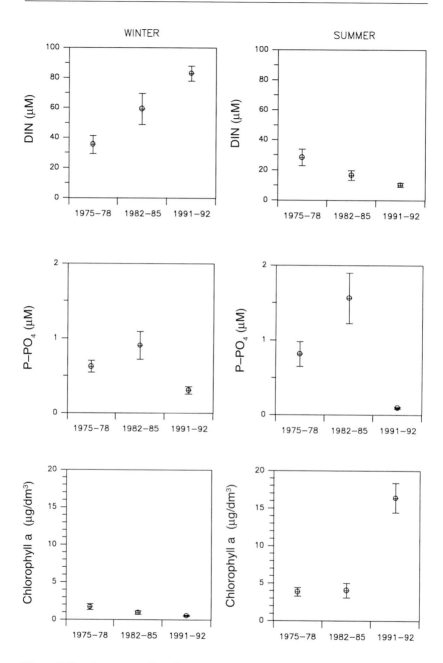

Figure 7.11 Averages and confidence limits of dissolved inorganic nitrogen (DIN), phosphate, chl *a* in winter and summer in 1975–78, 1982–85 and 1991–92

when data collected in the same area or in other basins of the Venice Lagoon were compared (Pavoni *et al.*, 1992; Sfriso *et al.*, 1992; Alberighi *et al.*, 1990), may be explained by the reduced amounts of these compounds in detergents from 1986–88.

(3) Phytoplankton cell counts, chl *a* (Figure 7.11), POC and PTN in summer increased between 1982–85 and 1991–92.

(4) Even within this relatively restricted area, various phytoplankton populations can develop in relation to the microhabitats: freshwater species in riverine areas, resuspended benthic diatoms in marshes and channels, and dinoflagellates in marginal, segregated areas.

(5) Zooplankton abundance was of the same order of magnitude from 1976 to 1992 but pronounced variations in species composition were observed in the inner areas. *Acartia margalefii* progressively decreased over the years while *A. tonsa* became dominant during the last sampling period. No significant changes were observed in the coastal zooplankton populations.

Comparison of tidal and annual variations

A comparison of the variability due to tidal dynamics with fluctuations on an annual time scale was done by calculating the coefficients of variation (CV, the standard deviation expressed as a percentage of the mean) for the data from this study and from the 1975–78 annual cycles (Table 7.4). The ratios identify the time scales (short or long) of the variations of a given parameter (Cadee, 1982). When the ratio is close to unity, then tidal and annual variations are of the same order: this was the case with TSM and zooplankton, since the former is linked to resuspension due to tidal currents or sudden events, and the latter because of its patchy distribution. Salinity, nutrients, POC, chl *a* and phytoplankton cell counts showed more pronounced variations on the annual scale, since they are associated to a seasonality, i.e. long-term changes in temperature, irradiance, rainfall, etc. The highest ratios were obtained for pH and dissolved oxygen, because of their wide fluctuations due to the nycthemeral cycle.

CONCLUSIONS

Dissolved nutrients in the Venice Lagoon derive mainly from land runoff and show a general inverse relationship with salinity, which is highly significant in late autumn but decreases in summer. In spring and summer nitrogen is removed from the dissolved phase and stored in macroalgae (spring) and in particulates (summer). Silicates show conservative behaviour (Franco, 1965; Bianchi *et al.*, 1990) that is modified only by diatom blooms. Phosphate concentrations are very low, suggesting a high exchange between bottom sediments, inorganic particulates and autotrophs (Degobbis *et al.*, 1986; Morris *et al.*, 1981; Rossi, 1979). Particulate matter derives from riverine

Table 7.4 Coefficient of variation (CV) for all stations

Variable	Tidal CV				Annual CV					RATIO
	1991–92				1975–76		1977–78			avg tidal/
	Palude della Rosa									avg annual
	Nov.	Apr.	Jul.	Avg	St. 7	St. 14	St. 7	St. 14	Avg	
Temperature	7	13	5	8.3	44	50	45	49	47.0	0.2
Salinity	12	9	8	9.7	8	25	11	33	19.3	0.5
pH	2	3	2	2.3	1	2	1	2	1.5	1.6
Relative oxygen	22	39	61	40.5	12	36	14	17	19.7	2.1
N-NH$_3$	37	41	57	45.0	119	108	139	81	111.8	0.4
N-NO$_2$	29	49	98	58.7	225	292	60	102	169.8	0.3
N-NO$_3$	30	77	73	60.0	166	220	150	158	173.5	0.3
DIN	28	53	54	45.0	83	116	102	111	103.0	0.4
Si-Si(OH)$_4$	24	23	62	36.3	80	58	62	57	64.3	0.6
P-PO$_4$	69	65	56	63.3	100	113	85	97	98.8	0.6
TSM	63	95	67	75.0			55	77	66.0	1.1
POC	49	57	52	52.7			53	71	62.0	0.8
PTN	49	51	55	51.7			71	79	75.0	0.7
Chlorophyll *a*	28	29	19	25.3			39	40	39.7	0.6
Total phytoplankton	2	12	11	8.5			15	14	14.6	0.6
Total zooplankton	24	27	10	20.7	15	20	18	24	19.0	1.1

inputs, resuspension of bottom sediments, decomposition of macroalgae and phytoplankton production. Resuspended and tidally-advected benthic diatoms play an important role in primary production.

Tidal and annual variations are frequently of the same order of magnitude. Suspended matter and zooplankton show even wider fluctuations on short time scales.

The trophic status of the area has changed in the last 15 years, mainly due to increased availability of nitrogen compounds which sustain blooms of both nitrophilous macroalgae and phytoplankton.

It is generally assumed that phytoplankton growth in the sea is strongly influenced by zooplankton grazing. It is difficult to demonstrate this in the Venice Lagoon since phytoplankton is often in excess and the zooplankton biomass is patchy and often scarce (Comaschi *et al.*, 1995).

Our results show that the Palude della Rosa is a source of organic matter which is transported to other parts of the Lagoon. Any reduction in water exchange hindering efflux may cause an accumulation of organic matter and can lead to dystrophic conditions.

ACKNOWLEDGEMENTS

We would like to thank the crew of R/V *U. D'Ancona* for their help; Idronaut (Italy) for CTD supply and assistance; A. Cesca, A. Locatelli, M. Marin and G. Penzo for logistic and sampling assistance; S. Tortato for drawings and Mrs. G. Walton for revision of the English text. We also thank the Magistrato alle Acque of Venice (Ministero dei Lavori Pubblici) for tide and rainfall data.

REFERENCES

Alberighi, L., Bianchi, F., Cioce, F. and Socal, G. (1992). Osservazioni durante un bloom di *Skeletonema costatum* in prossimità della centrale termoelettrica ENEL di Fusina Porta-Marghera (Venezia). *Oebalia*, suppl. **17**: 321–2

Andreoli, C. and Tolomio, C. (1988). Ciclo annuale del fitoplancton in una valle da pesca della laguna di Venezia (Valle Dogà). *Archo Oceanogr. Limnol.*, **21**: 95–115

Avanzi, C., Fossato, V., Gatto, P., Rabagliati, R., Rosa Salva, P. and Zitelli, A. (1979). Ripristino, conservazione ed uso dell'ecosistema lagunare veneziano. *Comune di Venezia*, pp. 197

Barillari, A., Bianchi, F., Boldrin, A., Cioce, F., Comaschi-Scaramuzza, A., Rabitti, S. and Socal, G. (1985). Variazione dei parametri idrologici, del particellato e della biomassa planctonica durante un ciclo tidale nella laguna di Venezia. *Atti VI Congr. AIOL*, 227–34

Battaglia, B., Datei, C., Dejak, C., Gambaretto, G., Guarise, G. B., Perin, G., Vianello, E. and Zingales, F. (1983). Indagini idrotermodinamiche e biologiche per la valutazione dei riflessi ambientali del funzionamento a piena potenza della Centrale (periodo dal 1979 al 1982). Relazione sintetica della Commissione tecnico-scientifica per la sperimentazione ed i controlli periodici sulla centrale termoelettrica dell'ENEL sita in località Fusina di Porto Marghera-Venezia. Regione Veneto, Venezia

Bianchi, F., Boldrin, A., Cioce, F., Rabitti, S. and Socal, G. (1987a). Variazioni stagionali dei nutrienti e del materiale particellato nella laguna di Venezia. Bacino Settentrionale. *1st.veneto Sci., Rapporti e studi XI*, 49–67

Bianchi, F., Boldrin, A., Cioce, F. and Socal, G. (1987b). Concentrazioni di nutrienti nella laguna di Venezia. Baciono Settentrionale. *Atti VII Congr. AIOL*, 155–64

Bianchi, F., Cioce, F., Comaschi-Scaramuzza, A. and Socal, G. (1990). Dissolved nutrient distribution in the central basin of the Venice lagoon. Autumn 1979. *Boll. Mus. civ. St. nat. Venezia*, **39**: 7–19

Bianchi, F., Socal, G., Alberighi, L. and Cioce, F. (1996). Cicli nictemerali dell'ossigeno disciolto nel bacino centrale della laguna di Venezia. *Biol. Mar. Medit.*, **3**(1): 628–30

Boldrin, A., Rabitti, S., Bianchi, F. and Cioce, F. (1987). Dinamica della sostanza sospesa nella laguna di Venezia. Bacino Settentrionale. *Atti VII Congr. AIOL*, 165–74

Cadee, G. C. (1982). Tidal and seasonal variations in particulate and dissolved organic carbon in the Ems-Dollart estury and the western Wadden Sea. *Netherlands Journal of Sea Research*, **15**: 228–49

Carrick, H. J., Alridge, F. J. and Schelske, C. L. (1993). Wind influences phytoplankton biomass and composition in a shallow productive lake. *Limnol. Oceanogr.*, **38**: 1179–92

Cavazzoni, S. (1973). Acque dolci nella laguna di Venezia. *Lab. Din. Gr. Masse, CNR, Tech. Rep. 64*, pp 40

Cioce, F., Comaschi-Scaramuzza, A., Lombardo, A. and Socal, G. (1979). Hydrological and biological data from the northern basin of the Venice Lagoon (1977–78). *Atti Ist. veneto Sci.*, **137**: 309–42

Cloern, J. E., Powell, T. M. and Huzzley, L. M. (1989). Spatial and temporal variability in South San Francisco Bay (USA). II. Temporal changes in salinity, suspended sediments, and phytoplankton biomass and productivity over tidal time scales. *Estuar. Coastal Shelf Sci.*, **28**: 599–613

Cole, B. E. and Cloern, J. E. (1987). An empirical model for estimating phytoplankton productivity in estuaries. *Mar. Ecol. Prog. Ser.*, **36**: 299–305

Comaschi-Scaramuzza, A. (1987). Studio di popolazioni di Copepodi planctonici nel bacino settentrionale della laguna di Venezia. Giugno 1977 – Giugno 1978. *Archo Limnol. Oceanogr.*, **21**: 1–17

Comaschi, A., Acri, F., Alberighi, L., Bastianini, M., Bianchi, F., Cavalloni, B. and Socal, G. (1994). Presenza di *Acartia tonsa* (Copepoda: Calanoida) nella laguna di Venezia. *Biol. Mar. Medit.*, **1**(1): 273–4

Comaschi, A., Bianchi, F. and Socal, G. (1995). Osservazioni sulla distribuzione e sul ciclo stagionale delle specie appartenenti al genere *Acartia* (copepoda: calanoida) presenti nella palude di Cona (bacino settentrionale della laguna di Venezia). *Ist. veneto Sci. Rappti e Studi*, **12**: 107–20

Comaschi-Scaramuzza, A. and Lombardo, A. (1977). Hydrological data from the northern part of the Venice lagoon, May 1975–July 1976. *Atti Ist. veneto Sci.*, **135**: 1–14

Degobbis, D., Gilmartin, M. and Orio, A. A. (1986). The relation of nutrient regeneration in the sediments of the Northern Adriatic to eutrophication, with special reference to the lagoon of Venice. *Science of the Total Environment*, **56**: 201-10

Demers, S., Therriault, T., Bourget, E. and Bah, A. (1987). Resuspension in the shallow sublittoral zone of a macrotidal estuarine environment: wind influence. *Limnol. Oceanogr.*, **32**: 327–39

Edler, L. (1979). Recommendations on methods for marine biological studies in the Baltic. Phytoplankton and chlorophyll. *BMP Publ.*, **5**: 1–38

Franco, P. (1965). Relazioni tra clorinità e concentrazioni del silicio in acque lagunari (laguna di Venezia). *Archo Oceanogr. Limnol.*, **14**: 139–50

Hansen, H. P., Grasshoff, K. (1983). Automated chemical analysis. In Grasshoff, K., Ehrhardt, M., Kremling, K. (eds) *Methods of seawater Analysis*. Verlag Chemie, Weinheim, 347–79

Hedges, J. I. and Stern, J. H. (1984). Carbon and nitrogen determination of carbonate-containing solids. *Limnol. Oceanogr.*, **29**: 657–63

Holm Hansen, O., Lorenzen, C. J., Holmes, R. W. and Strickland, J. D. H. (1965). Fluorometric determination of chlorophyll. *J. Cons. perm. int. Explor Mer,* **30**: 3–15

Lapointe, B. E. and Clark, M. W. (1992). Nutrient input from the watershed and coastal eutrophication in the Florida Keys. *Estuaries*, **15**: 465–76

Lara-Lara, J. R., Alvarez Borrego, S. and Small, L. F. (1980). Variability and tidal exchange of ecological properties in a coastal lagoon. *Estuar. Coastal Shelf Sci.*, **11**: 613–37

Loder, T. C. and Reichard, R. P. (1981). The dynamics of the conservative mixing in estuaries. *Estuaries*, **4**: 64–9

Lombardo, A. and Comaschi-Scaramuzza, A. (1977). Biological data from the northern Venice lagoon. May 1975–July 1976. *Atti Ist. veneto Sci.*, **135**: 133–48

Marra, J. and Heinemann, K. R. (1987). Primary production in the North Pacific Central Gyre: some new measurements based on ^{14}C. *Deep Sea Res.*, **34** (11): 1821–9

McEven, G. F., Johnson, M. W. and Folsom, T.R. (1954). A statistical analysis of the performance of the Plankton Sample Splitter, based upon test observations. *Arch. Met. Geophys. Bioklim.*, 7: 502–27

Morales Zamorano, L. A., Cajal-Medrano, R., Orellana-Cepeda, E. and Imenez-Perez, L. C. (1991). Effect of tidal dynamics on a planktonic community in a coastal lagoon of Baja California, Mexico. *Mar. Ecol. Progr. Ser.*, **78**: 229–39

Morris, A. W., Bale, A. J. and Howland, R. J. M. (1981). Nutrient distribution in an estuary: evidence of chemical precipitation of dissolved silicate and phosphate. *Estuar. Coast. Shelf Sci.*, **12**: 205–17

Morris, A. W., Bale, A. J. and Howland, R. J. M. (1982). Chemical variability in the Tamar Estuary, south-west England. *Estuar. Coast. Shelf Sci.*, **14**: 649–62

Nixon, S. W. (1982). Nutrient dynamics, primary production and fisheries yields of lagoons. *Oceanol. Acta, Proc. Int. Symp. on coastal lagoons, SCOR/IABO/ UNESCO, Bordeaux,* 357–71

Odum, E. P. (1959). *Fundamentals of Ecology* W.B. Saunders Company, Philadelphia and London, pp. 546

Paffenhofer, G. A. and Stearns, D. E. (1988). Why is *Acartia tonsa* (Copepoda: Calanoida) restricted to nearshore environments? *Mar. Ecol. Progr. Ser.*, **42**: 33–8

Pavoni, P., Marcomini, A., Sfriso, A., Donazzolo, R. and Orio, A. A. (1992). Changes in an Estuarine ecosystem. The Lagoon of Venice a case study. In Dunnette, D.A. and O'Brien, R.J. (eds). *The Science of Global Change*. American Chemical Society, Washington, DC, USA 287–305

Postma, H. (1969). Chemistry of coastal lagoons. In Castañares, A. A. and Phleger, F. B. (eds). *Lagunas Costeras, un Simposio. Mem. Simp. Lagunas Costeras., UNAM-UNESCO, Nov. 28–30, 1967, Mexico, D.f.,* 421–30

Powell, T. M., Cloern, J. E. and Huzzley, L. M. (1989). Spatial and temporal variability in South San Francisco Bay (USA). I. Horizontal distributions of salinity, suspended sediments, and phytoplankton biomass and productivity. *Estuar. Coast. Shelf Sci.*, **28**: 583–97

Redfield, A. C., Ketchum, B. H. and Richards, F. A. (1963). The influence of organisms on the composition of seawater. In Hill, M. N. (ed). *The Sea, Vol. 2*, John Wiley, London, 26–77

Reeve, R. R. and Walter, M. A. (1977). Observations on the existence of lower threshold and upper critical food concentrations for the copepod *Acartia tonsa* Dana. *J. Exp. Mar. Biol. Ecol.*, **29**: 211–21

Roman, M. R. (1978). Tidal resuspension in Buzzards Bay, Massachusetts. II. Seasonal changes in the size distribution of chlorophyll, particle concentration, carbon and nitrogen in resuspended particulate matter. *Estuar. Coast. Marine Sci.*, **6**: 47–53

Roman, M. R. (1980). Tidal resuspension in Buzzards Bay, Massachusetts. III. Seasonal cycles of nitrogen and carbon: nitrogen ratios in the seston and zooplankton. *Estuar. Coast. Marine Sci.*, **11**: 9–16

Roman, M. R. (1984). Utilization of detritus by the copepod *Acartia tonsa*. *Limnol. Oceanogr.*, **29**: 949–59

Roman, M. R. and Tenore, K.R. (1978). Tidal resuspension in Buzzards Bay, Massachusetts. I. Seasonal changes in the resuspension of organic carbon and chlorophyll *a*. *Estuar. Coast. Marine Sci.*, **6**: 37–46

Rossi, A. (1979). Distribuzione di clorofilla *a* e di carbonio e fosforo totali nel particellato sospeso in acque superficiali del bacino settentrionale della laguna veneta. *Tesi di Laurea, Università degli Studi di Padova, Dip. di Biologia, Padova*

Saggiomo, E., Magazzu, G., Modigh, M. and Decembrini, F. (1990). Produzione del fitoplancton. In Metodi nell'Ecologia del plancton marino. *Nova Thalassia*, **11**: 231–44

Sanders, J. G. and Kuenzler, E. J. (1979). Phytoplankton population dynamics and productivity in a sewage-enriched tidal creek in North Carolina. *Estuaries*, **2**: 87–96

Sfriso, A., Pavoni, B., Marcomini, A. and Orio, A. A. (1992). Macroalgae nutrient cycles and pollutants in the Lagoon of Venice. *Estuaries*, **15**: 517–28

Sfriso, A., Pavoni, B., Marcomini, A. and Orio, A. A. (1989). Macroalgae and phytoplankton standing crops in the central Venice lagoon: primary production and nutrient balance. *Science of the Total Environment*, **80**: 139–59

Sfriso, A. and Pavoni, P. (1994). Macroalgae and phytoplankton competition in the central Venice Lagoon. *Environmental Technology*, **15**: 1–14

Socal, G. (1981). Nota sulla distribuzione quantitativa del fitoplancton nel bacino settentrionale della laguna di Venezia. Giugno 1977 – Giugno 1978. *Ist. veneto Sci. Rappti. e Studi*, **8**: 105–19

Socal, G., Bianchi, F., Comaschi-Scaramuzza, A. and Cioce, F. (1987). Spatial distribution of plankton communities along a salinity gradient in the Venice Lagoon. *Archo. Oceanogr. Limnol.*, **21**: 19–43

Sokal, R. R. and Rohlf, F. J. (1969). *Biometry*. W.H. Freeman, San Francisco, USA, pp. 776

Solazzi, A., Orel, G. C., Chiozzotto, E., Scattolin, M., Curiel, D., Grim, F., Vio, E., Aleffi, F., Del Piero, D. and Vatta, P. (1991). Le alghe della laguna di Venezia. *Arsenale Editrice*, Venezia, pp. 119

Steeman-Nielsen, E. (1952). The use of radioactive (^{14}C) for measuring organic production in the sea. *J. Cons. perm. int. Explor. Mer.*, **18**: 117–40

Strickland, J. D. H. and Parsons, T. R. (1972). A practical handbook of seawater analysis. *Bull. Fish. Res. Bd. Canada*, **167**: pp. 311

UNESCO (1981). Coastal lagoon research, present and future. *UNESCO Technical Paper in Marine Sciences,* **32**: 51–79

Utermöhl, H. (1958). Zur Vervollkommung der quantitativen Phytoplankton-Methodik. *Mitt. int. Ver. Limnol.*, **9**: 1–38

Vatova, A. (1960). Primary production in the Northern Venice Lagoon. *J. Cons.*, **26**: 148–55

Volleweinder, R. A. (1974). *A manual on methods for measuring primary production in aquatic environments.* Blackwell Scientific Publications, Oxford

Voltolina, D. (1973). A phytoplankton bloom in the lagoon of Venice. *Archo Oceanogr. Limnol.* **18**: 19–37

Weiss, R. F. (1970). The solubility of nitrogen, oxygen and argon in water and seawater. *Deep Sea Res.*, **17**: 721–35

Zingales, F., Alessandrini, S., Bendoricchio, G., Marani, A., Pianetti, F., Rinaldo, A., Sartorio-Burotto, C. and Zanin, S. (1980). Inquinamento dovuto alle acque di un bacino agricolo sversante nella laguna di Venezia. *Inquinamento*, **12**: 25–31

CHAPTER 8

SEA SURFACE MICROLAYER CHEMISTRY

J.J. Cleary, L. Craboledda and G. Campesan

SUMMARY

Surface microlayer and subsurface water samples were collected at six stations in the Venice Lagoon and offshore in the Adriatic Sea between 10th and 21st May 1993 and analysed for concentrations of trace metals, organotins and polyaromatic hydrocarbons. The results indicate widespread pollution throughout the Lagoon system. The large variations in the concentrations of both metal and organic pollutants between duplicate samples masked any trends due to tidal differences or microlayer enhancement. The most polluted site was the CVE site (station 6) near the industrial complex at Marghera.

INTRODUCTION

The sea surface microlayer is an important area of concentration for both natural biogenic organic material and anthropogenic organic and metal pollutants. As a consequence, microlayer concentrations of PCBs, PAHs, toxic metals and organometals, such as TBT, are sometimes of several orders of magnitude greater than in subsurface waters (Hardy, 1982; Cleary, 1991; Cleary et al., 1993). The sea surface microlayer is also an important biological habitat and its typical inhabitants are microneuston such as bacteria, ciliates and algae which serve as food for copepods and larger organisms (Zaitsev, 1971). In addition, eggs and larvae of many fish and invertebrate species float to the surface due to their natural buoyancy and high lipid content and are therefore exposed to microlayer toxicants at the early stage in their life cycle. Chemical and biological water quality studies carried out on microlayer samples from the North Sea and British coastal waters using larval bioassay have shown that: (i) the microlayer is toxic to marine larval species, (ii) the microlayer contains enhanced concentrations of chemical toxicants, and (iii) that there is a strong correlation between larval mortality and chemical concentration (Hardy and Cleary, 1992; Cleary et al., 1993; McFadzen and Cleary, 1994).

Clearly, microlayer pollutants pose a potential threat to neustonic, pelagic and demersal organisms, particularly at the early life stages when they are most vulnerable. The reasons for the high incidence (5–26%) of embryo abnormality found in cod, whiting and plaice from the North Sea (Dethlefsen et al., 1985) and the even higher malformation rates (26–44%) in North Sea plaice, flounder and dab found by Cameron et al., (1992) have not been identified, but the authors suggest that the surface microlayer could be a contributory factor. It seems possible that the surface microlayer has a strong influence on survival and recruitment to fisheries in the North Sea.

The objectives of the present study were to provide a preliminary assessment of the spatial variability of chemical and biological water quality in the Venice Lagoon system (Figure 8.1), particularly with respect to the enhancement of pollutants in the surface microlayer. The extent of marine pollution in the surface microlayer compared with subsurface waters at selected mussel sites in the Venice Lagoon system was determined by analysing water samples for metals, organotins and polyaromatic hydrocarbons (PAHs). These contaminants arise from domestic and industrial sources as well as from marine traffic and atmospheric inputs, and are commonly found in coastal waters. Microlayer samples were also obtained and used for toxicity bioassays which are described elsewhere (see McFadzen Chapter 19).

MATERIALS AND METHODS

Sampling strategy

The original plan was to analyse both filtered and unfiltered water samples in order to assess the contribution of suspended matter to pollutant levels. But the filtering was so slow and the timetable so tight that the programme would not have been completed had we continued. Therefore, an alternative strategy was adopted and water samples were analysed without filtering, during ebb and flood conditions, to determine the effect of tidal state on water quality.

Microlayer sampling

Water samples were taken from established mussel sites in the Lagoon during ebb and flood tides, and from the platform (station 0) site 8 miles offshore in the Adriatic.

Duplicate water samples from surface microlayer and subsurface were collected in clean Pyrex bottles (1 l) for both chemistry and bioassay. Surface microlayer samples were collected using a Garrett screen (Garrett, 1965) consisting of a 60 cm^2 aluminium frame supporting a stainless-steel mesh (size 16) made from 0.5 mm wire. Samples were taken by lowering the frame vertically into the water, aligning it parallel to the surface and raising it through the surface layer, thereby removing discrete segments of the micro-layer as they become entrapped between the wires of the mesh. The screen was then tilted to drain water from the frame edges and, as the water film on the mesh began to break, it was tilted further and the sample collected in a glass bottle. The procedure was repeated, collecting 100 ml from each immersion of the screen, until a sample of 1 l was obtained. Metallic screens were used to sample for organics and organometals and screens made of nylon and perspex were used to sample for metals.

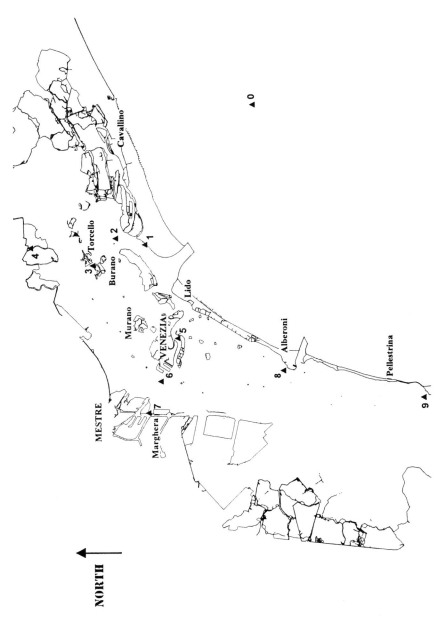

Figure 8.1 Map of the sampling stations in the Venice Lagoon

Subsurface samples were taken from 20 cm below the water surface by immersing a bottle by hand, removing the lid under water and allowing the bottle to fill. Sampling was carried out from a Boston whaler boat in the Lagoon and from a dinghy at the platform site. Water temperature was measured at the time of sampling and the salinity was measured at the CNR/IBM laboratory.

Sediment elutriates from S. Elena (station 12) and Carmini (station 22), prepared at the CNR/IBM laboratory, were analysed for metals, organotins and PAHs at the same time as the Lagoon samples (see McFadzen, Chapter 19 for more details).

Trace metals

Measured volumes (500 ml) of unfiltered water samples were pumped through micro Chelex-100 ion-exchange columns at 3–4 ml/min. Columns were washed with deionised water and trace metals eluted with 5 ml of 2 M nitric acid (Abdullah *et al.*, 1976). These extractions were carried out at the CNR/IBM laboratory using their standard technique for trace metal extraction from seawater. Analyses for copper, zinc, cobalt, manganese, nickel and cadmium were made at Plymouth Marine Laboratory using inductively-coupled plasma mass spectrometry (ICP-MS, VG Instruments Plasmaquad PQ 2) in the peak jumping mode.

Organotins and TBT

Unfiltered water samples (800 ml) were acidified with 4 ml conc. HCl and extracted with 50 ml hexane by shaking vigorously for 3 minutes. A second extraction was done with 20 ml of hexane. The extracts and the rinses were combined and reduced in volume to about 5 ml in a rotary evaporator and then transferred to storage vials pending analyses. The extractions were done at the CNR/IBM laboratory and the analyses, at the Plymouth Marine Laboratory.

This procedure results in a complete extraction of tributyltin (TBT) and a partial extraction of dibutyltin (DBT) but not of monobutyltin or inorganic tin. The hexane extract may be analysed directly or after a further concentration step to improve the detection limit. Treatment of the extracts with molar NaOH (3:1 v/v) removes the DBT from the hexane phase. Analyses were carried out by heated graphite furnace atomic absorption spectrometry using a graphite platform coated with tantalum pentoxide to increase the sensitivity (Cleary, 1991). Organotin and TBT concentrations are expressed in ng Sn/l.

Polyaromatic hydrocarbons

PAHs are readily removed from seawater by the hexane extraction described for organotins, giving complete extraction of naphthalene and chrysene. Therefore, the same extracts were analysed by fluorescence spectrometry for

2-ring and 4-ring PAH compounds, with naphthalene and chrysene as standards. Excitation and emission wavelengths for naphthalenes (Ex. 270 nm, Em. 330 nm) and chrysenes (Ex. 310 nm, Em. 360 nm) using 5 nm slit width were the same as those used to monitor light oil and crude oil fractions in the IGOSS Pilot Project on Marine Pollution (IOC, 1984; Law *et al.*, 1988). Analyses were carried out at the Plymouth Marine Laboratory.

Surface microlayer and subsurface samples from the CVE (station 6) and Salute (station 5) sites were analysed semi-quantitatively by HPLC at the CNR/IBM laboratory using their standard technique for PAH analysis. Although not part of the analytical programme, the opportunity to analyse individual PAHs ranging from 2-ring to 6-ring compounds provided additional information not available by fluorescence analysis.

RESULTS

Sampling conditions

Table 8.1 gives the salinity, temperature and tidal state at the time of the collections. The highest salinity and lowest temperature were measured at the platform site in the Adriatic.

Salinities during the ebb tide were lower than at the flood tide but salinity differences between surface microlayer and subsurface samples were small and did not show any systematic trend.

Trace metals

The trace metals Cu, Zn, Co, Mn and Ni were easily detected in both surface microlayer and subsurface waters but concentrations of Pb and Cd in most cases were less than blank values. Chelex column blank values were low and similar to acid blank values (Table 8 2).

The range in the mean concentrations (ppb) in surface microlayer and subsurface waters, respectively, were 2.8–0.7 and 1.6–0.4 for copper, 38.0–2.6 and 46.1–1.6 for zinc, 0.62–0.06 and 0.63–0.06 for cobalt, 7.6–1.1 and 7.3–1.2 for manganese and 1.6–0.4 and 1.5–0.5 for nickel. There was some indication of a microlayer enhancement of Cu and Zn, but the high concentration range found with some duplicate samples renders this interpretation difficult. At the CVE (station 6) site, Cu, Zn, Co and Ni occurred at consistently high concentrations in the microlayer and subsurface waters, during both flood and ebb conditions. Although some other sites had high concentrations of Zn, and to a lesser extent of Cu, they were in general less polluted than the CVE (station 6) site. Only at the Salute (station 5) site, where concentrations of Cu, Zn, Co, Mn and Ni were higher during the ebb tide, was there a consistent tidal pattern. No such pattern was seen at other sites. Results from a preliminary survey carried out between 2nd and 7th December 1991 in the Lagoon waters (V. U. Fossato, Chapter 6) also indicate little effect of tides on concentrations

Table 8.1 Salinity, temperature and tidal state at the sampling stations

Sample site	Date	Time	Tide	Salinity PSU		Temp. °C
				SMIC	SS	
0 Platform	11-5-93	1400–1500		36.3	36.3	17.3
2 Crevan	10-5-93	1500–1600	Flood	34.5	32.1	
	11-5-93	0900–1000	Ebb	28.1	29.3	
1 Lio Grande	10-5-93	1600–1700	Flood/Slack	35.2	34.8	
	11-5-93	1030–1130	Ebb/Slack	33.5	33.5	
6 CVE	12-5-93	1600–1700	Flood	30.3	30.6	21.5
	12-5-93	0900–0945	Ebb	27.4	27.6	21.5
5 Salute	12-5-93	1715–1800	Flood	33.5	33.3	19.5
	12-5-93	1000–1045	Ebb	31.6	31.5	20.5
8 Alberoni	13-5-93	1100–1200	Ebb/Slack	33.2	33.1	19.5
	17-5-93	1200–1300	Ebb		34.4	
9 Chioggia	17-5-93	1000–1100	Flood/Slack		33.6	

of metals. Samples from the platform (station 0) site in the Adriatic contained lower levels of metals than the Venice Lagoon samples.

Sediment elutriate samples indicate little release of Cu, Zn and Ni, but Co concentrations were markedly higher than in the purified water eluting agent. Very high release of Mn occurred from sediments at Carmini (station 22), giving an elutriate concentration of 123 ppb compared with a value of 6.48 ppb from the sediments at S. Elena (station 12).

Organotins and TBT

Organotins and TBT were found in both the surface microlayer and subsurface waters at all sites (Table 8.3). Higher concentrations occurred at the CVE (station 6) and Salute (station 5) sites and lower concentrations, at Crevan (station 2), Lio Grande (station 1) and platform (station 0) sites. Subsurface concentrations at Chioggia (station 9) were similar to those at CVE (station 6) and Salute (station 5) whereas those from Alberoni (station 8) agreed with the concentrations at Lio Grande (station 1) and Crevan (station 2). There was some indication of higher concentrations in the samples from CVE (station 6) and Salute (station 5) during ebb tide but the large variations between duplicate samples tended to mask such trends. Surface microlayer enhancement was most evident at Crevan (station 2) and Alberoni (station 8) but generally it was low and obscured by the large variations between replicates.

Maximum mean concentrations of organotins and TBT were 34.3 and 16.0 ng Sn/l, for the microlayer and 14.1 and 10.1 ng Sn/l for subsurface waters. These are similar to the values found in the North Sea in March 1990 (Hardy and Cleary, 1992) and the nearshore waters of southwest England in March 1993 (Cleary, unpublished data). TBT concentrations in most cases

Table 8.2 Trace metals in the surface microlayer (SMIC) and subsurface (SS) waters of the Venice Lagoon in May 1993

Sample site	Tide		*micrograms/litre*									
			Cu		*Zn*		*Co*		*Mn*		*Ni*	
			SMIC	SS	SMIC	SS	SMIC	SS	SMIC	SS	SMIC	SS
Platform	Flood	Mean	1.04	0.40	5.51	1.62	0.06	0.06	3.05	2.96	0.53	0.54
		Range	0.46	0.03	3.44	0.01	0.01	0.00	0.08	0.05	0.10	0.01
Crevan	Flood	Mean	2.78	0.83	16.49	11.61	0.28	0.22	7.61	7.32	1.59	0.69
		Range	1.19	0.04	3.26	0.47	0.01	0.01	0.54	0.55	0.40	0.02
	Ebb	Mean	1.06	0.46	2.64	1.64	0.23	0.24	6.03	4.10	0.83	0.80
		Range	0.51	0.01	1.15	0.22	0.04	0.01	0.46	0.44	0.13	0.04
Lio Grande	Flood	Mean	1.81	1.68	38.01	46.07	0.07	0.11	3.17	5.74	0.59	0.79
		Range	0.85	0.32	0.71	10.99	0.03	0.03	1.07	2.14	0.29	0.20
	Ebb	Mean	0.98	0.29	2.59	2.02	0.22	0.22	4.77	5.29	0.70	0.61
		Range	0.30	0.01	0.80	0.33	0.03	0.01	0.44	0.12	0.08	0.02
CVE	Flood	Mean	1.82	1.49	11.00	9.23	0.62	0.63	5.37	3.68	1.18	1.54
		Range	0.18	0.69	1.52	2.61	0.09	0.22	1.44	0.73	0.18	0.68
	Ebb	Mean	2.54	1.59	23.52	20.00	0.35	0.31	2.05	2.52	1.49	1.46
		Range	0.52	0.09	3.36	3.73	0.05	0.01	0.50	0.29	0.07	0.17
Salute	Flood	Mean	0.67	0.57	2.94	2.85	0.12	0.13	1.12	1.24	0.45	0.48
		Range	0.00	0.01	1.61	0.18	0.01	0.01	0.32	0.01	0.01	0.01
	Ebb	Mean	1.97	1.54	12.92	12.56	0.33	0.32	5.75	4.43	0.99	1.12
		Range	0.01	0.01	0.87	0.90	0.00	0.01	1.96	0.20	0.07	0.04
Alberoni	Ebb	Mean	1.06	0.69	6.35	5.77	0.23	0.34	2.13	3.61	1.06	0.97
		Range	0.01		0.27		0.01		0.21		0.06	

Continued on next page

Table 8.2 *Continued*

		Cu		Zn		*micrograms/litre* Co		Mn		Ni	
Sample site	Tide	SMIC	SS	SMIC	SS	SMIC	SS	SMIC	SS	SMIC	SS
Purified water	Mean		0.77		0.95		0.05		0.76		0.52
	Range		0.08		0.13		0.01		0.13		0.09
Sediment elutriate 12	Mean		0.48		1.38		0.43		6.48		1.03
	Range		0.02		0.13		0.02		0.30		0.03
Sediment elutriate 22	Mean		0.32		1.08		0.33		123.0		0.99
	Range		0.03		0.02		0.01		3.00		0.04
Method blank	Mean		0.03		0.10		0.00		0.05		0.03
	Range		0.01		0.05		0.00		0.01		0.01
Acid blank			0.04		0.04		0.00		0.05		0.03

Table 8.3 Organotins and polyaromatic hydrocarbons in the surface microlayer (SMIC) and subsurface waters (SS) of the Venice Lagoon in May 1993

Sample site	Tide		nanograms Sn/litre				micrograms/litre			
			Organotin		Tributyltin		Naphthalene		Chrysene	
			SMIC	SS	SMIC	SS	SMIC	SS	SMIC	SS
Platform		Mean	3.1	4.2	2.6	0.3	4.4	5.0	1.3	1.2
		Range	0.8	3.7	1.1	0.3	0.2	1.1	0.1	0.1
Crevan	Flood	Mean	7.0	0.6	2.7	0.4	10.7	19.4	3.4	5.0
		Range	0.6	0.6	0.7	0.4	2.3	3.9	0.8	0.6
	Ebb	Mean	6.2	1.0	2.1	0.1	10.0	39.4	3.3	11.1
		Range	1.7	0.9	0.8	0.1	2.6	34.5	0.4	9.5
Lio Grande	Flood	Mean	1.7	2.5	1.0	1.6	20.1	24.2	5.6	5.9
		Range	0.6	0.1	0.2	0.4	14.1	21.1	2.9	5.2
	Ebb	Mean	3.6	1.9	3.2	1.6	15.7	10.2	4.8	2.7
		Range	1.7	0.7	1.5	0.6	3.5	5.2	1.0	1.4
CVE	Flood	Mean	8.4	12.7	4.3	10.3	35.0	5.4	10.7	2.3
		Range	3.0	6.6	0.4	7.1	28.8	0.5	7.9	0.0
	Ebb	Mean	34.3	9.4	16.0	5.0	55.4	12.3	25.9	5.0
		Range	17.7	0.2	6.3	0.7	35.1	0.2	10.8	0.1
Salute	Flood	Mean	5.9	7.8	4.4	6.0	4.1	54.4	1.3	15.0
		Range	1.0	2.7	0.2	1.5	0.2	47.8	0.1	13.2
	Ebb	Mean	12.6	14.1	12.3	10.1	10.4	28.8	3.9	9.3
		Range	1.5	3.0	1.5	2.1	3.9	20.0	1.0	5.7
Alberoni (1)	Ebb	Mean	3.3	1.5	2.5	0.0	18.6	38.4	5.1	8.3
		Range	0.6	0.5	0.5	0.0	0.0	6.8	0.2	2.5
Alberoni (2)	Ebb	Mean		0.4		0.0		8.1		2.5
		Range		0.4		0.0		1.3		0.4

Continued on next page

Table 8.3 *Continued*

Sample site	Tide		*nanograms Sn/litre*				*micrograms/litre*			
			Organotin		*Tributyltin*		*Naphthalene*		*Chrysene*	
			SMIC	SS	SMIC	SS	SMIC	SS	SMIC	SS
Chioggia	Flood	Mean		8.6		7.9		34.6		6.4
		Range		0.4		0.1		3.0		0.6
Purified water		Mean		0.6		0.2		3.3		0.7
		Range		0.4		0.2		2.0		0.5
Sediment elutriate 12		Mean		14.5		10.4		4.3		1.4
		Range		0.7		0.2		0.3		0.1
Sediment elutriate 22		Mean		6.7		3.6		14.5		5.9
		Range		2.0		0.1		1.4		0.1
Blank		Mean		1.6		0.4		0.5		0.1
		Range		1.5		0.4		0.3		0.1

exceeded the UK EQS value of 2 ng TBT/l (equivalent to 0. 8 ng Sn/l) for marine waters.

Organotin and TBT concentrations (14.5 and 10.4 ng Sn/l, respectively) in sediment elutriates from S. Elena (station 12) were more than twice those from Carmini (station 22) (6.7 and 3.6 ng Sn/l respectively). Partition studies have shown that TBT has an affinity for sediment and that the sediment can act as a secondary source of TBT to the water column. In this case it is quite possible that mooring, cleaning and traffic of boats might have contributed a greater amount of TBT to the sediment at S. Elena (station 12) than at Carmini (station 22), even though other characteristics suggest that Carmini is more polluted. Besides, significant concentrations of organotins have been found in municipal wastewaters and sewage sludge in Zurich. Such sources may also contribute to organotin concentrations in elutriates.

Polyaromatic hydrocarbons

PAHs were detected at all sites and naphthalenes were present in greater amounts than chrysenes. Blank values were an order of magnitude lower than in samples from the platform (station 0) site where the lowest concentrations were found (Table 8.3). The overriding feature of the results from the Lagoon was the vast difference between duplicate samples which obscured any trends that might exist between sites, ebb and flood conditions, or microlayer and subsurface water. Maximum and minimum concentrations for naphthalenes were 102.3 and 3.1 ppb and for chrysenes, 36.8 and 0.7 ppb. Such a range is not unusual in harbours and fishing ports near Plymouth in southwest England where microlayer enhancement commonly occurs. In the shallow waters of the Venice Lagoon, high values were found in both the microlayer and sub-surface waters, perhaps caused by the greater vertical mixing and distribution of suspended particulates.

PAH concentrations in sediment elutriates were higher at Carmini (station 22) than at S. Elena (station 12). This agrees with the high Mn in elutriate 22 but differs from organotin and TBT values which are greater in elutriate 12, suggesting that the two sites are subjected to different inputs with different characteristics.

The UV fluorescence method used here to determine hydrocarbon concentrations in seawater is suitable as a first-line screening technique. It is not specific for oil since it also detects PAHs derived from fossil fuel combustion. Therefore, when the UVF shows high concentrations of hydrocarbons in the extracts, the latter should ideally be further investigated by more sophisticated techniques to identify the individual components. This was possible for only a few samples.

Surface microlayer and subsurface samples from the CVE (station 6) and Salute (station 5) sites analysed by HPLC contained PAHs ranging from 2-

Table 8.4 Polyaromatic hydrocarbons (PAHs) in surface microlayer (SMIC) and subsurface (SS) waters at the CVE (station 6) and Salute (station 5) sites in the Venice Lagoon

Compound	Salute (station 5)		CVE (station 6)	
	SMIC	SS	SMIC	SS
i) Determined by HPLC (ng/l)				
Naphthalene	522	114	86	9
Acenaphthalene	606	1277	4584	nd
Fluorene	1183	222	180	14
Acenaphthene				
Phenanthrene	441	21	148	nd
Anthracene	nd	nd	nd	nd
Fluoranthene	93	110	206	15
Pyrene	138	69	185	11
Benz(a)anthracene	112	186	275	5
Chrysene	221	94	144	9
Benzo(b)fluoranthene	502	227	255	23
Benzo(k)fluoranthene	254	94	107	15
Benzo(e)pyrene	1090	226	202	26
Dibenzo(a, h)anthracene	2081	575	294	34
Indeno(1, 2, 3)pyrene	3173	507	514	71
Benzo(ghi)perylene				
Total PAH	10416	3722	7180	232
ii) Determined by fluorescence spectrometry (ng/l)				
Naphthalenes	14260	8810	20320	12150
Chrysenes	4930	3680	15110	4960
Total PAH	19190	12490	35430	17110

ring to 6-ring compounds, most of which exhibited microlayer enhancement (Table 8.4).

Microlayer concentrations at Salute (station 5) were only slightly greater than at CVE (station 6), but subsurface waters at Salute were much more polluted than at CVE and contained 10 times more of total PAH (Table 8.4). The more highly condensed PAHs (4, 5 and 6 ring compounds) were predominant in both the microlayer and subsurface at both sites, but lower molecular weight PAHs (containing 2 and 3 ring compounds) also occurred in high amounts, particularly in the CVE (station 6) microlayer, although concentrations in subsurface waters there were low.

Total PAH values measured by HPLC were less than those measured by fluorescence analysis, but this is not unexpected since the latter technique may include a range of substituted PAH compounds that would not be detected by HPLC. PAH enhancement in the microlayer occurred at both sampling sites, but it was greater at CVE (station 6) than at Salute (station 5) due mainly to the much lower subsurface concentrations. Enhancement varied with individual PAH compounds but mean enhancement factors (i.e. SMIC conc./SS conc.) were 30.9 at CVE (station 6) and 2.8 at Salute (station 5) for HPLC measurements.

CONCLUSIONS

It is clear that concentrations of metals, organotins and polyaromatic hydrocarbons were higher in Venice Lagoon waters but much lower at the platform (station 0) site in the Adriatic, with PAH and TBT levels similar to those found previously in Adriatic waters (Dujmov and Sucevic, 1989; Chiavarini *et al.*, 1992).

The continuous discharge of pollutants into the semi-enclosed waters of the Lagoon leads to an accumulation of toxic species in water and sediment, and produces an excessive growth of algae. Over the last decade, the blooms of macroalgae have frequently attained biomass exceeding 10 kg/m^2 of water surface (Zonta *et al.*, 1992). It seems likely that the strong tidal currents in the Lagoon, coupled with the busy traffic and shallow waters, prevents settling of fine-size particles. Unfiltered water samples, therefore, are likely to be loaded with suspended particulate material and possibly with macroalgae. This could account for the large variations between duplicate samples, found for both metal and organic pollutants, which masks any trends that may be due to tidal differences or surface microlayer enhancement. Individual samples at Salute (station 5) and CVE (station 6), however, have enhanced PAH microlayer concentrations. Cu also shows a similar trend. The high Zn concentrations which occur in both microlayer and subsurface water throughout the Lagoon are indicative of urban and industrial pollution. The CVE (station 6) site near the industrial complex at Marghera appears to be the most polluted, with the highest organotin, TBT and PAH values as well as high Zn concentrations. PAH concentrations in the Lagoon waters are probably derived from both petroleum and combustion sources, since PAH pollution in the Adriatic comes from marine traffic (47%), urban waste and industrial sources (32%), and atmospheric inputs and other sources (21%) (Dujmov and Sucevic, 1989) but their range of concentrations is similar to those found in harbours and ports in southwest England. Organotin and TBT results indicate widespread low level pollution, in agreement with previous data from Venice Lagoon (Chiavarini *et al.*, 1992). Nevertheless microlayer levels exceed the UK Environmental Quality Standard of 0.8 ng Sn/l (as TBT) for marine waters and, therefore, must be regarded with some concern.

The limitations of the present initial study could be overcome by a more extensive programme of work. A clearer picture of the importance of the surface microlayer in the distribution of pollutant chemicals in the water column would be obtained by: (i) analysing filtered and unfiltered samples; (ii) using a drum sampler which collects a more concentrated microlayer

(50 μm thickness) instead of the Garrett screen (250–300 μm thickness); and (iii) extending the range and sophistication of chemical analysis.

Although chemical exchange processes between the surface microlayer and subsurface waters are not clearly understood, the ubiquitous nature of the surface microlayer and its ability to accumulate chemical contaminants provide a ready means for monitoring pollution in aquatic environments which may act as an early warning system for the bulk of the water body. Despite being small in mass balance terms, the surface microlayer could act as a sensitive indicator of water quality for the whole water column, as well as a specific indicator for surface water.

REFERENCES

Abdullah, M., El-Rayis, O. and Riley, J. (1976). Re-assessment of chelating ion-exchange resins for trace metal analysis of sea-water. *Anal. Chim. Acta.*, **84**: 363–8

Cameron, P., Berg, J., Dethlefsen, V. and Westernhagen, H. Von. (1992). Developmental defects in pelagic embryos of several flatfish species in the southern North Sea. *Neth. J. Sea Res.*, **29**: (1–3): 239–56

Chiavarini, S., Cremisini, C., Ferri, T., Morabito, R. and Ubaldi, C. (1992). Liquid–solid extraction of butyltin compounds from marine samples. *Appl. Organomet. Chem.*, **6**: 147–53

Cleary, J. J. (1991). Organotin in the marine surface microlayer and subsurface waters of southwest England: relation to toxicity thresholds and the UK Environmental Quality Standard. *Mar. Env. Res.*, **32**: 213–22

Cleary, J. J., McFadzen I. R. B. and Peters, L. (1993). Surface microlayer contamination and toxicity in the North Sea and Plymouth near-shore waters. *ICES paper CM 1993/E. 28, International Council for the Exploration of the Sea*

Dethlefsen, V., Cameron, P. and Westernhagen, H. Von. (1985). Untersuchungen uber die Haufigkeit von Mibbildungen in Fischembryonen der sudlichen Nordsee. *Inf. Fischwirtsch*, **32**: 22–7

Dujmov, J. and Sucevic, P. (1989). Contents of polycyclic aromatic hydrocarbons in the Adriatic Sea determined by UV-fluorescence spectroscopy. *Mar. Poll. Bull.*, **20**, (8): 405–9

Fent, K., Hunn, J., Renngli, D. and Siegrist, H. (1991). Fate of tributyltin in sewage sludge treatment. *Mar. Env. Res.*, **32**: 223–31

Garrett, W. D. (1965). Collection of slick-forming materials from the sea surface. *Limnol. Oceanogr.*, **10**: 602–5

Hardy, J. T. (1982). The sea surface microlayer, biology, chemistry and anthropogenic enrichment. *Prog. Oceanogr.*, **11**: 307–28

Hardy, J. T. and Cleary, J. J. (1992). Surface microlayer contamination and toxicity in the German Bight. *Mar. Ecol. Prog.* Ser., **91**: 203–10

Law, R. J., Fileman T. W. and Portmann, J. E. (1988). Methods of analysis of hydrocarbons in marine and other samples. In *Aquatic environment protection: Analytical Methods, No. 2..* MAFF, Directorate of Fisheries Research

McFadzen, I. R. B. and Cleary, J. J. (1994). Toxicity and chemistry of the sea-surface microlayer in the North Sea using a cryopreserved larval bioassay. *Mar. Ecol. Prog. Ser.*, **103**: 103–9

UNESCO, (1984). Manual for monitoring oil and dissolved/dispersed petroleum hydrocarbons in marine waters and on beaches. Procedures for the Petroleum Component of the IOC Marine Pollution Monitoring System (MARPOLMON-P). *IOC Guide No. 13*

Zaitsev, Y. P. (1971). Marine neustonology. In Vinogradov, K. A. (ed.), *Israel Programmes for Scientific Translations*. Jerusalem

Zonta, R., Costa, F., Ghermandi, G., Pini, R., Perin, G., Traverso, P., Vazzoler, S. and Argese, E. (1992). Geochemical and chemical–physical characterization of a polluted mud flat in the Venice lagoon. *Bull. de l'Inst. Oceanogr., Monaco,* special **11**: 207–25

CHAPTER 9

PELAGIC NITROGEN FLUXES IN THE VENICE LAGOON

P. Morin, P. Lasserre, C. Madec, P. Le Corre,
E. Macé and B. Cavalloni

SUMMARY

The Venice Lagoon is a shallow coastal ecosystem which, since the thirteenth century, has been subjected to large anthropogenic influence. During the past 40 years, an excessive enrichment with nitrogen and phosphorus has led to large phytoplankton blooms and a proliferation of macroalgae (mainly *Ulva rigida*). Measurement of nitrogen uptake rates by phytoplankton was thus the focus of this study. Measurements were carried out using ^{15}N as a tracer in the northern basin of the Venice Lagoon in 1992, along salinity and nitrogen gradients extending from the the Lido entrance to the Silone Channel (low salinity–high nitrogen region).

Nitrogen uptake rates were extremely high, and ammonium and nitrate uptake rates were among the highest reported from other similar temperate coastal environments. Maximum uptake rates were measured in summer. Ammonium uptake rates were generally higher than those for nitrate in all seasons, except in summer in the low-salinity waters of Silone Channel and Palude della Rosa. Ammonium was the the major nitrogen source for phytoplankton, satisfying more than 70% of their N requirements. On an annual basis, nitrogen taken up by phytoplankton was about 36–49% of that taken up by *Ulva*.

INTRODUCTION

The Venice Lagoon is a shallow coastal system along the western part of the northern Adriatic Sea. This Lagoon has been highly influenced by human activities since the thirteenth century, and during the past decades the anthropogenic pressure has increased significantly, resulting in problems of eutrophication and chemical pollution (Sfriso *et al.*, 1987, 1988a, b, 1989; Pavoni *et al.*, 1990). In numerous shallow coastal areas around the world, excessive enrichment with N and P compounds has progressively changed the trophic relations, leading to a reduction of species diversity and to abnormal blooms of a few species of phytoplankton and macroalgae (Fonselius, 1978; Bach and Josselyn, 1979; Nixon and Pilson, 1983; Larsonn *et al.*, 1985; Valliela and Costa, 1988). Such conditions have been observed over the last forty years in the Venice Lagoon where a huge proliferation of macroalgae has led to dystrophic conditions accompanied by mass mortality of benthic

143

organisms (Sfriso *et al.*, 1988a, Marcomini *et al.*, 1992). Nitrogen is a major factor limiting primary production in coastal marine waters (Ryther and Dunstan, 1971; Eppley *et al.*, 1972; Thomas *et al.*, 1974; Thayer, 1974; Goldman, 1976). Since nitrogen may account for a major part of the high biomass and productivity in the Venice Lagoon, it was considered essential to obtain a better knowledge of the N dynamics of this coastal system. The present paper is an attempt in this direction and reports on the seasonal distribution of different forms of N (ammonium, nitrite and nitrate) and the rates of their uptake by phytoplankton. It may be mentioned here that direct measurements of N uptake have not been made in the Venice Lagoon until now.

Perspectives for future work

Possible perspectives for future work are:

(1)   To compare the relative importance of the N taken up by phytoplankton and macroalgae (*Ulva*) on a seasonal basis.
(2)   To determine the forms of N taken up by *Ulva* and by phytoplankton in order to evaluate whether there is competition for N between these two primary producers.
(3)   To quantify the importance of water column regeneration rates through microheterotrophic activities (both as direct excretion and remineralization through the microbial loop).
(4)   To assess the importance of the internal and external inputs of N to the total N demand of primary producers, including phytoplankton and *Ulva*.
(5)   To compare the relative importance of N regeneration in the water column and at the benthic level.

GENERAL PELAGIC CONDITIONS IN THE VENICE LAGOON

Physical and geographical properties

The Lagoon extends over 549 km^2 (52 km long and 8–14 km wide) and is divided into three main hydrological (Lido, Malamocco and Chioggia) basins. The Lido basin, located in the northern part, is the largest (276 km^2), followed by the Malamocco (112 km^2) and Chioggia (111 km^2) basins. The mean depth is around 1 m and 75% of the Lagoon has a depth of less than 2 m. Only 5% of the total surface of the Lagoon has a depth greater than 5 m (Fossato, 1990).

The Lagoon waters

The Lagoon waters are directly influenced by freshwaters advected by the rivers and seawater exchanges with the Adriatic Sea through the three port entrances.

Freshwater inputs

The freshwater flow into the Lagoon comes mainly from small rivers (average discharge 15–17 m³/s) permanently draining a basin of 1850 km² (Veneto Region, 1988) and partially from Sile, Brenta and Bacchiglione rivers that drain an external basin (the Sile, Brenta and Piave rivers which formerly entered the Lagoon have been partially diverted to flow directly into the Adriatic Sea). The present rate of discharge of the Sile River through the Silone Channel is 7–10 m³/s (Veneto Region, 1988), the same as that of the Brenta and Bacchiglione. The combined base flow is 30–40 m³/s (Bernardi *et al.*, 1986; Appi *et al.*, 1989; Vazzoler *et al.*, 1987) and the storm flow may attain 600 m³/s (Cavazzoni, 1973). The mean residence time of freshwater in the whole Lagoon is around 15–30 days (Battiston *et al.*, 1983).

Lagoon–sea exchanges

The volume of water exchanged between the Adriatic Sea and the Venice Lagoon is 111×10^9 m³/year (Marcomini *et al.*, 1992): 40, 47 and 24×10^9 m³/year, respectively, through the Lido, Malamocco and Chioggia entrances. These estimates agree with those obtained using a unidimensional model (Consorzio Venezia Nuova, 1988): a total of 140×10^9 m³/year, with 56, 56 and 28×10^9 m³/year for the three entrances. The volume exchanges in a 12 h tidal cycle of $3.1–5.4 \times 10^8$ m³ (Sfriso *et al.*, 1991), corresponding to 3.5×10^8 m³ during flood and 2.87×10^8 m³ during ebb tides. The volume of water exchanged between the Lagoon and the sea is thus about two orders of magnitude higher than that of freshwaters flowing into the Lagoon.

Nutrients

Nutrient inputs

Nutrients entering the Venice Lagoon originate from the rivers draining the intensively cultivated hinterland, treated and untreated sewage effluents from the cities of Venice, Mestre, Chioggia and other islands (approximately 350,000 inhabitants), industrial wastes mainly from the Porto Marghera area (about 40,000 inhabitants) and lagoon–sea exchanges through the three port entrances.

The total direct and indirect nutrient loading in the Venice Lagoon has been evaluated at 8200–10,600 tons N/year and 1100–1900 tons P/year (Marcomini *et al.*, 1992). The two main sources of nutrients are the river inputs–urban sewage and the lagoon–sea exchanges (Table. 9.1), with the atmospheric fallout and harbour activities representing only a few percent of the total inputs.

Table 9.1 Nitrogen and phosphorus inputs in the Venice Lagoon

Origin	Nitrogen inputs		Phosphorus inputs		References
	(tons/year)	*% of total inputs*	*(tons/year)*	*% of total inputs*	
River inputs	3400	24–25	790	23–26	ENEA, 1990
					Vazzoler et al., 1987
Harbour activities	94–187	0.6–1.4	42–85	1.2–2.8	ENEA, 1988
Urban sewage	4268	30–31	1658	49–55	Bendoricchio et al., 1985
Atmospheric fallout	560	4–7	23	0.6–2.2	Cossu & Giuliano, 1980
	870		66		ENEA, 1988
Lagoon–sea exchanges	27600	39–40	555	16–28	Magistrato alle Acque, 1985
	5450		850		ENEA, 1988
Total	13700–14100		3000–3400		

Nutrient concentrations

Nutrient concentrations inside the Lagoon are at least an order of magnitude higher than those observed in the Gulf of Venice. The annual average nitrite + nitrate concentrations range from 10 to 50 μmol/l in the Lagoon and from 3 to 5 μmol/l in the Gulf of Venice (Marcomini *et al.*, 1992). In the central part of the Lagoon, the area that has been intensively studied, the average nitrate and phosphate concentrations range, respectively, from 50 to 70 μmol/l and from 5 to 25 μmol/l (Sfriso *et al.*, 1992). Facco *et al.* (1986) reported, for the same part of the Lagoon, an average total dissolved inorganic (ammonium + nitrite + nitrate) nitrogen (DIN) concentration of 106 μmol/l and reactive P concentration of 14 μmol/l. In the northern part of the Lagoon, similar high levels of nutrients have been reported by Bianchi *et al.* (1987, 1990).

The large range of nutrient concentrations (Table 9.2) may be partly explained by the rapid regional and seasonal variations that are difficult to quantify with the common sampling techniques. Nevertheless, a general decreasing trend in dissolved nutrients, especially for ammonium, has been observed during the last decade (Sfriso *et al.*, 1992). In the industrial zone around the Porto Marghera area, ammonium decreased from 360–2, 500 μmol/l in the early 1970s to 10–200 μmol/l in the late 1970s and to < 3–50 μmol/l in the second half of the 1980s. Such a distinct decrease has not been observed for other nutrients (nitrate, nitrite and phosphate – Sfriso *et al.*, 1992).

Seasonal differences in the predominant N forms have also been reported (Sfriso *et al.*, 1988a): nitrate is predominant in winter whereas ammonium may represent 40 to 97% of DIN from April to December. Nitrite is < 3% of DIN from October to May but becomes a non-negligible part (> 30 %) of it from July to September.

Table 9.2 Nitrogen concentrations in the waters (μM) and sediments (mg/g dry wt) of the Venice Lagoon, Gulf of Venice and the Emilia-Romagna shoreline

	Venice Lagoon	*Gulf of Venice*	*Emilia-Romagna shoreline*
Water (μM)			
References	Facco *et al.*, 1986	Sfriso *et al.*, 1992	Sfriso *et al.*, 1988
DIN	106		6.5
NO_3	50–70	3–5	3
NO_2	0–3		0.35
NH_4	3–50		2.9
PO_4	5–25		0.16
Sediments (mg/g dry weight)			
References	Orio & Donazzolo, 1987		
Nitrogen	2.5		
Phosphorus	0.36–0.70		

As in other estuarine and nearshore marine environments, and in contrast with what it is generally observed in oceanic waters, maximum phosphate concentrations occurs in summer (Riley, 1941; Nixon and Lee, 1981). This may be attributed to remineralization of organic matter from the spring phytoplankton bloom and to release of phosphate from sediments. In recent years, no significant changes in the nutrient concentrations in the sediments have been observed (Orio and Donazzolo, 1987; ENEA, 1988, 1990). Average N in the sediment was 2.5 mg/g dry weight and average P ranged from 0.36 to 0.7 mg/g dry weight (Table 9.2).

Nutrient fluxes

The only nutrient fluxes that have so far been measured are the inputs from anthropogenic sources and release from sediments. The contribution of the former has been estimated at 3.7 mmol $DIN/m^2/day$ and 0.38 mmol $P/m^2/day$. Phosphorus release from the sediments is of the same order, ranging from 0.26 (Sfriso *et al.*, 1988) to 0.29 $mmol/m^2/day$ (Orio and Donazollo, 1987) but DIN release ranges from 8.36 (Orio and Donazzolo, 1987) to 10.5 $mmol/m^2/day$ (Sfriso *et al.*, 1988). These estimates are averages since there are strong seasonal variations. Until now, no direct measurements of N uptake by primary producers (macroalgae and phytoplankton) have been made.

Macroalgae and phytoplankton

Coastal Lagoons have often been considered as highly productive systems (Vanucci, 1969). More recent studies (Nixon, 1980, 1981) have shown that the amount of carbon fixed in lagoon areas does not appear to be significantly higher than that measured in estuaries or in productive coastal waters.

In these shallow coastal systems, phytoplankton, benthic microflora and macrophytes account for the major part of total primary production (Nixon, 1981) but in varying proportions as a function of the area under investigation. For example, production by phytoplankton and benthic microflora in the Wadden Sea were of the same order (100 $gC/m^2/year$, Cadée and Hegeman, 1974a and b) whereas macrophyte production was four times higher than that of phytoplankton in Bogue Sound, North Carolina (Dillon, 1971). In the Venice Lagoon, Marcomini *et al.* (1992) estimate that macroalgae (with *Ulva rigida* as the dominant species) account for at least 80% of total primary production. During the last forty years, a change in the diversity and in the populations of macroalgae has been noted in the central part of the Lagoon with a general increase of Chlorophyceae (from 20 to 34%) and a simultaneous decrease of Rhodophyceae (from 65 to 51%). This change was accompanied by a decrease in the diversity of macroalgae, with *U. rigida* becoming one of the dominant species.

Macroalgae

According to Sfriso *et al.* (1988), the annual net production of macroalgae in the central part of the Lagoon is 1,155,000 tons wet weight, equivalent to 3,026 to 4,034 tons of N and 269 to 359 tons of P. Macroalgal biomass ranging from 0 to 12.6 kg wet weight/m², and during peak growth its production can be as high as 250–1506 g wet weight/m²/day (Sfriso *et al.*, 1989). At this time, N is the major factor limiting the primary production (Sfriso *et al.*, 1987). On an annual basis, macroalgal biomass production is 14.7–20 kg wet weight/m², equivalent to 475–647 gC/m²/year, 51.4–69.9 gN/m²/year and 4.57–6.22 gP/m²/year.

In the 1970s and 1980s, peak macroalgal blooms were generally observed during spring–summer, with possible sudden decreases during anaerobic conditions (Sfriso *et al.*, 1989). Before the 1970s, the production cycle was quite different: it began at the end of February and was maximal from April to May (Sfriso *et al.*, 1987). In summer, biomass fluctuated and declined to become negligible in autumn and winter. The changes observed in the 1970s and 1980s have been attributed to different factors, including changes in water circulation patterns leading to poor water exchange over vast areas, and the establishment of hypertrophic conditions (Sfriso and Cavolo, 1983; Albertotanza and Piergallini, 1986; Sfriso *et al.*, 1987).

Recently, a marked decrease in peak biomass has been observed in the central part of the Lagoon. Between 1989 and 1991, the maximum biomass decreased by 50%, from 617–1833 to 388–928 g wet weight/m²/year, and the duration of *U. rigida* proliferation decreased from 4 to 2 months in some places (Sfriso *et al.*, 1992).

Phytoplankton

Earlier studies on phytoplankton were on species composition (Socal *et al.*, 1987) and biomass (chl *a*). Phytoplankton diversity is generally lower in the Venice Lagoon than in the adjacent Adriatic Sea (Voltolina, 1973). The spring blooms are generally dominated by diatoms (*Skeletonema costatum*, with concentrations up to $60,000 \times 10^3$ cells/l, representing 87–97% of total cell counts – Fossato, 1990). At other times, nanoplankton and flagellates may contribute a significant (up to 80%) part of the phytoplankton biomass (Sfriso *et al.*, 1988b).

Relatively few measurements of the phytoplankton production of the Venice Lagoon were made with the classical ^{14}C technique. The earliest measurements were made by Vatova (1960) during an annual cycle in 1959. The results showed a general increase during spring, a marked maximum in August (107.4 mgC/m³/day), a rapid decrease in September and a minimum in November–December. Battaglia *et al.* (1983) also measured minimum rates of production (0.6–16 mgC/m³/h) in autumn–winter and the maximum (17–580 mgC/m³/h) in summer. These values are higher than those measured by Degobbis *et al.* (1986) in April in the industrial area (0.4–186 mgC/m³/h), at the historical centre (1–42 mgC/m³/h) and at the Lido entrance (0.8–20 mgC/m³/h), but are of the same order as the averages found in the Northern Adriatic Sea (0-143 mgC/m³/h) by Smodlaka (1986) during the 1966–1981 period.

Seasonal changes in macroalgal and phytoplankton production appear to be closely related. Sfriso *et al.* (1988b) observed that during the bloom of macro-algae in April–May, phytoplankton biomass (chl *a*) remained relatively low (< 5 μg/l). When the macroalgae began to decay after the bloom, chl *a* attained very high (> 100 μg/l) concentrations, which are rare even for highly productive coastal systems. In June–July, when macroalgal biomass increased again, chl *a* decreased to < 2.2 μg/l. After the total decomposition in late July, chl *a* concentrations rose to about 10 μg/l.

Pelagic nitrogen uptake

From the earlier studies it is apparent that maximum phytoplankton development occurs when the collapse of the macroalgal bloom leads to rapid decomposition and substantial nutrient regeneration. During decomposition of organic matter, N is regenerated in different steps, beginning with the reduced form (ammonium–NH_4^+), through successive oxidized (nitrite – NO_2^-; nitrate – NO_3^-) forms. The use of the ^{15}N technique permits direct measurements of the uptake of these different forms and a distinction between the uptake of reduced forms (those arising from regeneration in summer) and oxidized forms (those derived mainly from river inputs or exchange at the sediment–water interface). Until now, no such measurements have been carried out in the Venice Lagoon. Since N appears to be the major factor limiting primary production, measurements of its flux, rather than that of carbon, would be more useful. Such a study will provide new insights into the conditions of development of phytoplankton populations, their responses to N enrichment of the Lagoon waters, the preferred form of N nutrient and its uptake rate, and the possible source of N assimilated in different seasons (external inputs by river discharge and urban sewage or internal inputs through regeneration within the Lagoon). Such data will give a better knowledge of the key processes of the N cycle that should be addressed in future studies, especially in view of the current levels of eutrophication, to determine what can be done to safeguard the ecosystem from further deterioration.

GENERAL ASPECTS OF N DYNAMICS IN MARINE ENVIRONMENTS

Many studies in the last few years were mainly focused on measuring seasonal variations in concentrations or biomass in different compartments of the ecosystem. Nutrient sources (rivers inputs, sewage effluents, exchange at the sediment–water interface) and biomass (phytoplankton, zooplankton and macroalgae) were generally considered, but few studies dealt with measuring fluxes between these different compartments. Our objective in this study is to assess the importance of uptake flux of different forms of N by phytoplankton on a seasonal basis.

Nitrogen is distributed among different compartments (boxes in Figure 9.1) of the ecosystem and is exchanged (arrows in Figure. 9.1) between them.

The main compartments in marine ecosystems are:

(1) *The N sources*, where N is present under different dissolved organic and inorganic forms. The major inorganic forms are molecular N_2, ammonium, nitrite and nitrate, and the organic forms are urea and amino acids. In coastal environments, inputs through rivers and sewage, atmospheric fall-out and sediment–water exchange constitute the sources of N;
(2) *The autrotrophs*, which include phytoplankton, phytobenthos and macroalgae;
(3) *The heterotrophs*, which include bacteria, microzooplankton;
(4) *The detritus*, which corresponds to zooplankton fecal pellets and dead organic matter which are transferred from the surface towards the bottom and the sediment.

The main exchange rates between these compartments are:

(1) Autotrophic uptake;
(2) Grazing by herbivorous heterotrophs (e.g. grazing of phytoplankton by herbivorous microzooplankton, macrozooplankton and nekton);
(3) Grazing of bacteria by bacterivorous heterotrophs (ciliates, nano-flagellates);
(4) Predation on (herbivorous and carnivorous) heterotrophs;
(5) Excretion of ammonium and urea by heterotrophs;
(6) Production of detrital organic matter by all organisms;
(7) Recycling of part of the detritus by microheterotrophs (bacteria, ciliates and flagellates) through the microbial loop (Azam *et al.*, 1983) and export of the rest towards sediments.

The present study is restricted to precise measurements of N uptake by phytoplankton on a seasonal basis.

MATERIAL AND METHODS

Field surveys

Four field surveys, on a seasonal basis, were carried out in 1992 and the samples were obtained using the research vessels *Umberto D'Ancona*, *Mysis* and *Orata*.

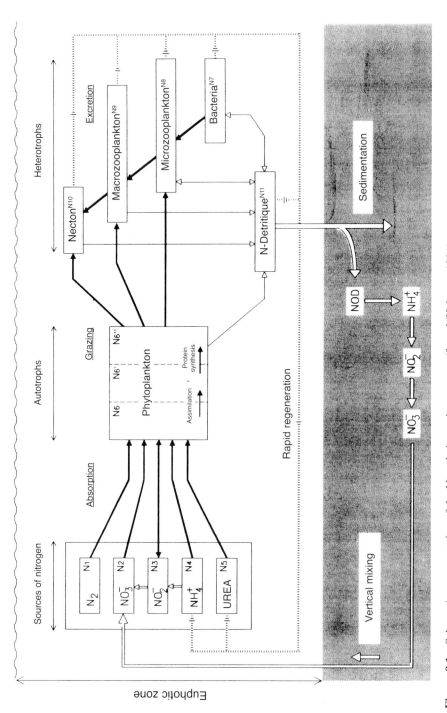

Figure 9.1 Schematic representation of the N cycle in marine systems (from L'Helguen, 1991)

152

Sampling area

Five stations, along salinity and N (mainly ammonium and nitrate) gradients, from the Silone Channel to the Lido entrance (Figure 9.2), were sampled. Station 1 was located at the Lido entrance and characterized coastal marine waters of the Adriatic Sea which enter the Lagoon. Station 2 was located near Crevan Island in the Burano Channel and had an intermediate salinity (around

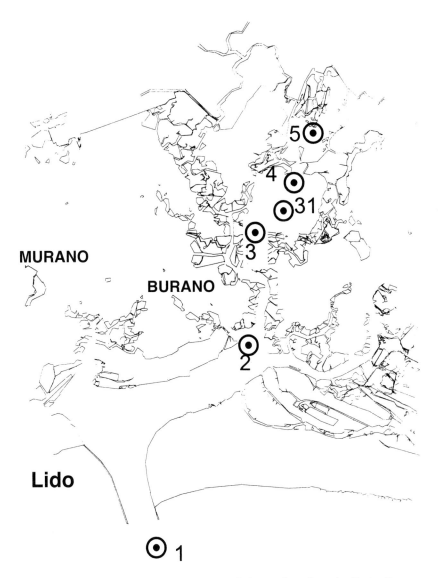

Figure 9.2   Position of stations along the salinity gradient from the Venice Lagoon entrance (Porto di Lido) to the Silone Channel

33 PSU) between the Adriatic waters and the low-salinity waters of the palude areas. Stations 3, 31 and 4 were in the Palude della Rosa. Station 3 was located in the Torcello Channel which is the main seawater input/output channel into the Palude della Rosa (near Torcello Island). Station 31 was at the centre of the palude where the IBM/CNR group carried out intensive studies (see Bianchi *et al.*, Chapter 7). Station 4 was located in the Silone Channel at the main freshwater inlet of the Palude della Rosa.

Two additional stations (5 and 6) upstream in the Silone Channel were sampled to characterize the low salinity waters (station 5 had an average salinity of 5 PSU and station 6, 1 PSU).

The precise locations of the five stations on the salinity and N gradients were fixed after a reconnaissance survey one day prior to actual sampling. The collections were always made at a period when the tide had a well-marked two component semi-diurnal cycle. Strong short-term variations of hydrological, chemical and biological characteristics occur during the tidal cycle (see Bianchi *et al.*, Chapter 7). These variations correspond to a horizontal advection of the Lagoon waters and, in order to avoid them, samples were obtained at a constant reference salinity for each station (15 PSU at Silone Channel, 25 PSU at Palude della Rosa, 27 PSU at Torcello Channel, 31 PSU at Crevan and 34 PSU at Porto di Lido). Nitrogen uptake measurements at the five stations were made on five successive days.

Methods

Sampling

Vertical profiles of temperature ($\pm 0.01°C$) and salinity (± 0.005 PSU) were obtained using an Idronaut CTD system. Surface temperatures were measured with Richter and Wiese reversing thermometers (precision: $\pm 0.01°C$). Salinity of surface samples was measured in a Guildine Autosal 8400 salinometer (precision: ± 0.005 PSU).

Surface water samples for nutrients, chlorophyll and particulate organic matter were collected with 7 l Niskin bottles and stored at 6°C on board pending analysis later in the day at the IBM/CNR shore laboratory.

Chemical and biological measurements

Concentrations of nitrite, nitrate, phosphate and silicate were measured in a Technicon II AutoAnalyser following the procedures given by Tréguer and Le Corre (1975). Ammonium concentrations were measured by the indophenol blue method (Koroleff, 1969). Chlorophyll was measured fluorometrically (Yentsch and Menzel, 1963) in a Turner Design's fluorometer. Particulate organic C and N concentrations were measured in a Perkin Elmer model 240 elemental analyser.

Nitrogen uptake measurement

Nitrate and ammonium uptake rates were measured using the ^{15}N tracer method (Dugdale and Goering, 1967). The samples were pre-filtered with a 200 μm mesh nytex net and ^{15}N-labelled nitrate ($Na^{15}NO_3$, 97.5%) or ammonium ($^{15}NH_4Cl$, 95%) was added at about 10% of the ambient concentrations (Figure 9.3). Half of each sample was filtered immediately after the inoculation so as to have a zero-time ^{15}N enrichment of particulate nitrogen. The remaining samples were incubated in polycarbonate bottles under natural light (simulated *in situ*) for 3–4 h. The temperature of the incubator was maintained nearly constant with a continuous flow of surface lagoon water. At the end of the incubation, samples were filtered on Whatman GF/F glass-fibre filters (47 mm dia, 0.7 μm pore size). The filters were then dried at 60°C for several hours and stored in a plastic vial until further analyses. The atom percentage excess of ^{15}N in the particulate organic matter (POM) on the filters was measured in a SOPRA GS1 emission spectrometer after converting the particulate organic nitrogen (PON) to gaseous N.

Specific uptake rates (V/h) were calculated as the ratio of atom percentage excess of ^{15}N in particulate N, to dissolved N, divided by incubation time. Absolute uptake (transport) rates ($\mu mol/l/h$) were calculated by multiplying the specific uptake rates by the PON concentration ($\mu mol/l$).

RESULTS

Winter situation

The winter survey was in late January 1992, when biological activity is at its minimum and remineralization has been completed. At this time, advection of nutrients by rivers is also at its maximum and the nutrient stock measured represents the maximum that would be available for spring growth of phytoplankton.

Hydrological characteristics along the Silone Channel – Palude della Rosa – Porto di Lido transect

A well-marked salinity gradient was observed along the transect (Figure 9.4b). Salinity ranged from 12 PSU in the Silone Channel (at the northern entrance of the Palude della Rosa) (which is also the seasonal minimum at which N uptake was measured) to 36.3 PSU at the Lido entrance, where salinity is similar to that of Adriatic coastal waters. Intermediate values were observed at the Torcello Channel (27 PSU) and Crevan (33 PSU) areas.

The Lagoon waters were relatively cold, with temperatures ranging from 1.98°C at the Torcello Channel to 8.79°C at Porto di Lido (Figure 9.4a). The temperature gradient decreased steadily from Porto di Lido to the Torcello Channel but an inversion was observed in the low-salinity waters of the Silone

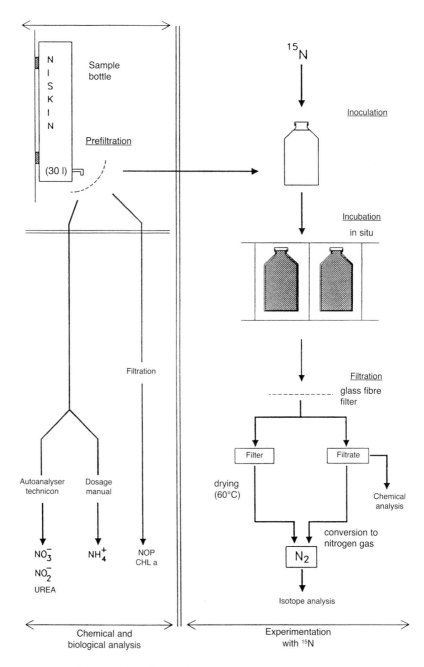

Figure 9.3 ^{15}N experimental procedure for the determination of the N uptake rates (from L'Helguen, 1991)

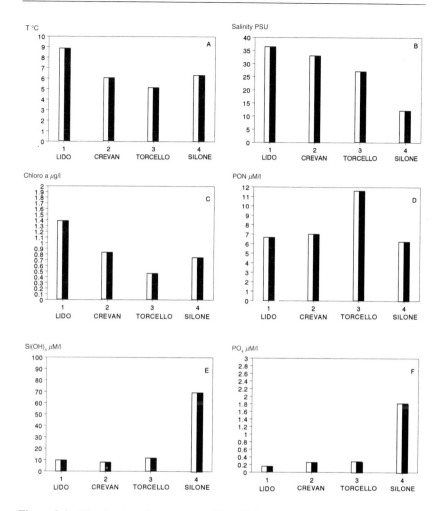

Figure 9.4 Distribution of temperature (A), salinity (B), chl *a* (C), PON (D), silicate (E) and phosphate (F) along the Porto di Lido–Silone Channel transect in winter

Channel which were slightly warmer than the Crevan and Torcello waters. The reasons for this inversion are not known.

Chemical and biological characteristics

The N gradient was well-marked, with an increase in ammonium and nitrate concentrations from the Lido entrance (0.5 and 11 μmol/l, respectively) to the Silone Channel (2.4 and 140 μmol/l) (Figures 9.5a,e). Higher nitrate concentrations (> 235 μmol/l) were measured upstream in very low-salinity (2 PSU) waters. Nitrite concentrations increased from 1.4 to 2.4 μmol/l along the same

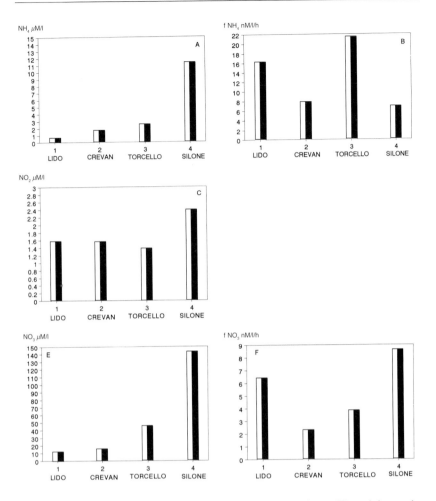

Figure 9.5 Distribution of ammonium (A), nitrite (C) and nitrate (E), and the uptake rates of ammonium (B) and nitrate (F) along the Porto di Lido–Silone Channel transect in winter

transect from Porto di Lido (Figure 9.5c), with a weakly-marked gradient, which is not surprising given that it is only an intermediate form between ammonium and nitrate in the N cycle and thus is transient in occurrence. As with ammonium and nitrate, concentrations of phosphate and silicate also showed horizontal gradients (Figures. 9.4e, f), with values ranging respectively, from 0.16 and 2.4 μmol/l to 1.82 and 69.5 μmol/l from Porto di Lido to the Silone Channel.

There was a well-marked conservative distribution of nutrients in winter (Figures 9.6a–d), with strong negative correlations between nutrients and salinity:

$$NH_4^+ \ (\mu mol/l) \quad = \ -0.504 \times S + 18.64 \qquad r = 0.977 \qquad n = 17$$
$$NO_2^- \ (\mu mol/l) \quad = \ -0.039 \times S + 2.70 \qquad r = 0.955 \qquad n = 27$$
$$NO_3^- \ (\mu mol/l) \quad = \ -5.859 \times S + 214.43 \qquad r = 0.984 \qquad n = 27$$
$$PO_4^{3-} \ (\mu mol/l) \quad = \ -0.064 \times S + 2.39 \qquad r = 0.982 \qquad n = 27$$
$$Si(OH)_4 \ (\mu mol/l) = \ -2.785 \times S + 99.54 \qquad r = 0.975 \qquad n = 27$$

The highly significant correlations indicate that nutrient distribution at this time can, therefore, be interpreted in terms of physical mixing (two-component dilution model) between two waters: marine waters, typical of Adriatic coastal waters, with a salinity above 36 PSU and nitrate concentrations around 11 $\mu mol/l$, and low-salinity waters, which are highly influenced by land runoff which leads to a decrease in salinity and an increase in nutrients.

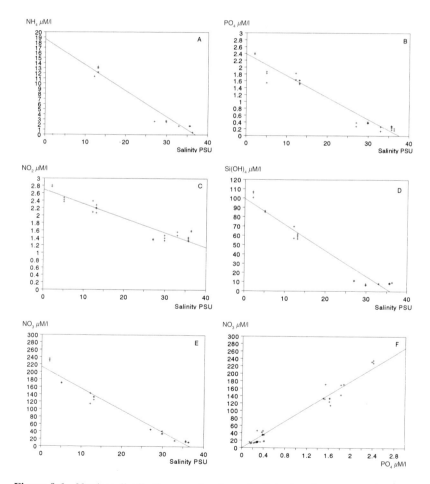

Figure 9.6 Nutrient distribution as a function of salinity in winter: ammonium (A), phosphate (B), nitrite (C), silicate (D), nitrate (E) and nitrate–phosphate relationship (F)

The mean nitrate/phosphate concentration ratio from 31 measurements was 94.1 (Figure 9.6f), much higher than the usual 16 calculated for oceanic waters (Redfield *et al.*, 1963). This is due to excessive enrichment of lagoon waters with N, compared with P, through river inputs.

Chlorophyll concentrations are generally at their minimum in winter. Except at the Torcello Channel station (Figure 9.4c), chlorophyll concentrations were relatively low (<1.5 μg/l), with values increasing from Porto di Lido (0.79 μg/l) to the Silone Channel station (1.5 μg/l). During ^{15}N experiments, chl *a* was exceptionally high (>5.4 μg/l) at the Torcello Channel station. This is rather unusual and may arise from particularly early phytoplankton development in this part of the Lagoon. The PON maximum (>12 μmol/l, Figure 9.4d) corresponded with this chl maximum (Figure 4d). Elsewhere at the other three stations, PON concentrations were of the same order, ranging from 6 to 7 μmol/l.

N uptake

As we anticipated only a low uptake in winter, we measured only ammonium and nitrate. The uptake rates of ammonium and nitrate were very low (< 10 nmol/l/h for nitrate$^-$ and < 30 nmol/l/h for NH_4^+) and of the same order as measured in winter in other coastal environments (Table 9.3) of Northern America (Garside, 1981; Pennock, 1987; Glibert *et al.*, 1991) or of Western Europe (Paasche and Kristiansen, 1982; L'Helguen 1991). Neither ammonium nor nitrate uptake rates (Figures 9.5b,f) followed N and salinity gradients. The maximum ammonium uptake rate (> 20 nmol/l/h) was at the Torcello Channel station and corresponded with the very high chlorophyll and PON concentrations. The minimum ammonium uptake rates (< 8 nmol/l/h) were at the Crevan and Silone Channel stations and corresponded with the low chl (< 1.5 μg/l) and PON (< 7 μmol/l) concentrations. At these two sites phytoplankton production was at its minimum level of the year. The ammonium uptake rate at the oceanic station of Porto di Lido was intermediate

Table 9.3 Ammonium and nitrate uptake rates during winter in different coastal environments of Northern America and Northwestern Europe compared to this study

Areas	Nitrate uptake rates nmol/l/h	NH$_4$ uptake rates nmol/l/h	References
New York Bight	10–30	10–30	Garside, 1981
Delaware Estuary	10–30	10–30	Pennock, 1987
Chesapeake Bay Plume	10–25	10–30	Glibert *et al.*, 1991
Oslofjord	10–20	10–40	Paasche and Kristiansen, 1982
Western English Channel	0–5	0–5	L'Helguen, 1991
Venice Lagoon	0–10	0–22	This study

(15 nmol/l/h) but corresponded with a minimum value in chlorophyll (0.79 μg/l). Hence, the chlorophyll-specific ammonium uptake rate (ammonium uptake rate divided by chlorophyll concentration) was five times higher than at the Torcello Channel station. The phytoplankton populations at this oceanic station were, therefore, more productive and healthier than those in the Torcello Channel low-salinity waters.

Nitrate uptake rates increased regularly along the transect from the Crevan to Silone Channel stations with increasing nitrate concentrations (Figure 9.5f). The uptake rates were maximum in the low salinity–high nitrate waters of the Silone Channel, but were still low (< 9 nmol/l/h). As with ammonium, nitrate uptake rates were at intermediate levels at the Lido entrance, thus confirming the presence of a healthy and productive phytoplankton population in this area.

Uptake rates of ammonium were generally higher than those of nitrate (Figures 9.5b, f). Though nitrate accounted for > 80% of DIN, its uptake was only between 15 and 30% of total N uptake. This indicates a preferential uptake of ammonium over nitrate by phytoplankton for its winter N nutrition. The inhibition of nitrate uptake by ammonium has been observed previously in other coastal environments (MacIsaac and Dugdale, 1972; MacCarthy *et al.*, 1975, 1977; Paasche and Kristiansen, 1982), particularly when its concentrations are relatively high (>1 μmol/l). The situation was different in the low salinity waters of the Silone Channel, where both ammonium and nitrate were taken up at similar rates (7 nmol/l/h for ammonium and 8 nmol/l/h for nitrate). This lack of inhibition of nitrate uptake by ammonium even though ammonium concentration was >12 μmol/l may partly be explained by the presence of different phytoplankton populations more adapted to low salinity waters.

Spring situation

Measurements in late April were done at the same stations and along the same transect as in winter. An additional station (station 31) in the centre of Palude della Rosa, where intensive studies were carried out by Bianchi *et al.* was also sampled. Intensive proliferation of macroalgae in these shallow waters (depth < 1 m) occurred at this time.

Hydrological characteristics along the Silone Channel – Palude della Rosa – Porto di Lido transect

Lagoon waters were much warmer than in winter: temperatures ranged from a minimum of 13.6°C at the the Lido entrance to a maximum of 20.77°C at the Crevan station (Figure 9.7a). The temperatures measured between Crevan Island and the Silone Channel were rather similar. This can be explained by uniform spring heating of the water masses due to the uniform bathymetry

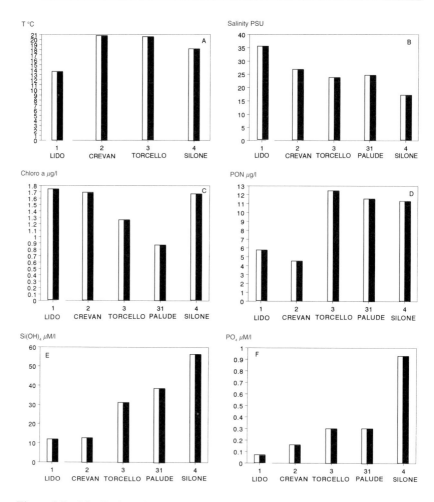

Figure 9.7 Distribution of temperature (A), salinity (B), chl *a* (C), PON (D), silicate (E) and phosphate (F) along the Porto di Lido–Silone Channel transect in spring

between these points (depths between 2 and 5 m). At the Lido entrance, where the depth is >10 m, spring heating is far less important and is slower than in the shallower waters.

The horizontal salinity gradient between the Silone Channel and Porto di Lido stations in spring was lower than in winter (ΔS = 18.4 PSU in spring, and ΔS = 24.1 PSU in winter). Salinity ranged from 15.7 PSU in the Silone Channel to 33.3 PSU at the Porto di Lido (Figure 9.7b), where salinity values were lower than those measured in winter at approximatively the same state of the tide. This lower salinity at the entrance to the Lagoon can probably be explained by an increased inflow of low salinity waters over a major part of the Lagoon at this time.

Chemical and biological characteristics

Well-marked gradients were observed between the Porto di Lido and the Silone Channel stations with values ranging, respectively, for ammonium, nitrite and nitrate from 0.89, 0.34 and 4.0 μmol/l to 5.55, 1.42 and 39.7 μmol/l (Figures 9.8a,c,e). As with salinity, the horizontal N gradients were also lower in spring ($\Delta NH_4^+ = 3.65$ μmol/l, $\Delta NO_3^- = 41.3$ μmol/l) than in winter ($\Delta NH_4^+ = 10.8$ μmol/l, Δnitrate$^- = 129$ μmol/l). This decrease by a factor of three may be explained either by a decrease in advection by river flow in spring (but unlikely since low-salinity waters were observed at the entrance to the Lagoon)

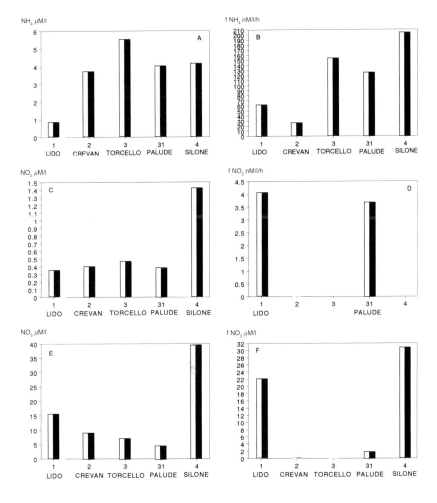

Figure 9.8 Distribution of ammonium (A), nitrite (C) and nitrate (E), and their respective uptake rates (B, D and F) along the Porto di Lido–Silone Channel transect in spring

or more probably by a decrease of N inside the Lagoon corresponding to an increase in N assimilation by primary producers (phytoplankton and macroalgae) which are known to develop intensively at this time of the year.

The N gradient was interrupted in the Palude della Rosa area where ammonium and nitrate concentrations were lower (respectively from 1.99 and 4 μmol/l at the Palude centre station to 4.08 and 7 μmol/l at the Torcello Channel station) than the theoretical concentrations expected at these salinity levels. According to the winter dilution model, the salinity values of 24 and 27 PSU in the Palude area should correspond, respectively, to ammonium and nitrate concentrations ranging from 5 and 56 μmol/l to 6.5 and 74 μmol/l, which are well above the concentrations observed at these two stations. This deficit is also evident on the N salinity diagrams (Figures 9.9a–e) which show

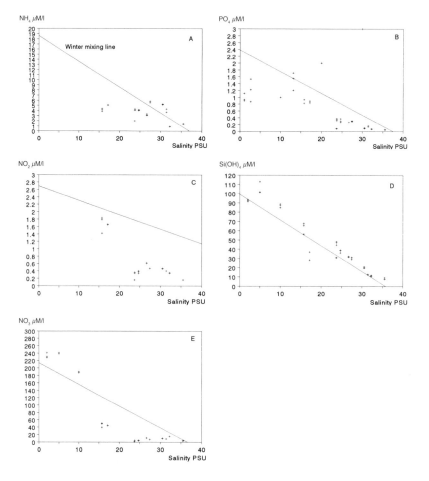

Figure 9.9 Nutrient distribution as a function of salinity in spring: ammonium (A), phosphate (B), nitrite (C), silicate (D) and nitrate (E)

that, in the salinity range above 15 PSU, the N concentrations observed were well under the winter mixing line indicating a general depletion of N in the Lagoon. While phosphate changes also showed a depletion, silicate behaved conservatively, indicating an absence of its consumption and consequently, a reduced development of diatoms (which are the main silicate consumers), a fact also supported by the low chlorophyll levels (< 2 μg/l).

The N and P deficits may be attributed to spring proliferation of macroalgae, and more particularly to *Ulva*, which was very abundant in large areas of the northern part of the Lagoon during our field sampling. At this time, chlorophyll concentrations were rather uniform (1.5 μg/l, Figure 9.7c) along the salinity gradient, except in the Palude della Rosa area (stations 3 and 31) where minimum chl concentrations were observed. The distribution of PON along the transect (Figure 9.7d) showed relatively low concentrations at the Lido and Crevan stations compared to those observed in the Torcello, Palude and Silone stations. If the PON/chl ratios are considered, then the values observed at the two downstream stations (Lido and Crevan) are comparable to those of the coastal area of Narragansett Bay (Furnas, 1983). The PON/chl ratios were higher (7.4 to 13.3) at the three other upstream stations, indicating either the presence of detrital particulate material or a high N content in the phytoplankton cells corresponding to an intracellular accumulation of N.

N uptake

Measurements of nitrite uptake rates were included in spring. Ammonium and nitrate uptake rates were higher in spring than in winter. As in winter, variations in uptake rates of different forms of N did not follow the salinity gradient, but appeared rather to be related to the concentrations of the same form of N at the stations along the transect. The N uptake rates were at intermediate levels between the lower ones measured in the coastal area of the western English Channel (L'Helguen, 1991) and in the Oslofjord (Paasche and Kristiansen, 1982) and the higher ones measured in the North American coastal environments of the Chesapeake Bay Plume (Glibert *et al.*, 1991) and New York Bight (Garside, 1981) (Table 9.4). They were of the same order of magnitude as those measured in the Delaware estuary (Pennock, 1987) where the study was carried out along a salinity gradient.

Maximum ammonium uptake rates (140 to 210 nmol/l/h) were measured at the three upstream stations (Torcello Channel, Palude della Rosa and Silone Channel, Figure 9.8b), where ammonium concentrations were also relatively high (> 4 μmol/l). At these three stations the changes in ammonium uptake followed remarkably well those in *in situ* ammonium concentrations. Lower uptake rates (25 to 65 nmol/l/h) were measured at the Porto di Lido and Crevan stations. Ammonium concentrations at the Crevan station were rather high (> 3.5 μmol/l) and chlorophyll levels were in the same range as at other

Table 9.4 Ammonium and nitrate uptake rates during spring in different coastal environments of Northern America and Northwestern Europe

Areas	Nitrate uptake rates nmol/l/h	NH_4 uptake rates nmol/l/h	References
New York Bight	310–480	430–750	Garside, 1981
Delaware Estuary	0–100	80–200	Pennock, 1987
Chesapeake Bay Plume	100–500	100–800	Gilbert et al., 1991
Narragansett Bay	0–150	0–720	Furnas et al., 1986
Oslofjord	0–25	0–10	Paasche and Kristiansen, 1982
Western English Channel	0–15	0–20	L'Helguen, 1991
Venice Lagoon	0–30	0–200	This study

stations in the transect, yet the uptake rates were particularly low. This is rather surprising and we could not find any specific reason.

Nitrite uptake rates, measured at the stations at the Lido entrance and Palude della Rosa centre, were very low (< 5 nmol/l/h, Figure 9.8d) and were the lowest of the three N forms, indicating that nitrite was a negligible N source for phytoplankton at this time of the year. These rates are in the same range as those measured in the Chesapeake Bay Plume (0–15 nmol/l/h, Glibert and Garside, 1992) and in the coastal area of the western English Channel (3–4 nmol/l/h, L'Helguen, 1991) at the same period of the year.

As with ammonium, changes in nitrate uptake rates were related to its concentrations along the salinity gradient (Figure 9.8f): decreasing nitrate concentrations between the Lido entrance and the Palude station corresponded to decreasing trends in nitrate uptake, with very low values (<2 nmol/l/h) at the Crevan, Torcello Channel and Palude della Rosa stations. In these areas, an intense development of macroalgae (and more particularly *Ulva*) was observed. Since nitrate concentrations decreased between winter and spring, and chlorophyll concentrations and nitrate uptake were at low levels, the nitrate depletion observed may be attributed to intense development of macroalgae in these areas.

The highest nitrate uptake rates were measured at the Porto di Lido and the Silone Channel stations (respectively, 22 and 31 nmol/l/h) where nitrate concentrations were also significantly high (respectively, 15 and 50 μmol/l). At these two stations, where nitrate concentrations represented 91% and 87% of DIN, nitrate uptake represented only 26% and 13% of the total N taken up. At the other three stations in the Crevan and Palude areas, it represented only a small percentage (< 5%) of the total N uptake.

Ammonium was the main form of N taken up by phytoplankton at all stations in spring, with a 70–98% share in total N nutrition. Nitrate uptake was significant only in areas where nitrate concentrations were high (the Porto di Lido and Silone Channel stations). Nitrite was a negligible N source for phytoplankton nutrition at this time of the year, as observed in spring in other coastal environments (L'Helguen, 1991; Glibert and Garside, 1992).

Summer situation

The summer study was carried out in late July.

Hydrological characteristics along the Silone Channel – Palude della Rosa – Porto di Lido transect

The horizontal salinity gradient between the Porto di Lido and the Palude della Rosa area was not very marked ($\Delta S = 7$ PSU, Figure 9.10b) and was not uniform along the transect, as found in winter and spring. It was very marked only between Palude della Rosa and Silone Channel ($\Delta S = 14$ PSU). This

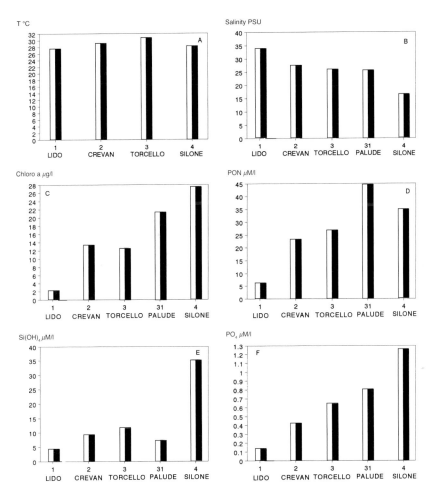

Figure 9.10 Distribution of temperature (A), salinity (B), chl *a* (C), PON (D), silicate (E) and phosphate (F) along the Porto di Lido–Silone Channel transect in summer

situation may be explained by the diminished influence of freshwater inputs by rivers in the northern part of the Lagoon in summer. Freshwater inputs are thus confined to the Silone Channel in the northern part of the Palude della Rosa area.

The temperature of the Lagoon waters was at its maximum (27.61 to 30.8°C, Figure 9.10a) and increased slightly upstream from Porto di Lido (27.61°C) to the Torcello Channel (30.8°C) and decreased in the freshwaters of the Silone Channel (28.23°C).

Chemical and biological characteristics

As in winter and spring, nitrite and nitrate gradients were well-marked between the Porto di Lido and Silone Channel stations (Figure 9.11c,e), with the concentrations increasing upstream, respectively, from 0.11 μmol/l and 1.9 μmol/l to 2.25 μmol/l and 54.3 μmol/l. These gradients were interrupted in the Palude della Rosa centre where nitrite and nitrate concentrations were lower than in the surrounding areas. Nitrite and nitrate levels in the low salinity waters of the Silone Channel were high in summer and were in the same range as in spring. Input by rivers was, therefore, still important but its influence was confined spatially to the immediate vicinity of the Palude della Rosa area.

Ammonium concentrations were high (> 2 μmol/l) at all stations. Its distribution along the Porto di Lido – Silone Channel transect (Figure 9.11a) was completely different from those of the other two N forms and did not follow the salinity gradient. Maximum concentrations were at Crevan station (> 6 μmol/l) and they decreased steadily from Crevan to the Silone Channel. Since ammonium concentrations were lower in the Silone Channel than in the Palude and Crevan areas, the high concentrations observed in the Palude and Crevan areas certainly did not come from river inputs, as was the case in winter and spring. This enrichment occurs within the Lagoon itself and derives probably from an important *in situ* degradation of organic matter that has been produced earlier in the seasonal biological cycle. A substantial part of the degrading organic matter is probably the macroalgal biomass, particularly *Ulva*, that underwent intensive growth in spring. Another source of enrichment may be exchange at the sediment–water interface. Ammonium represented an important part (35–64%) of DIN in summer.

As with nitrite and nitrate, phosphate and silicate gradients were well marked, with increasing values along the transect between Porto di Lido and the Silone Channel (Figures 9.10e,f). Again, as with the two N forms, the silicate gradient was interrupted in the Palude della Rosa centre station where lower concentrations (7.4 μmol/l) occurred. The phosphate gradient, however, was not interrupted in the Palude area. Moreover, for the same salinity range, phosphate concentrations were at the same levels in summer and winter (Figure 9.12d), as though there was no assimilative loss between winter and summer. This characteristic summer phosphate maximum has also been

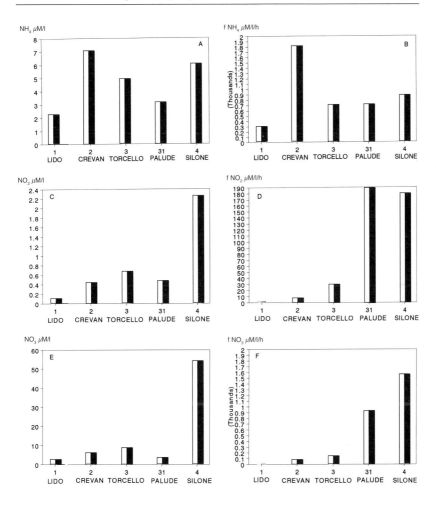

Figure 9.11 Distribution of ammonium (A), nitrite (C) and nitrate (E), and their respective uptake rates (B, D and F) along the Porto di Lido–Silone Channel transect in summer

observed in other lagoons (Riley, 1941; Smayda, 1957; Taft and Taylor, 1976; Nixon and Lee, 1981; Postma, 1981) and has been attributed to the release of phosphate from sediments during anoxic events (Taft and Taylor, 1976) and to remineralization of organic matter produced during the spring development of phytoplankton and macroalgae. The high ammonium and phosphate concentrations seem to indicate that a substantial remineralization of organic matter occurs in the Venice Lagoon during the summer period, leading to enhanced availability of nutrients for subsequent development of phytoplankton and macroalgae.

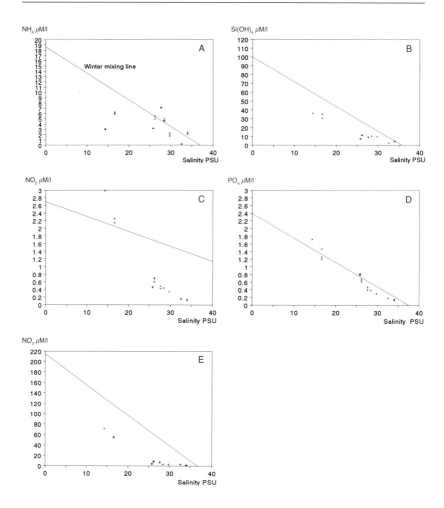

Figure 9.12 Nutrient distribution as a function of salinity in summer: ammonium (A), phosphate (D), nitrite (C), silicate (B) and nitrate (E)

The maximum development of phytoplankton was observed at this time. Apart from the Porto di Lido station, significantly high chlorophyll concentrations (>10 μg/l) were measured in the northern part of the Lagoon (Figure 9.10c), with maximum values in the Palude centre station (>20 μg/l) and in the Silone Channel (>27 μg/l) where nutrient levels were also high. Changes of chl concentrations followed the same increasing gradient as nutrients (except for ammonium) along the transect from Porto di Lido to the Silone Channel. The same trend was also observed for PON, with values increasing from 6.37 μmol/l at the entrance to 44.78 μmol/l at the Palude centra station (Figure 9.10d).

N uptake

The maximum uptake rates of the three forms of N appear to be much higher than those reported from other similar coastal temperate systems of North America and Western Europe (Table 9.5). As in spring, the uptake rates of the different forms of N appeared to be more closely related to concentrations of the same N form in the water column rather than to the salinity gradient.

Maximum ammonium uptake rates (>1800 nmol/l/h) were obtained at the Crevan station (Figure 9.11b) and these corresponded with the high ammonium concentrations (> 6 μmol/l) in the water column. This rate is significantly high for this type of coastal environment since it was two times higher than the maximum reported by Garside (1981) in the New York Bight (one of the highest uptake rates reported in the literature) during a summer phytoplankton bloom. Chlorophyll concentrations (>13 μg/l) were also relatively high, confirming that phytoplankton populations were developing intensively in this area. High ammonium uptake rates (700–880 nmol/l/h) were also observed in the Palude della Rosa and Silone Channel along with high chl (>20 μg/l). The minimum ammonium uptake rate was at the Lido entrance, but the chl-specific uptake rate was in the same range as at the Crevan station. The ammonium sources at the Crevan and Torcello Channel stations come mainly from regeneration within the Lagoon (through degradation of organic matter produced earlier), since the ammonium concentrations were higher at these two stations than at the Silone Channel, which had a maximum amount of ammonium advected by the rivers. This result emphasizes, once again, the importance of regeneration as an internal source of N contributing to summer phytoplankton bloom.

Nitrite uptake rates increased in the same way as did its concentrations along the Porto di Lido–Silone Channel transect, with minimum values at the Lido entrance and maximum values in the Silone Channel (Figure 9.11d). Maximum uptake rates were at the Silone Channel and Palude centre stations (respectively, 180 and 190 nmol/l/h) and were ten times higher than those observed in the Chesapeake Bay Plume (Glibert *et al.*, 1991) during the

Table 9.5 Ammonium and nitrate uptake rates during summer in different coastal environments of Northern America and Northwestern Europe

Areas	Nitrate uptake rates nmoles/l/h	NH$_4$ uptake rates nmoles/l/h	References
New York Bight	0–460	0–850	Garside, 1981
Delaware Estuary	0–110	100–300	Pennock, 1987
Chesapeake Bay Plume	0–60	0–500	Glibert *et al.*, 1991
Narragansett Bay	0–300	0–374	Furnas *et al.*, 1986
Oslofjord	0–130	60–300	Paasche and Kristiansen, 1982
Western English Channel	0–50	0–40	L'Helguen, 1991
Venice Lagoon	0–1575	300–1800	This study

summer period. At the Silone Channel station, the high nitrite uptake rates also corresponded with high nitrite concentrations originating from river inputs, which are probably the major sources of nitrite in this area. However, at the Palude centre station, nitrite concentrations were lower; the high uptake might have exceeded the nitrite input from rivers and *in situ* regeneration, resulting in a progressive depletion of nitrite in the water column. At other stations, nitrite uptake rates were in the same range (3–30 nmol/l/h) as those reported in the Chesapeake Bay Plume.

Nitrate uptake rates also increased along the transect, with maximum values at the Silone Channel station (>1600 nmol/l/h, Figure 9.11f). As with ammonium, this maximum value was significantly very high and three times higher than that reported by Garside (1981) during a summer phytoplankton bloom in the New York Bight. At this station, the high nitrate concentrations arose mainly from the river inputs (NO_3^- > 50 μmol/l). It is remarkable that, despite relatively high ammonium concentrations (> 3.6 μmol/l), there was apparently no suppression of nitrate uptake by phytoplankton, as usually observed in coastal and oceanic waters. This is similar to the findings of Pennock (1987) in the Delaware estuary: when nitrate concentrations are high, there is no suppression of its uptake by high ammonium concentrations. High nitrate uptake rates were also recorded at the Palude centre station (>930 nmol/l/h) where nitrate concentrations were low (< 3.5 μmol/l). As observed for nitrite, nitrate assimilation by phytoplankton was probably higher than nitrate inputs at this station, resulting in progressive nitrate depletion.

Nitrate was the major source of N for phytoplankton at the Silone Channel and Palude centre stations, respectively, with 64% and 57% of the total N taken up. Along the transect from Porto di Lido to the Torcello Channel where nitrate uptake rates were lower (15 to 140 nmol/l/h), ammonium still constituted the main source of N and represented 81–95% of the total N uptake. At these downstream stations, regeneration through bacterial nitrification and microheterotrophic activities were certainly the main N sources for phytoplankton nutrition.

Remarkable high N uptake rates characterized the summer period. These were among the highest reported from different coastal environments around the world in recent years. Ammonium constituted the main source of N over a major part of the Venice Lagoon (from Porto di Lido to the Torcello Channel) and nitrate was the major source in the low-salinity waters of the Silone Channel and in Palude della Rosa. At this time of the year, apart from the Silone Channel where N inputs by rivers were still significantly high, regeneration constituted an important N source inducing a phytoplankton bloom in the northern part of the Venice Lagoon. The regeneration was sustained by degradation of organic matter produced in spring, a major part of which certainly comes from macroalgae, most probably from *Ulva*, which underwent intense growth in spring. In short, it seems likely that *Ulva* indirectly induced the summer phytoplankton bloom.

Autumn

Sampling in autumn was done in October and in the same manner as in the other seasons, except that the station located in the centre of Palude della Rosa was not sampled.

Hydrological characteristics along the Silone Channel – Palude della Rosa – Porto di Lido transect

The horizontal salinity gradient between the Lido entrance and the Torcello Channel was very low ($\Delta S < 3$ PSU, Figure 9.13b) indicating a quite uniform distribution of the water masses in this part of the Lagoon. Salinity values were relatively low at the Lido (32.8 PSU), indicating an influence of the freshwater inputs down to the Lagoon entrance. As noted above, in summer, the salinity gradient was very marked only between the Torcello Channel and the Silone Channel ($\Delta S = 12.8$ PSU). During this autumn study, low-salinity waters were still confined to the approaches of the Silone Channel area.

Temperature decreased from 15.4 to 12.8°C (Figure 9.13a) along the transect between Porto di Lido and the Silone Channel. The temperature decrease between summer and autumn was very sharp (the mean decrease is around 15°C). The range in autumn temperatures was similar to that of spring.

Chemical and biological characteristics

Gradients of all N forms were well-marked (Figures 9.14a,c,e). N concentrations increased steadily from the Lido entrance to the Torcello Channel, with a rapid increase in the Silone Channel where the maximum N concentrations ($NH_4^+ > 15.0$ μmol/l, $NO_2^- > 4.4$ μmol/l and $NO_3^- > 179$ μmol/l) were found. For ammonium and nitrite, these concentrations were much higher than those observed during winter, probably due to a N enrichment from significant bacterial regeneration at this time of the year. Nevertheless, nitrate still represented the major form of N in autumn, ranging from 70 to 86% of DIN.

Like the N compounds, silicate and phosphate concentrations increased steadily from the Lido ($Si(OH)_4 = 14.6$ μmol/l) to the Toricello Channel ($Si(OH)_4 = 24.2$ μmol/l), with the same rapid increase in the Silone Channel, where the maximum concentrations of these two compounds were measured ($Si(OH)_4 = 125$ μmol/l, $PO_4^{3-} = 1.7$ μmol/l, Figures 9.13e,f). These concentrations were in the same range, or even higher for silicate, than those observed during winter.

Chlorophyll concentrations were relatively low (1.02–1.76 μg/l, Figure 9.13c). Their changes along the transect did not follow the salinity and N gradients: increasing values were observed between the Crevan and Silone Channel stations, with intermediate values at the Lido entrance. Changes of PON followed approximately the same trend with relatively low values (< 6.0 μmol/l, Figure 9.13d).

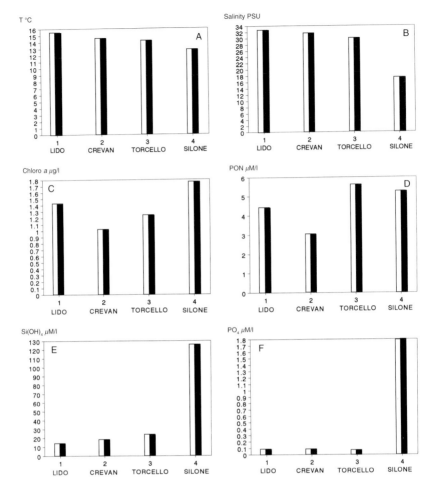

Figure 9.13 Distribution of temperature (A), salinity (B), chl *a* (C), PON (D), silicate (E) and phosphate (F) along the Porto di Lido–Silone Channel transect in autumn

N uptake

Ammonium uptake rates were within the range of those reported from other coastal temperate systems of North America and Western Europe (Table 9.6). A general decrease in the rates was observed along the transect from Porto di Lido to the Silone Channel (Figure 9.14b) corresponding to an increase in ammonium concentrations. The maximum rates, observed at the Lido entrance (106.61 nmol/l/h), were similar to those reported in the Delaware estuary (Pennock, 1987) and in Narragansett Bay (Furnas *et al.*, 1986) in autumn. The minimum rate measured at the Crevan station also corresponded to a minimum in chlorophyll and PON concentrations. Though ammonium represented

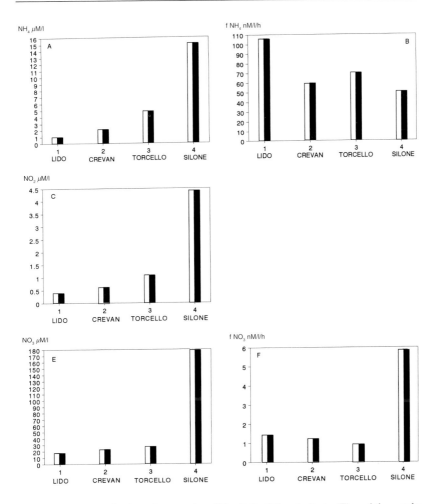

Figure 9.14   Distribution of ammonium (A), nitrite (C) and nitrate (E), and the uptake rates of ammonium (B) and nitrate (F) along the Porto di Lido–Silone Channel transect in autumn

only 11 to 16% of DIN in autumn, its uptake accounted for 90 to 99% of the total uptake. During autumn, ammonium uptake rates no longer related to ammonium concentrations, as observed in summer.

Nitrate uptake rates were particularly low (< 2 nmol/l/h) with a decreasing trend from Porto di Lido to the Torcello Channel (Figure 9.14f). The maximum rates (around 6 nmol/l/h) were measured in the nitrate-enriched low salinity waters of the Silone Channel. These values are in the same range as those reported for the New York Bight (Garside, 1981) but are much lower than those observed in similar coastal environments during autumn. Nitrate uptake represented only 1 to 11% of total N uptake.

Table 9.6 Ammonium and nitrate uptake rates during autumn in different coastal environments of Northern America and Northwestern Europe

Areas	Nitrate uptake rates nmol/l/h	NH_4 uptake rates nmol/l/h	References
New York Bight	0–10	20–30	Garside, 1981
Delaware Estuary	0–30	20–90	Pennock, 1987
Narragansett Bay	0–20	0–140	Furnas *et al.*, 1986
Oslofjord	14–24	16–200	Paasche and Kristiansen, 1982
Western English Channel	0–20	0–30	L'Helguen, 1991
Venice Lagoon	0–6	0–110	This study

The autumn period was characterized by relatively low uptake rates. Ammonium still represented the main source of N over a major part of the northern Venice Lagoon. Phytoplankton biomass was relatively low and autumn can be considered as a transition period leading to the winter situation when nutrient concentrations are at their highest levels and chlorophyll and N uptake rates at their lowest levels of the year.

Nitrite uptake rates in autumn were less precise and hence are not reported here.

Ammonium and nitrate uptake by *U. rigida*

Ammonium and nitrate uptake rates of *U. rigida* were measured in summer, in co-operation with the group of Prof. Ravera (University of Venice at La Celestia).

Samples of *Ulva* (area = 32 cm²) were introduced in 300 ml glass bottles containing natural Lagoon seawater and inoculated with ^{15}N-labelled ammonium and nitrate. Initial ambient N concentrations were close to those observed in the northern basin of the Lagoon during summer (NH_4^+ = 2 μmol/l, NO_2^- = 0.21 μmol/l and nitrate$^-$ = 6.5 μmol/l). Incubation was carried out at constant temperature (25.5°C) and light (11,500 to 11,800 lux). Nitrogen uptake rates were measured after different incubation periods (1, 2, 4 and 8 h).

Preliminary results were quite satisfying since it was the first time that our group had measured uptake rates of N by macroalgae using the ^{15}N method. Uptake of nitrate was three times higher than that of ammonium (Figure 9.15): at the end of 8 h of incubation, 24 μmol of nitrate and 8.7 μmol of ammonium had been taken up by 1 g wet weight biomass of *U. rigida*. From these results, it appears that *U. rigida* prefers nitrate over ammonium as a N source. These results explain, in part, the depletion of nitrate in the Palude della Rosa area during spring, when the growth of *U. rigida* was intense. In this same area, phytoplankton populations were using ammonium rather than nitrate for spring development. So, a question can be asked: is there competition between

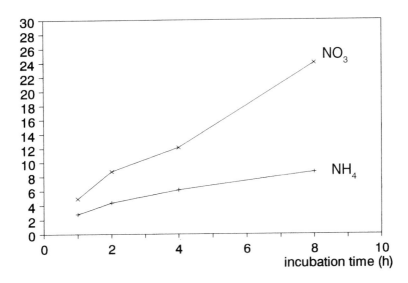

Figure 9.15 Ammonium and nitrate uptake (μmol N/g wet weight) by *U. rigida* during summer

macroalgae and phytoplankton for N? Future work is necessary to assess the relative importance of the different forms of N for the development of macroalgae and phytoplankton in the Venice Lagoon.

DISCUSSION AND CONCLUSIONS

The seasonal changes of hydrological, chemical and biological characteristics and N uptake rates are summarized in Figures 9.16 and 9.17. They indicate the main trends observed during our study.

Temperatures of the Lagoon waters have a well-marked seasonal cycle with values ranging from 2.0°C in winter to 30.8°C in summer ($\Delta T = 29$°C) in the Torcello Channel (Figure 9.16a). The annual mean salinities at the study stations were: 33.75 PSU at the Lido entrance, 31.05 PSU in Crevan, 27.33 PSU in the Torcello Channel, 25.25 PSU in Palude della Rosa and 15.74 PSU in the Silone Channel (Figure 9.16b).

Maximum nutrient concentrations were observed in winter and autumn, with notable differences according to nutrient. The concentrations generally increased from the Lido entrance to the Silone Channel. Ammonium and nitrite concentrations were at their highest levels during autumn (Figures 9.17a,c) as already observed in the Lagoon in several studies (Marcomini *et al.*, 1992; Sfriso *et al.*, 1992). Nitrate concentrations were similar during winter and autumn (Figure 9.17e) but decreased markedly in spring and

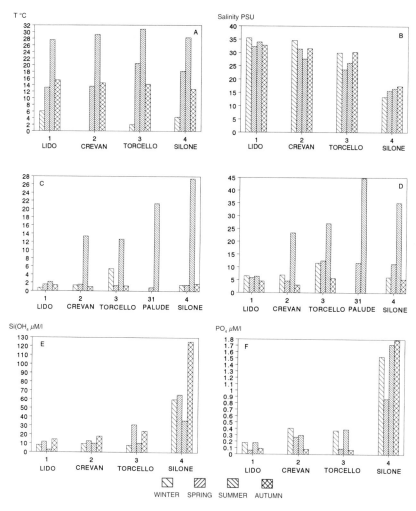

Figure 9.16 Seasonal changes of temperature (A), salinity (B), chl *a* (C), PON (D), silicate (E) and phosphate (F) along the Porto di Lido–Silone Channel transect

summer, corresponding to utilization of this N compound by primary producers. Unlike nitrate, the minimum silicate concentrations were only observed in summer (Figure 9.16e). The seasonal changes in phosphate were characterized by a summer maximum which is in the same range as in winter (Figure 9.16f). This remarkable pattern has been also reported in other coastal lagoons (Riley, 1941, Smayda, 1957, Nixon, 1981).

The seasonal changes in chlorophyll *a* are marked by the absence of a spring phytoplankton bloom and an intense development in summer with chl concentrations ranging from 12 to 28 µg/l (Figure 9.16c). Macroalgal growth

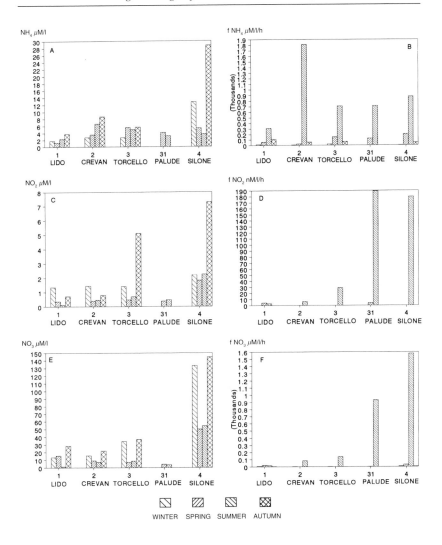

Figure 9.17 Seasonal changes in ammonium (A), nitrite (C) and nitrate (E), and their respective uptake rates (B, D and F) along the Porto di Lido–Silone Channel transect

was intense in spring and this may explain the absence of the classical phytoplankton spring bloom (Chl < 2 μg/l). Chlorophyll concentrations at the Lido entrance remained at relatively low levels (< 2 μg/l) throughout the year. Seasonal changes of PON (Figure 9.16d) were similar to those of chl a.

Seasonal changes in N uptake rates were characterized by a well-defined summer maximum with exceptionnally high values of ammonium and nitrate uptake rates (respectively 1.81 μmol/l/h at Crevan and 1.575 μmol/l/h in the Silone Channel, Figures 9.17b,f) corresponding to high chl concentrations.

179

These uptake rates are among the highest reported in the literature for coastal temperate systems (Table 9.5). During the other three seasons, N uptake rates were in the same range as in other coastal environments (Tables 9.3, 9.4 and 9.6).

During the four seasonal studies, ammonium uptake rates were generally higher than nitrate uptake rates (Figures 9.17b,f). Nitrate uptake exceeded ammonium uptake only in summer in the low-salinity waters of Palude della Rosa and Silone Channel where nitrate concentrations were higher. Nitrite uptake rates were negligible (Figure 9.17d) and were generally an order of magnitude lower than ammonium uptake rates. Ammonium, therefore, represented the main source of N for phytoplankton nutrition and contributed more than 70% of the N demand. Nitrate was the second most important N source (< 30%) for phytoplankton. Nitrite accounted only for a few percent and was a negligible source of N for phytoplankton (Table 9.7).

Except in a few cases (during summer, in low salinity waters), ammonium uptake was proportionally greater than its presence in the ambient N pool (Figure 9.18a) and reciprocally, nitrate uptake was proportionally lower than its proportion (Figure 9.18b). Ammonium was, therefore, generally taken up preferentially over nitrate in the Venice Lagoon. Though the proportion of ammonium did not exceed 50% of DIN (Figure 9.19a), its uptake represented more than 70% of total N uptake (Figure 9.19b), thus confirming its importance for nitrogenous nutrition of phytoplankton in the Venice Lagoon throughout the year. The major ammonium sources were river inputs during spring and probably autumn and *in situ* regeneration in summer. It would be worthwhile in future studies to quantify exactly the N regeneration rates in the Venice Lagoon using ^{15}N methodology in order to determine the relative importance of external inputs (river inputs) and internal inputs (regeneration) of ammonium into the Venice Lagoon.

An estimate of the total amount of N taken up by phytoplankton during an annual cycle was calculated for the salinity range of 15–35 PSU measured

Table 9.7 Mean seasonal salinities and mean percentages of ammonium, nitrite and nitrate uptake rates *vs.* total inorganic N uptake rates

Station	Mean seasonal salinity	% NH_4 uptake rate	% NO_3 uptake rate	% NO_2 uptake rate
Porto di Lido	33.75	0.83	0.15	0.02
Crevan	31.05	0.92	0.08	0.01
Torcello Channel	27.33	0.92	0.08	0.01
Palude della Rosa	25.25	0.68	0.29	0.03
Silone Channel	15.74	0.64	0.36	0.01
Area weighted average		0.80	0.19	0.01

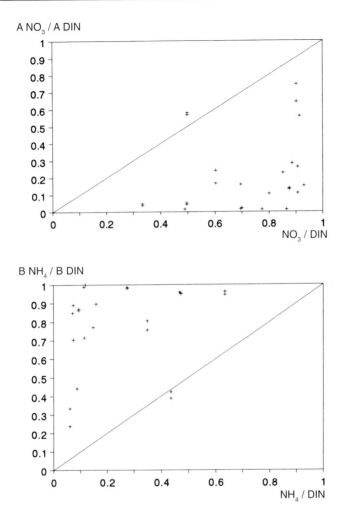

Figure 9.18 Uptake rates of nitrate (A) and ammonium (B) as percentages of total N uptake *vs* percentage nitrate (A) and ammonium (B) content in the DIN pool in the Venice Lagoon

over a major part of the Lagoon. An average estimate of annual N utilization was made taking into account the different seasons and stations, and taking 1 m as the mean water column depth. This amounted to 25.3 gN/m^2/year, equivalent to 167.7 gC/m^2/year (with a C/N ratio of 6.625). Compared with the amount of N fixed by *Ulva*, estimated at 51.4–69.9 gN/m^2/year (Sfriso *et al.*, 1987), this represents 36% to 49% of the amount fixed by *Ulva* and is, therefore, far from negligible. These estimates of N uptake by phytoplankton give new insights on N dynamics in the Lagoon. These must be taken into

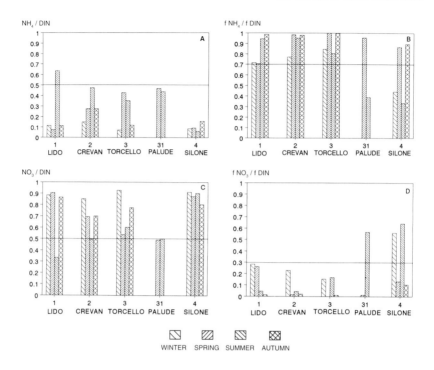

Figure 9.19 Relative importance of ammonium ($< 50\%$, A) and nitrate ($> 50\%$, C) concentrations and of ammonium ($>70\%$, B) and nitrate ($< 30\%$, D) uptake rates vs concentrations and uptake rates of total inorganic N

consideration when establishing a budget of the yearly rates of organic matter in the Venice Lagoon.

In conclusion, the salient results of this study are:

(1)   The unsuspected importance of N uptake rates by phytoplankton: for ammonium and nitrate, they are among the highest reported in similar coastal temperate environments;

(2)   The existence of the maximum rates of N uptake during summer. During the four seasonal studies, ammonium uptake rates were generally higher than those of nitrate, except in summer in the low-salinity waters of the Silone Channel and Palude della Rosa;

(3)   Ammonium represents the main N source ($> 70\%$ of total N uptake rates) for phytoplankton nitrogenous nutrition;

(4)   On an annual basis, N uptake by phytoplankton has been evaluated to be 36–49% of the nitogen uptake by *Ulva*;

(5)   The existence of an important summer phytoplankton bloom, and the absence of a spring phytoplankton bloom (chl < 2 μg/l) when the proliferation of macroalgae (*U. rigida*) was intense;

(6)   The macroalgal cycle in the Venice Lagoon in 1992 was comparable to that observed in the 1960s, with an intense spring proliferation and reduced summer development replaced by an intense phytoplankton bloom. This trend, if confirmed in future years, could indicate a significant decrease in eutrophication of the Venice Lagoon.

REFERENCES

Albertotanza, L. and Piergallini, G. (1986). Biomasse algale nella laguna di Venezia. *Ambiente Risorse Salute*, **51**: 45–7

Appi, A., Basili, M., Bendoricchio, G., Berbenni, P., Bortolussi, R., Fontanive, A., Franco, M., Marchesi, R., Perrucio, L. and Prokopowicz, J. (1989). Agriculture run-off pollution in the Venice lagoon. Consorzio Venezia Nuova (ed.), Soc. Coop. Tipogr., Padova

Azam, F., Fenchel, T., Field, J. G., Gray, J. S., Meyer Reil, L. A. and Thingstad, F. (1983). The ecological role of water-column microbes in the sea. *Mar. Ecol. Progr. Ser.*, **10**: 257–63

Bach, S. D. and Josselyn, M. N. (1979). Production and biomass of *Cladophora prolifera* (Chlorophyta, Cladophorales) in Bermuda. *Botan. Mar.*, **22**: 163–8

Battaglia, B., Datei, C., Dejak, C., Gambaretto, G., Guarise, G.B., Perin, G., Vianello, E. and Zingales, F. (1983). Hydrothermodynamic and biological investigations to determine the environmental consequences of the functionality of the ENEL thermoelectric plant in Porto Marghera. Regione Veneto, Vol. 4, Venezia

Battiston, L., Giommoni, A., Pilan, L. and Vincenzi, S. (1983). Salinity exchange induced by tide in the Venice lagoon. *Proc. Conf. On five centuries of water management in the Venice Territory*, **2**: 2–23

Bendoriccho, G., Agostinello, L. and Alessandrini, S. (1985). Nutrients of non-point sources from the river catchment (In Italian). *Acqua Aria*, **2**: 177

Bernardi, S., Cecchi, R., Costa, F., Ghermandi, G. and Vazzoler, S. (1986). Freshwater and pollutant transfer in the lagoon of Venice. *Inquinamento*, **28**: 46–55

Bianchi, F., Boldrin, A., Cioce, F., Rabitti, S. and Socal, G. (1987). Variazioni stagionali dei nutrienti e del materiale particellato nella laguna di Venezia. Bacino settentrionale. *Ist. Ven. Sci. Lett. Arti*, **11**: 49–67

Bianchi, F., Cioce, F., Comaschi-Scaramuzza, A. and Socal, G. (1990). Dissolved nutrient distribution in the central basin of the Venice lagoon. Autumn 1979. *Boll. Mus. civ. St. nat. Venezia*, **39**: 7–19

Cadée, G. C. and Hegeman, J. (1974a). Primary production of the benthic microflora living on tidal flats in the Dutch Wadden Sea. *Neth. J. Sea Res.*, **8**: 260–91

Cadée, G. C. and Hegeman, J. (1974b). Primary production of phytoplankton in the Dutch Wadden Sea. *Neth. J. Sea Res.*, **8**: 240–59

Cavazzoni, S. (1973). Acque dolci della laguna di Venezia. *Rapporto n.64 CNR, Laboratorio delle Grandi Masse*

Consorzio Venezia Nuova (1988). Distribution of the contaminant concentrations and of current speeds at different sections of the port entrances. *Internal Report, Study* 1.3.3

Cossu, R. and Giuliano, M. (1980). L'apporto di nutrienti dall'atmosfera alla laguna di Venezia. *Ingegneria Ambientale*, **9**: 283–90

Degobbis, D., Gilmartin, M. and Orio, A. A. (1986). The relation of nutrient regeneration in the sediments of the Northern Adriatic to eutrophication, with special reference to the lagoon of Venice. *Science Total Environment*, **56**: 201–10

Dillon, C. R. (1971). A comparative study of the primary productivity of estuarine phytoplankton and macrobenthic plants. Ph.D Thesis, Univ. North Carolina, Chapel Hill, pp. 112

Dugdale, R. C. and Goering, J. J. (1967). Uptake of new and regenerated forms of nitrogen in primary productivity. *Limnol. Oceanogr.*, **12**: 196–206

Dugdale, R. C. and Wilkerson, F. P. (1986). The use of ^{15}N to measure nitrogen uptake in eutrophic oceans; experimental considerations. *Limnol. Oceanogr.*, **31**: 673–89

ENEA (1988). Integrative elements on the polluting inputs in the lagoon. *Final report, Consorzio Venezia Nuova*

ENEA (1990). Final report on the present conditions of the lagoon ecosystem. *Report 1.3.9, Consorzio Venezia Nuova*

Eppley, R. W., Carlucci, A. F., Holm-Hansen, O., Kiefer, D., McCarthy, J. J. and Williams, P. M. (1972). Evidence for eutrophication in the sea near southern California coastal sewage outfalls. – July 1970. *Cal. Coop. Oceanic Fish. Inv. Rpt.*, **16**: 74–83

Facco, S., Degobbis, D., Sfriso, A. and Orio, A. A. (1986). Space and time variability of nutrients in the waters and sediments of the Venice lagoon. In Wolfe, D. A. (ed.). *Estuarine variability*, Academic Press, NY, 307–18

Fonselius, S. H. (1978). On nutrients and their role as production limiting factors in the Baltic. *Acta Hydrochim. Hidrobiol.*, **6**: 329–39

Fossato, V. (1990). Lagoon waters. In Ente zona industriale di Porto Marghera e associazione degli industriali della provincia di Venezia, (eds). *Porto Marghera, Venice, and its environment*, Canova, Treviso, 81–112

Furnas, M. J. (1983). Nitrogen dynamics in lower Narragansett Bay, Rhode Island. I. Uptake by size-fractionated phytoplankton populations. *J. Plankt. Res.*, **5**: 657–76

Furnas, M. J., Smayda, T. J. and Deason, E. A. (1986). Nitrogen dynamics in lower Narragansett Bay. II. Phytoplankton uptake, depletion rates of Nous nutrient pools, and estimates of ecosystem remineralization. *J. Plankt. Res.*, **8**: 755–69

Garside, C. (1981). Nitrate and ammonium uptake in the apex of the New York Bight. *Limnol. Oceanogr.*, **26**: 731–9

Glibert, P. M. and Garside, C. (1992). Diel variability in nitrogenous nutrient uptake by phytoplankton in the Chesapeake Bay plume. *J. Plankt. Res.*, **14**: 271–88

Glibert, P. M., Garside, C., Fuhrman, J. A. and Roman, M. R. (1991). Time-dependent coupling of inorganic and organic nitrogen uptake and regeneration in the plume of the Chesapeake Bay estuary and its regulation by large heterotrophs. *Limnol. Oceanogr.*, **36**: 895–909

Goldman, J. C. (1976). Identification of nitrogen as a growth-limiting nutrient in wastewaters and coastal marine waters through continuous culture algal assays. *Wat. Res.*, **10**: 97–104

Koroleff, F. (1969). Direct determination of ammonia in natural waters as indophenol blue. *Int. Rep. Cons. int. Explor. Mer*, **3**: 19–22

Kveder, S., Revelante, N., Smodlaka, N. and Skrivanic, A. (1971). Some characteristics of phytoplankton and phytoplankton productivity in the Northern Adriatic. *Thalassia Jugosl.*, **7**: 151–8

Larsonn, U., Elmgren, R. and Wulff, F. (1985). Eutrophication and the Baltic Sea: causes and consequences. *Ambio*, **14**: 9–14

L'Helguen, S. (1991). Absorption et régénération de l'azote dans les écosystèmes pélagiques du plateau continental de Manche Occidentale. Relations avec le régime de mélange vertical des masses d'eaux; cas du front thermique d'Ouessant. Thèse Doct. Chimie Marine, Université de Bretagne Occidentale, Brest, pp. 212

MacCarthy, J. J., Taylor, W. R. and Taft, J. L. (1975). The dynamics of nitrogen and phosphorus cycling in the open waters of the Chesapeake Bay. In Church, T. M. (ed.). *Marine Chemistry in the coastal environment, A.C.S. Symp. Ser.,* **18**: 664–81

MacCarthy, J. J., Taylor, W. R. and Taft, J. L. (1977). Nitrogenous nutrition of the plankton in the Chesapeake Bay. I. Nutrient availability and phytoplankton preferences. *Limnol. Oceanogr.,* **22**: 996–1011

MacIsaac, J. J. and Dugdale, R. C. (1972). Interactions of light and inorganic nitrogen in controlling nitrogen uptake in the sea. *Deep-Sea Res.,* **19**: 209–32

Magistrato alle Acque (1985). Stato delle Conoscenze sullo Inquisnamento della Laguna di Venezia. *Consorizio Venezia Nuova,* 4 vol.

Marcomini, A., Sfriso, A., Pavoni, B. and Orio, A. A. (1992). Eutrophication of the lagoon of Venice: nutrient loads and exchanges. In McComb, A. J. (ed.). *Eutrophication in shallow estuaries and lagoons.* CRC Press, Boca Raton, Florida, USA

Nixon, S. W. (1980). Between coastal marshes and coastal waters – A review of twenty years of speculation and research on the role of salt marshes in estuarine productivity and water chemistry. In Hamilton, E. P. and McDonald, K. B. (eds). *Estuarine and wetland processes.* Plenum Publishing Corp., NY, 437–525

Nixon, S. W. (1981). Remineralization and nutrient cycling in coastal marine ecosystems. In Nelson, B. J. and Cronin, L. E. (eds). *Estuaries and nutrients.* Humana Press, NY, 111–38

Nixon, S. W. and Lee, V. (1981). The flux of carbon, nitrogen and phosphorus between coastal lagoons and offshore waters. In *Coastal lagoon research, present and future.* UNESCO, Div. Marine Sci., Paris, 325–48

Nixon, S. W. and Pilson, M. Q. (1983). Nitrogen in estuarine and coastal marine ecosystems. In Carpenter, E. J. and Capone, D. G. (eds). *Nitrogen in the Marine Environment.* Academic Press, NY, 565–648

Orio, A.A. and Donazzolo, R. (1987). Toxic and eutrophicating substances in the lagoon and the Gulf of Venice. *Ist. Ven. Sci. Lett. Arti,* **11**: 149–215

Paasche, E. and Kristiansen, S. (1982). Nitrogen nutrition of the phytoplankton in the Oslofjord. *Est. Coast. Shelf Sci.,* **14**: 237–49

Pavoni, B., Sfriso, A., Donazzolo, R. and Orio, A. A. (1990). Influence of waste waters from the city of Venice and the hinterland on the eutrophication of the lagoon. *Science Total Environment,* **96**: 235–52

Pennock, J. R. (1987). Temporal and spatial variability in phytoplankton ammonium and nitrate uptake in the Delaware Estuary. *Est. Coast. Shelf Sci.,* **24**: 841–57

Postma, H. (1981). Exchange of materials between the North Sea and the Wadden Sea. *Mar. Geol.,* **40**: 199–213

Redfield, A. C., Ketchum, B. H. and Richards, F. A. (1963). The influence of organisms on the composition of sea-water. In Hill, M. N. (ed.). *The Sea.* Intercience, NY, London, **2**: 26–7

Riley, G. A. (1941). Plankton studies. III. Long Island Sound. *Bull. Bingham Oceanogr. Coll.,* **7**: 1–93

Ryther, J. H. and Dunstan, W. M. (1971). Nitrogen, phosphorus, and eutrophication in the coastal marine environment. *Science,* **117**: 1008–13

Sfriso, A. and Cavollo, S. (1983). La situazione della alghe nella laguna di Venezia. *Ambiente Risorse Salute,* **16–17**: 38–9

Sfriso, A., Marcomini, A. and Pavoni, B. (1987). Relationships between macroalgal biomass and nutrient concentrations in a hypertrophic area of the Venice lagoon. *Mar. Environ. Res.*, **22**: 287–312

Sfriso, A., Pavoni, B., Marcomini, A. and Orio, A. A. (1988a). Annual variations of nutrients in the lagoon of Venice. *Mar. Poll. Bull.*, **19**: 54–60

Sfriso, A., Pavoni, B., Marcomini, A. and Orio, A. A. (1988b). Macroalgal production and nutrient recycling in the lagoon of Venice. *Ingegneria Sanitaria*, **5**: 255–66

Sfriso, A., Pavoni, B., Marcomini, A. and Orio, A. A. (1992). Macroalgae, nutrient cycles and pollutants in the lagoon of Venice. *Estuaries*, **15**: 517–28

Sfriso, A., Marcomini, A. and Pavoni, B. (1989). Macroalgae and phytoplankton standing crops in the central Venice lagoon: primary production and nutrient balance. *Science Total Environment*, **80**: 139–59

Sfriso, A., Marcomini, A., Pavoni, B. and Orio, A. A. (1991). Species composition, biomass and net primary production in shallow coastal waters: the Venice lagoon. *Biores. Technol.*, **28**: 365–84

Smayda, T. J. (1957). Phytoplankton studies on lower Narragansett Bay. *Limnol. Oceanogr.*, **2**: 342–59.

Smodlaka, N. (1986). Primary production of organic matter as an indicator of eutrophication in the Northern Adriatic Sea. *Science Total Environment*, **56**: 211–20

Socal, G., Bianchi, F., Comaschi-Scaramuzza, A. and Cioce, F. (1987). Spatial distribution of plankton communities along a salinity gradient in the Venice lagoon. *Archo Oceanogr. Limnol.*, **21**: 19–43

Taft, J. L. and Taylor, W. R. (1976). Phosphorus dynamics in some coastal plain estuaries. In Wiley, M. (ed.). *Estuarine Processes*, Vol.1, Academic Press, NY, 79–89

Thayer, G. W. (1974). Phytoplankton production and the distribution of nutrients in a shallow unstratified estuarine system near Beaufort, NC. *Oecologia*, **14**: 75–92

Thomas, W. H., Seibert, D. L. R. and Dodson, A. N. (1974). Phytoplankton enrichment experiments and bioassays in natural coastal seawater and in sewage outfall receiving waters off southern California. *Est. Coast. Shelf Sci.*, **2**: 191–206

Tréguer, P. and Le Corre, P. (1975). Manuel d'analyse des sels nutritifs dans l'eau de mer (utilisation de l'AutoAnalyser II Technicon) 2ème édition. *Université de Bretagne Occidentale*, pp. 110

Valliela, I. and Costa, J. E. (1988). Eutrophication of Buttermilk Bay, a Cape Cod coastal embayment: concentrations of nutrients and watershed nutrient budgets. *Environ Management*, **12**: 539–53

Vanucci, M. (1969). What is known about production of coastal lagoons. *Lagunas costeras, un simposio, Mem. Simp. Inter. Lagunas Costeras.* UNAM-UNESCO, Paris. 457–78

Vatova, A. (1960). Primary production in the high Venice lagoon. *J. Cons. Perm. Explor. Mer*, **26**: 75–84

Vazzoler, S., Costa, F. and Bernardi, S. (1987). Laguna di Venezia: trasferimento di acqua dolce e inquinanti. *Rapporti e studi, Ist. ven. Scienze*, **11**: 81–124

Veneto Region (1988). Rapporto sul disinquinamento della laguna di Venezia e piano degli interventi. pp. 121

Voltolina, D. (1973). A phytoplankton bloom in the lagoon of Venice. *Archo. Oceanogr. Limnol.*, **18**: 19–37

Yentsch, C. S. and Menzel, D. W. (1963). A method for the determination of phytoplankton chlorophyll and phaeophytins by fluorescence. *Deep-Sea Res.*, **10**: 221–31

Section 3

Benthic Studies

CHAPTER 10

SEDIMENT BIODEGRADATIVE POTENTIAL OF THE VENICE LAGOON (PALUDE DELLA ROSA)

N. Sabil, D. Tagliapietra, Y. Aissouni and M.-A. Coletti-Previero

SUMMARY

Within a shallow-water area, the Palude della Rosa, of the Venice Lagoon, insoluble bacterial enzymes (cellulase, phosphatase and urease) were found in the sediment phase, while no detectable enzyme activity was present in the aqueous phase, even after centrifugation. Urease was strongly bound and unaffected by all procedures tested, whereas phosphate and some cellulase could be removed by phosphate anions. Cellulase was also solubilized by a non-ionic detergent. After release, phosphatase and cellulase were purified by chromatographic procedures.

These enzymes, immobilized on the inorganic component of the sediment, were probably of bacterial origin but were able to outsurvive the organisms from which they originated. They showed marked heat stability, increased resistance to environmental changes and a lifespan prolonged by their insolubility. As a consequence, they resist conditions where similar soluble enzymes would be rapidly inactivated, thus they have evolved a protective relay system when a natural or anthropogenic event lowers the density of microorganisms in the Lagoon. They are useful as diagnostic factors of the ecosystem, since their presence is related to waste products.

INTRODUCTION

The Palude della Rosa area of the Venice Lagoon was studied in the framework of the UNESCO-MURST Venice Lagoon Project as a model ecosystem of the northern Lagoon. As with all lagoons, it can be roughly subdivided into three phases, gaseous (air), aqueous (water) and solid (sediments) (Figure 10.1). Biodegradation occurs in each of these compartments and at the interphases between them. Figure 10.1 represents the vertical profile of a sedimentary ecosystem with an ideal subdivision of the degradation events. Distinct populations of microorganisms decompose the organic matter in the sequence shown in Figure 10.1 toward terminal acceptors which are, successively, carbonate, nitrate, sulfate and oxygen. The molecular events underlying this degradation are the actions of catalytic systems (enzymes) produced by the microorganisms, and this is the driving force in the transformations accompanying decomposition.

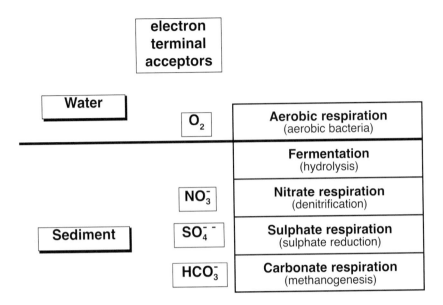

Figure 10.1 Schematic representation of the vertical profile of a sedimentary ecosystem

Biological degradation has been around for millions of years and any compound present in the environment can be degraded by one or other organism. However, xenobiotic compounds may or may not be biodegradable. They include many products arising from combustion of natural elements which are often, at best, slow to degrade. Substantial evidence exists to show that enzymes, evolved for the degradation of biogenic compounds, can be recruited to degrade xenobiotics and that microorganisms are also able to modulate their enzyme synthesis to transform large quantities of new compounds. While abundant literature is available on the pollution of two, of the air and water layers of the habitat, relatively little is known about the sediments. This is surprising, since, of the three components, it is the least subject to changes such as those induced by winds and tides/currents. Only studies on bacterial development are available, and these, although interesting, overlook an important parameter: the 'immobilized enzymatic potential'. We have focused our efforts on the insoluble phase of the Palude by undertaking large-scale monitoring of its enzymatic activity and by evaluating its biodegradative potential and the relative importance of anthropogenic influences.

From previous studies on anaerobic sediments from septic tanks and sludges, we were able to show that the enzymatic activities responsible for degradation of phosphate derivatives, proteins, cellulose, lipids and urea were mostly linked to the insoluble material. Since this insolubility protects the enzymatic activity, it is of interest to determine whether the same situation occurs in other biological systems that include both a solid and liquid phase and to ascertain whether such enzymatic processes are ubiquitous. We also investigated *in vitro* changes in sediment enzyme activities with temperature, pH, time and seasonal variation, in order to evaluate their responses to environmental changes, even extreme ones, and compared them with the behaviour of the same catalysts, when solubilized.

EXPERIMENTAL

Samples

The *Palude della Rosa* is a round-shaped pond 2 km in diameter and with average depths of 80 cm in the centre and 10 to 50 cm at the edges, with reference to mean sea level. Thirty-eight sediment cores were collected near the centre and along the boundaries of the Palude (Figure 10.2) from 10 to 30 cm under the Lagoon bed. The cores were collected using 5 cm-diameter core tubes, stored at −18 °C to avoid losses in enzymatic activities, and analysed within a week of sampling. A comparison of the results from some samples analysed before freeze-drying and after storage showed no significant decay in storage, even after 20 days. The sediment in all cores was a relatively uniform mix of 80% silt clay (ø < 62 μm) and sand, with some shell fragments which were normally removed. This composition was similar to that (clay and silted clay at 75 to 90%) reported by Froelich (1980) for the Palude della Rosa sediments. The pH of the samples varied from 7.7 to 8.4 and the local water

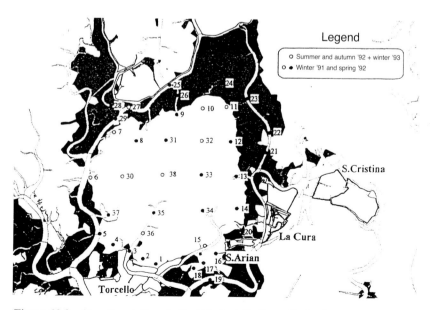

Figure 10.2 Sampling stations in the Palude della Rosa, Venice Lagoon

temperatures at the time of sampling were 5–7 °C (winter), 9–11 °C (spring), 25–30 °C (summer) and 14–16 °C (autumn).

Enzyme activities

The insoluble material was centrifuged for 30 min at 10,000 rpm and washed twice with an isotonic solution. The enzyme activities were measured in triplicate on suitable quantities of the centrifuged sediment (from 0.1 to 1 g in the same volume) and of the combined supernatants. The sediment samples were suspended in the buffer required for each assay and the kinetics measured as a function of time (Figure 10.3).

(1)   Protease activity: haemoglobin (10 mg/ml) was hydrolyzed at 37 °C, pH 8 (0.2 M Tris/HCl buffer) (Kunitz, 1947).
(2)   Phosphatase activity: *p*-nitrophenyl phosphate (PNPP) was hydrolyzed to *p*-nitrophenol at 37 °C, pH 5.5 (0.2 M maleate buffer) and the reaction was followed at 400 nm (Bonmati *et al.*, 1985).
(3)   Cellulase activity: carboxymethyl cellulose (CMC) was incubated at 37 °C, pH 4.5 (0.2 M acetate buffer) and the digestion was followed by analysis of liberated reducing sugars (Park and Johnson, 1949).
(4)   Lipase activity: tributyrin was incubated at 25 °C, pH 7.5 and the digestion was followed by continuous neutralization of the mixture with 0.01 M NaOH using an automatic titrimeter pHstat (Norbert *et al.*, 1989).
(5)   Urease activity: urea was incubated in the same experimental conditions as (4) above but using 0.01 M HCl as neutralizing solvent (Maunoir *et al.*, 1991).

Desorption

Washing in isotonic solution and centrifugation for 30 min at 10,000 rpm did not release any enzymatic activity. The sediments were therefore treated with the following solutions: 0.5 M NaCl, 0.5 M $NaHCO_3$, 0.5 M phosphate buffer at pH 8 and 1% Triton X-100. A suitable quantity of each sediment (0.1 to 1.0 g) was suspended in each solution (5 ml total volume) and allowed to react for 15 min at room temperature. After centrifugation, the activity was determined in the supernatant.

Temperature and pH

The sediments were heated to different temperatures (from 37 ° to 100 °C) for 15 min and then rapidly cooled to room temperature. The different residual activities were determined as described above.

The enzymatic reactions were determined at pH values ranging from 3 to 9.5.

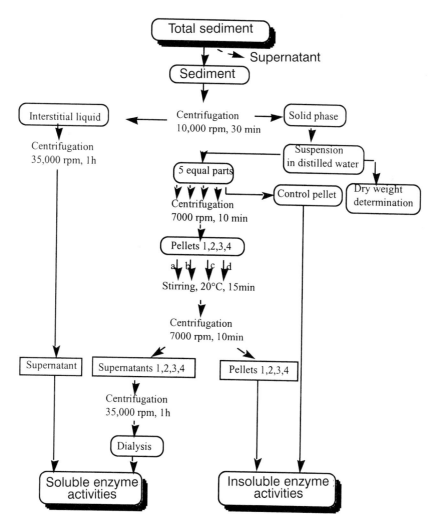

Figure 10.3 Sample procedure for enzyme determination: a) 0.5 M NaCl; b) 0.5 M NaHCO$_3$; c) 0.5 M Phosphate buffer (pH 8), and d) 1% Triton X-100

Seasonal variations

The activities in the sediments of Palude della Rosa were measured in the four seasons over the year (December 1991, March, July and October 1992).

Isolation of phosphatase and cellulase

The enzymes were released from the sediment by phosphate buffer (0.5 M, 15 min) and were first chromatographed on Sephadex G 75, which separates by molecular sieving. The active fractions were then charged in the same buffer on a Mono Q column (basic ion exchange) where they eluted only under the effect of increasing salt concentration.

ENZYMATIC ACTIVITIES

It is usually assumed that enzymatic activity is confined to the soluble phase or to the pore-water phase of the upper layer of the sediment column. However, the aqueous phase of the sample cores was devoid of measurable enzymatic activities showing that the concentration of enzymes was, at best, undetectable.

By contrast, the insoluble phase showed distinct and sometimes unexpectedly strong enzymatic activities responsible for the degradation of urea, cellulose and phosphate derivatives, mostly linked to the insoluble material. Protease and lipase activities were undetectable: they are probably present in the ecosystem but at a dilution that is impossible to determine with any degree of certainty. Table 10.1 summarises the situation.

Representation of the enzyme activities on the geographical map of the Palude shows distinct areas of activity (Figure 10.4). These patterns were obtained by joining the sample positions with nearly-identical to identical activity by lines, hereafter called iso-enzymatic (lines). The shaded areas represent zones of higher activity while the striped lines are zones of lowest activity. The curves take into account the nearest neighbouring enzymatic parameter (Sabil *et al.*, 1993).

The form of binding between enzymes and the insoluble material is an essential parameter. If only Van der Waals forces were responsible for adsorption of the enzymes onto the insoluble matrix, then $NaHCO_3$, $NaCl$ or any other salt solution should desorb them; whereas if the enzymes were entrapped in whole or fragments of membranes of bacteria or other microorganisms, the non-ionic detergent Triton X-100 should partly desorb them. Three approaches, as detailed below, were used to differentiate between molecules bound to bacterial membranes and those bound to the inorganic part of the sediment.

Table 10.1 Enzymatic activities of the different components of the sediment from the Palude della Rosa (spring survey, averages from 30 experiments)

		Phosphatase[1]	*Cellulase*[2]	*Urease*[1]	*Protease*	*Lipase*
Supernatant	Control	–	–	–	–	–
	Palude d. R					
Interstitial liquid	Control	trace	trace	trace	–	–
	Palude d. R	0.24 ± 0.04	0.11 ± 0.03	0.075 ± 0.005	–	–
Solid phase	Control	0.82 ± 0.12	2.26 ± 0.18	2.44 ± 0.09	trace	trace
	Palude d. R	9.14 ± 0.82	9.35 ± 0.55	7.39 ± 0.57	–	–

[1]μmoles/min/g, [2]mg CMC/h/g

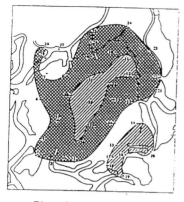

Phosphatase activity

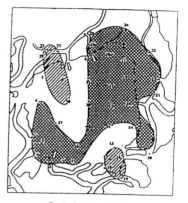

Cellulase activity

Urease activity

Figure 10.4 Enzymatic maps of the Palude della Rosa sediments. Striped zone = lower activity; grid zone = higher activity

(1) Some active sediment samples were analysed by scanning electron microscopy. This did not show any intact bacteria but, since only a small part of the sediment samples was taken for this analysis, the presence of bacteria cannot be completely ruled out. In any case, they may not occur in quantities sufficient enough to account for the intensity of the enzymatic activities measured. This is a first indication suggesting that most of the activity detected is from enzymes present in an insoluble form.

(2) The centrifuged sediments were submitted to differential extraction tests. Treatment with 0.5 M NaHCO$_3$ and 0.5 M NaCl at pH 8 did not release any urease and phosphatase activities and less than 10% of the cellulase activity was extracted. However, when a phosphate solution of the same

195

molarity (0.5 M, pH 8) was used, most of the phosphatases and a significant part of the cellulase activity were extracted. This confirms that the ionic strength was not *per se* responsible for the release of the enzymes from the sediment and that the form of ion was essential. This specific effect of phosphate anions on enzymes linked to inorganic material was previously reported in studies from our laboratory on biotechnological procedures that include immobilized enzymes (Pugnière *et al.*, 1986; Coletti-Previero and Previero, 1989).

(3)   A fraction of immobilized cellulase activity could be liberated using a 1% solution of Triton X-100, a non-ionic detergent often used to solubilize membrane enzymes.

Urease activity remained insoluble under all treatments. This enzyme has a tendency to adsorb to inorganic supports and, because of its stability, was one of the first insoluble enzymes to be used for therapeutical purposes. Table 10.2 summarizes the results.

Table 10.2   Release of enzymatic activities from 1 g dry weight of sediment samples with different extracting solutions and 15 min extraction time. $NaHCO_3$ did not release any activity. Extraction of urease activity was negligible in all samples

Sample (No)	NaCl 0.5 M $NaHCO_3$ 0.5M	Phosphate 0.5M buffer pH 8	Triton 1%
	Phosphatase activity (μmoles PNPP/min)		
4	tr	3.5	0.1
8	0	3.0	0.3
13	0	5.0	0.2
19	0	1.0	0.0
20	0	2.0	0.2
21	tr	3.0	0.4
23	0	0.5	0.2
25	0	0.5	0.2
29	0	2.0	0.1
30	0	0.5	0.1
33	0	0.2	0.4
35	0	2.0	0.1
	Cellulase activity (μmoles reducing sugars/h)		
1	0.1	2.0	2.5
4	0.1	2.0	3.0
8	0.2	1.5	3.5
12	0.1	1.5	2.0
14	0.1	1.0	2.5
22	0.0	2.0	2.0
23	0.5	2.0	2.0
24	0.0	0.5	1.5
25	0.1	1.5	2.0

Since phosphate anions were present in the Palude della Rosa at concentrations far below 0.5 M, we studied the effect of lower concentrations on the release of the enzymatic activity (Figure 10.5A). As can be seen from this Figure, the desorption was proportional to the concentration of phosphate anions. The effect of time of contact (Figure 10.5B) showed that, given time, the lower concentrations were effective in releasing the enzyme activities (phosphatase and cellulase). The increase of phosphate concentration in the ecosystem (0.2 to 0.3 mM), due to its increased use in surrounding agricultural lands, probably stimulated phosphatase synthesis in microorganisms. However, this effect was offset by the ability of phosphate anions to wash out the enzymatic activity from the sediment, even though they had promoted the enzyme synthesis.

It is well known that soluble enzymatic activities are easily destroyed even in a relatively mild environment. They can be affected by a number of factors, such as salt concentration, solvent composition, temperature, proteolytic enzymes, etc. Insolubility has a well-known protective effect on enzymes. Thus the enzymatic activity of the bio-sediment of the Palude della Rosa should have a high resistance to changes in the environment.

This ability to resist environmental changes is particularly evident when the effect of temperature is examined. The sediment samples were heated and residual activity tested after cooling. The temperature dependence of the sediment enzymatic activities was examined from 4 to 100 °C. This allowed us to calculate the activities at temperatures which were similar to those *in situ* and examine the resistance of the immobilized enzymes to increasing temperatures. As can be seen from Figure 10.6A, the phosphatase activity was low, but not absent, at 4 °C, and increased up to 80 °C *in situ* (Figure 10.6B). The

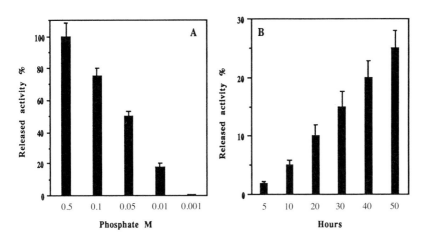

Figure 10.5 Relationship between phosphate concentration (A), contact time (B) and the release of enzymatic activity

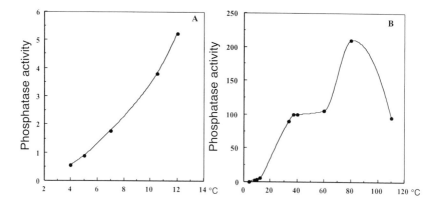

Figure 10.6 Variations of phosphatase activity with increasing temperature. (A) from 4 to 12 °C; (B) from 4 to 110 °C

sharp decrease between 80 ° and 100 °C, without loss of activity, even after 15 min, could be attributed to denaturation of the enzyme itself. This implies that the starting activity is much higher than measured and is hindered by an unknown component, whose presence became apparent only upon heating.

Cellulase activity was partly denatured by increasing temperatures but its resistance was higher than the corresponding soluble enzyme (Figure 10.7A). Urease activity at lower temperature was higher than expected: at higher temperatures the enzyme behaved like phosphatase, giving a sharp increase at around 80 °C (Figure 10.7B). These results are in agreement with the hypothesis that these enzymes are linked to the solid support since resistance to temperature increase (conditions under which the free enzymes are totally denaturated) is a typical feature of immobilized enzymes.

When phosphatase activity of a given sample was released from the insoluble support and heated (Figure 10.8), it showed the typical behaviour of a soluble enzyme, thus refuting the hypothesis of thermostable catalysts produced by thermostable bacteria.

In conclusion, while part of the insoluble cellulase activity was trapped in bacterial membrane fragments released by the detergent and sensitivity to temperature increase, the insoluble phosphatase and urease activities of the Lagoon sediment could be attributed to enzymes linked to the inorganic matrix. It is difficult at this point to ascertain the kind of linkage that binds the enzyme to the sediment but it is probably dependent on the structure of the enzymes and sediment composition. Previous work from our laboratory has shown that clay spontaneously linked hydrolytic enzymes *in vitro* (Maunoir *et al.*, 1991). Since clay is the main component of the sediment of the Palude (Froelich, 1980), the immobilization of these enzymes seems to be a reason-

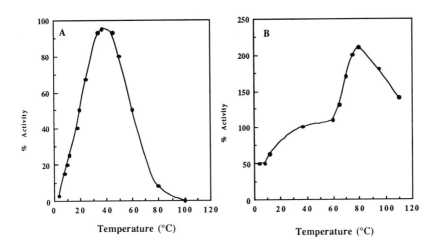

Figure 10.7 Variations of cellulase (A) and urease (B) activities with increasing temperature

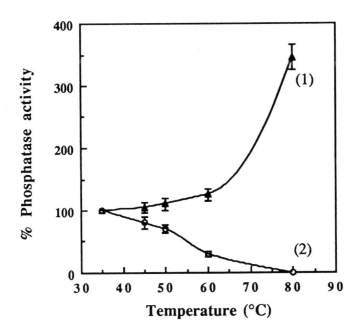

Figure 10.8 Temperature effect on phosphatase activity of immobilized (1) and released (2) enzyme. The samples were heated for 15 min and after cooling the activities were measured at 37 °C. The original activity at 37 °C was taken as 100%

able consequence of the affinity reported previously. Phosphate ions were shown to have a conflicting influence on the degradative efficiency of the Lagoon sediment. Even diluted phosphate solutions were able to release an important part of the insolubilized activities when they were allowed to interact with the sediment for a sufficiently long time. This could explain the lower enzyme concentration at the edges than at the centre of the Palude (Figure 10.4). In the case of cellulase, phosphate anions should only induce the release of the enzymatic activity, and the decrease of enzyme concentration in the most likely area of phosphate arrival was evident on the map (Figure 10.4).

The pH dependence of phosphatase and cellulase-insoluble enzymatic activities (Figure 10.9A and B) shows two optimum pH values. The basic pH zone is the value found at the Palude della Rosa and is considered the relevant value.

The pH dependence of urease activity (Figure 10.10) could be determined only on the immobilized enzyme since all our attempts to release this activity were unsuccessful.

In order to allow an evaluation of the biodegradative capacity of the inorganic part of the sediment, the following Table (Table 10.3) translates the kinetic values to more common orders of magnitude. The enzymatic values from distant sites in the Venice Lagoon (Figure 10.11) were taken as blanks and the mean value for each activity was subtracted from the values given in Table 10.3. The formula used for calculating the activity of the whole Palude is:

$$A = a \times M_s \times S \times q$$

where A = activity in the whole Palude (4 km², 1 cm depth),
a = enzymatic activity,
M_s = substrate molecular weight,
S = surface area of the Palude (4 km²) and
q = sediment weight in 1 m², to a depth of 1 cm

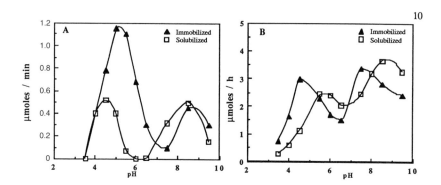

Figure 10.9 Enzymatic activities of phosphatase (A) and cellulase (B) as a function of pH from the Palude della Rosa sediment. Both immobilized and solubilized activities are shown

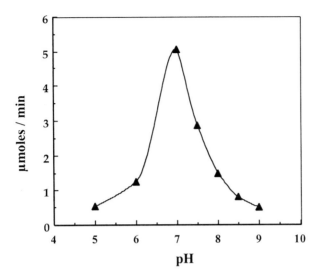

Figure 10.10 Enzymatic activity of immobilized urease as a function of pH

Table 10.3 Enzymatic activities of the sediment from the Palude della Rosa. The bold values are for the *whole* surface of the Palude

Activity	Winter	Spring	Summer	Autumn
Phosphatase[1]	0.686 ± 0.046	3.132 ± 0.256	7.236 ± 0.426	2.52 ± 0.2
	54 582 kg/min	246 902 kg/min	578 862 kg/min	194 036 kg/min
Cellulase[2]	2.61 ± 0.02	7.09 ± 0.37	4.08 ± 0.52	1.29 ± 0.19
	2 680 kg/min	9 453 kg/min	5 440 kg/min	1 717 kg/min
Urease[3]	0.008 ± 0.0005	0.29 ± 0.02	0.47 ± 0.04	0.21 ± 0.02
	624 kg/min	23 760 kg/min	37 728 kg/min	16 464 kg/min

[1]mg PNPP/min/g, [2]mg CMC/h/g, [3]mg Urea/min/g

The strength of the enzymatic activity is clearly remarkable, even if it depends on the availability of dissolved substrate in the Lagoon. This situation, where the enzyme remains in a latent form until a sufficient quantity of substrate diffuses, is common. These data, however, show that enough enzyme activity is present to digest the substrates even if they are at low concentrations.

We also analysed the phosphatase, cellulase and urease activities at the lowest temperature found in the Palude and at the relevant pH. We complemented these data with a similar determination, measured after centrifugation (to mimic eventual packing of the sediment). Even under these unfavourable conditions, the quantity of urea, phosphate derivatives and cellulose transformed per unit time (Table 10.4) will be substantial.

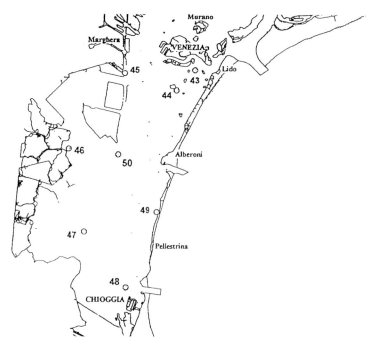

Figure 10.11 Positions in the Venice Lagoon from where blank samples were obtained

Table 10.4 Enzymatic activities of the *whole* Palude at 4 °C without mixing

Activity at optimum pH	At 4 °C without centrifugation (kg/min)				At 4 °C after centrifugation (kg/min)			
	Winter	Spring	Summer	Autumn	Winter	Spring	Summer	Autumn
Phosphatase								
pH 5.5	300	1358	3183	1067	210	948	2229	747
pH 8.0	100	455	1060	356	70	316	743	249
Cellulase								
pH 4.5	70	246	141.5	44.6	20.9	74	42.4	13.4
pH 8.0	117.6	411.6	236.6	74.8	52.5	184.8	106.4	33.7

The seasonal variations of phosphatase, cellulase and urease are shown in Figures 10.12A, B and C. As expected, all activities increase from winter to spring, while only phosphatase and urease activities were at their maximum in summer (Sabil *et al.*, 1993).

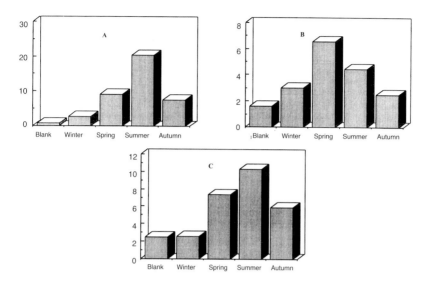

Figure 10.12   Seasonal variations of phosphatase (A), cellulase (B) and urease (C) activities (μmol/min/g) in the Palude della Rosa sediment

ISOLATION OF PHOSPHATASE AND CELLULASE

Phosphatase and cellulase, once extracted from the sediment of the Palude della Rosa, were further examined to demonstrate their presence, characterize their efficiency as catalysts, and gain insight as to which microbial entities are responsible for their synthesis.

The sediment was extracted with phosphate buffer, as described above, and the supernatant was reduced in volume and chromatographed by molecular sieve on a column of Sephadex G 75 (Pharmacia). Both eluted to the volume of the column, corresponding to a molecular weight between 55 and 70 kD. The molecular weight of cellulase was estimated to be of 57 kD and of the phosphatase, 69 kD (Figure 10.13). The active fractions were then charged on an ion exchange column (Mono Q) where the active material required a positive salt gradient to elute. This shows that both enzymes are acidic, with cellulase being slightly more acidic than phosphatase. As seen in Figure 10.14, a satisfactory separation was achieved, which confirmed the ability of phosphate anions to release these enzymes from the sediment of the Palude della Rosa. The two active peaks were apparently homogeneous in gel electrophoresis and an analysis of their amino acids confirmed purity up to about 90%.

CONCLUSIONS

Large amounts of material, including dielectric fluids, coolants, lubricants, pesticide, etc, are released into the Palude della Rosa environment. An

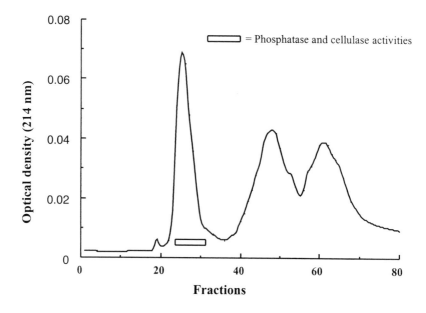

Figure 10.13 Molecular sieve of extracted enzymatic activity from the sediment of the Palude della Rosa

increasing need to assess the safety of waste repositories and the fate of agricultural/industrial chemicals has arisen in the last few years, mainly because of the fact that they accumulate in land, ponds, lakes, rivers, groundwater, soils and in the atmosphere. Predictions about the migration of contaminants are regularly tested but they generally fail: even the little we know of the ecosystem is usually far too complex for extrapolation. The way forward is still through good empiric research, unless and until we can soundly describe most of the variables. One of the factors, the degradation of chemicals by enzymes accumulated by immobilization on the inorganic part of the sediment, had completely escaped our attention and was never considered.

What are the conclusions this new finding suggests and what are the new questions it raises? Even though an initial choice of enzymatic activity to be measured had to be made, thereby limiting our study to initial degradative events, some of the activities found are strong enough not only to be measured with great precision but also to allow the study of their release without having to deal with the problems that could have arisen from the need to treat a large quantity of sediment.

(1) The activities in the sediment from different locations of the Palude della Rosa are significantly different enough to provide a clear distinction among areas of activity, through iso-enzymatic lines. The presence of an

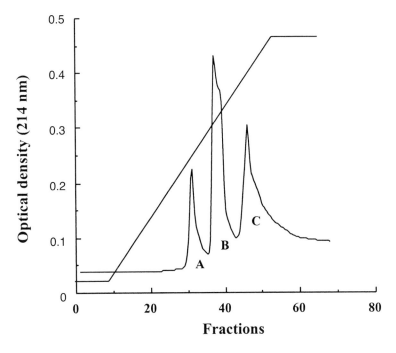

Figure 10.14 Ion exchange chromatography of the active fractions from G 75 Sephadex. A: Phosphatase activity; B: Inactive peak; C: Cellulase activity

'enzymatic map' in a Lagoon sediment had never been suspected and therefore its delineation had never been attempted.

(2) The relation between the bio-molecules and the insoluble material could be inferred from the results on the release of the enzymes by chosen solutions. Part of the cellulase was trapped in bacterial membranes, while urease and phosphatase were linked to the inorganic material of the sediment with different affinities depending on the enzyme.

(3) A new understanding of the action of phosphates on the degradative potential of the Lagoon sediment begins to unravel. In fact, even very low concentrations of phosphate ions can wash out the insoluble enzymatic activity if they are allowed to stay in contact long enough.

(4) This biodegradative capacity appears to be able to survive without the microorganisms from which it was generated and represents an important, and perhaps, vital share of the degradative power of the ecosystem. Moreover, these activities can be considered, up to a certain point, of diagnostic importance since they are directed towards compounds which are well known to be the leading contaminants of the Venice Lagoon (urea and phosphate derivatives from industry and agriculture, cellulose from the degradation of algal cell walls).

(5) The enzymatic activity, if present, is found in the insoluble part of the system. This is worth considering since it is well known that soluble enzymatic activities are easily destroyed even in a protected environment. This insolubility has a protective effect, giving to the enzymatic activity of a bio-sediment the possibility to resist otherwise denaturating conditions.

(6) The biodegradative potential of any ecosystem is a function of the environment. We have shown here that it is also a function of the previous history of the sediment, since immobilization of enzymes allows their accumulation and prolongs their lifespan for months (Aissouni *et al.*, in press). When a natural or anthropogenic event lowers the density of a microorganism in the Lagoon, the insoluble enzymatic activity remaining can take over while the population of the microorganism develops again to its optimum density. If, however, the negative effect goes beyond enzyme breakpoint, the two-step relay collapses, sometimes dramatically.

(7) The regenerating power of the sediment can be put at risk by contaminants which might not be particularly dangerous *per se* but could simply poison these activities, thus initiating an irreversible chain of events. Mercury, which inactivates all enzymes with -SH groups in their active centre is an example of such a selective and specific interaction. Obviously, if the inactivated enzyme is at a metabolic crossroads the consequences will be far more visible than if the enzyme has an activity which can be overtaken by other similar ones (Sabil *et al.*, submitted). At the moment, however, it is impossible to foresee whether the effect will be positive or negative.

(8) The sediment composition is an essential parameter of this immobilization of enzymes and close attention should be paid to any interference that tends to change the inorganic composition of the sediment. For instance, new channels from the sea, especially if they are deep, can facilitate an increase in sand level as compared to clay and/or silted clay (the main components of Venice Lagoon sediments). It can be predicted that the recent replacement of phosphate by zeolite in commercial washing powders will, given time, affect the sediment composition and hence its enzymatic capacity both quantitatively and qualitatively.

ACKNOWLEDGEMENT

The preliminary results were presented as a poster, Activités et rôle biologique des sédiments de la lagune de Venice (N. Sabil, A. Cherqui, D. Tagliapietra and M-A. Coletti-Previero) at the Forum Environment (May 1992) meeting. Two publications were submitted to international scientific Journals. The first appeared in *Water Research* (Sabil *et al.*, 1994). The second appeared in *Environmental Technology* (Sabil *et al.*, 1993). These results were also presented in a plenary lecture, Enzymes linked to inorganic supports: preparation and natural occurrence, by M.-A. Coletti-Previero, by invitation of the Max-Delbrück Zentrum für Molekulare Medizin in June 1993. M.-A. C.-P. has also expanded this research within a different framework, Programma di verifica sperimentale di tecniche di arresto ed inversione del processo di degrado della laguna di Venezia, where the enzymatic activity of the Palude della Rosa sediment was studied at superficial and deeper levels for another year.

REFERENCES

Aissouni, Y., Sabil, N., Tagliapietra, D. and Coletti-Previero, M.-A. (1995). Biodegradative power of the sediment Venice lagoon area, Palude della Rosa. *Venice Lagoon Ecosystem* (UNESCO Paris) 1–15

Bonmati, M., Pulola, J., Sana, M. and Soliva, M. (1985). Chemical properties, populations of oxidizers, urease and phosphatase activities in sewage sludge-amended soils. *Plant and Soil.* **84**: 79–91

Coletti-Previero, M.-A. and Previero, A. (1989). Alumina–phosphate complexes for immobilization of biomolecules. *Analytical Biochemistry*, **180**: 1–10

Froelich, P. N. (1980). Analysis of organic carbon in marine sediments. *Limnol. Oceanogr.*, **25**: 564–72

Kunitz, M. (1947). Crystalline soybean trypsin inhibitor. *J. Gen. Physiol.*, **30**: 291–310

Maunoir, S., Sabil, N., Rambaud, A., Philip, H. and Coletti-Previero, M.-A. (1991). Role of insoluble enzymes in anaerobic wastewater treatment and enzyme-bioactivator interactions. *Env. Technol.*, **12**: 313–23

Norbert, W. T., Astles, J. R. and Shuey, D. F. (1989). Lipase activity measured in serum by a continuous monitoring pH-stat technique – an update. *Clin. Chem.*, **35**: 1688–93

Park, J. T. and Johnson, M. J. (1949). Submicrodetermination of glucose. *J. Biol. Chem.*, **181**: 149–51

Pugniére, M., Skalli, A., Coletti-Previero, M.-A. and Previero, A. (1986). Peptide and ester synthesis in organic solvents catalyzed by seryl proteases linked to alumina. *PROTEINS: Struct. Funct. Gen.*, **1**: 134–8

Sabil, N., Cherqui, A., Tagliapietra, D. and Coletti-Previero, M.-A. (1994). Immobilized enzymatic activity in the Venice lagoon sediment. *Wat. Res.*, **28**: 1: 77–84

Sabil, N., Tagliapietra, D. and Coletti-Previero, M.-A. (1993). Insoluble biodegradative potential of the Venice lagoon. *Env. Technol.*, **14**: 1089–95

Sabil, N., Aissouni, Y., Coletti-Previero, M.-A., Donazzolo, R., D'Ippoloto, R. and Pavoni, B. (1995). Immobilized enzymes and heavy metals in the sediments of Venice. *Env. Technol.* **16**: 765–74

CHAPTER 11

SEASONAL VARIATIONS OF THE MACROBENTHIC COMMUNITY IN THE PALUDE DELLA ROSA

D. Tagliapietra, M. Pavan, C. Targa and C. Wagner

SUMMARY

Fluctuations in the macrozoobenthic community were followed, by monthly sampling, during 18 campaigns performed from December 1991 to July 1993. The sampling was executed at five stations located in a shallow water basin (Palude della Rosa, Venice Lagoon – Italy). The observed fluctuations of benthic community structure closely followed the seasonal cycle of the green macroalga *Ulva rigida* C.Ag. and its overgrowth. Significant differences in community composition and diversity were found between stations more influenced by *Ulva rigida* presence and those less influenced. Sediment enrichment in organic matter, due to such an algal overgrowth, greatly modified the physico-chemical conditions of the Palude, leading to a prolonged anoxic crisis over the summer. All species were strongly affected by the dystrophic conditions and almost all of them disappeared at this time. This was then followed by a phase of recolonization of the denuded bottom in the following winter. The community was finally composed mostly of opportunistic species that were able to take advantage of strong enrichment with organic matter.

INTRODUCTION

This study was conducted in a shallow water basin called the Palude della Rosa (Figure 11.1) located in the northern part of the Venice Lagoon. This basin was influenced by the presence of both continental freshwater (Dese River and a branch of the Sile River) and tidal saltwater. The peculiar topography of this semi-enclosed shallow basin ('Palude' in Italian), and nutrient load (see Morin *et al.*, Chapter 9), allowed the growth of huge amounts of the green macroalga *Ulva rigida* Agardh (up to 5 kg m^{-2} Bendoricchio, Chapter 26). Therefore, in our case, the environmental conditions were subject to fluctuations far beyond the normal range.

In this context, the aim of the present research was to detect and follow the changes in the macrozoobenthic communities throughout an annual cycle of the green macroalga.

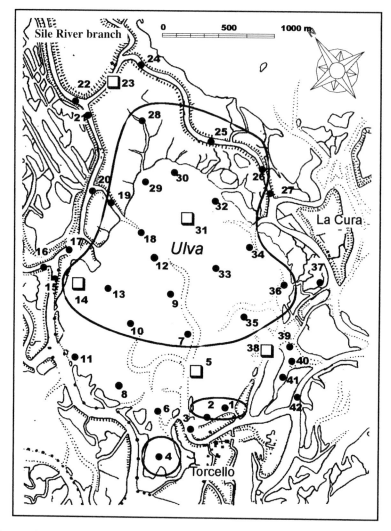

Figure 11.1 Map of the Palude della Rosa showing locations of the sampling stations and *Ulva rigida* distribution. The five monthly sampling stations are labelled with a '□'

MATERIALS AND METHODS

Choice of the sampling stations

After an initial survey of 42 stations in the Palude and the canals around it in December 1991, five of them (5, 14, 23, 31 and 38) were retained for monthly studies. The data reported in the present paper refer exclusively to these five sampling stations. In addition, another ten stations (5, 9, 10, 11, 17, 23, 29, 30,

210

31, 38) were sampled seasonally for a comparative study with the meiobenthos (see Chapter 12). The sampling stations were located along a gradient of *Ulva* presence and confinement: station 38 on the edge of the Palude close to a canal, station 5 was close to a canal end but inside the basin, station 31 was located in a shallow water area covered by a dense mat of *Ulva*, station 14 was in an internal cul de sac but with a lower *Ulva* cover. Finally, station 23 was located in a peripheral canal, with low hydrodynamics and influenced by continental waters (Figure 11.1). Stations 14, 23 and 31, located in the northern part of the Palude della Rosa, were subject to the direct influence of *Ulva rigida*, consequently called in this paper 'Ulva Stations', while stations 5 and 38, located in the southern reach, were only indirectly affected by the macroalga presence, because of their position close to the incoming canals conveying tidal seawater and therefore called 'Non-*Ulva* Stations'. In terms of faunal assemblages, these stations were representative of the whole area.

All collections were done at high water of each spring tide, in order to maintain a consistency in tidal conditions and have easy access to the very shallow areas of the basin. Samples for benthic macrofauna, one at each station, were obtained with a 0.1 m² Van Veen grab, washed through a box sieve of 1 mm mesh and preserved in 4% formaldehyde buffered with seawater. The animals were sorted in the laboratory, identified to species level, when possible, and counted. After fixing in formalin, the specimens were preserved in 70% ethanol. The preserved organisms were dried at 105°C for 24 h, weighed and then combusted in a muffle furnace at 450°C (overnight) to estimate ash content. The weights mentioned in this paper are ash-free dry weights (AFDW). Some data tables are also available (Tagliapietra *et al.*, 1997). During each sampling, some parameters such as water temperature, salinity (hand refractometer) and dissolved oxygen (Winkler method) were also measured.

Data analysis

Abundance and biomass values were analysed using univariate and multi-variate techniques. Similarity matrices based on the Bray–Curtis similarity index were calculated on fourth root transformed abundance data. These similarity matrices were the basis on which to perform Hierarchical Agglomerative Clustering and 2D and 3D Multidimensional Scaling Ordination (MDS) to allow the identification of similar community structures and to follow temporal trends (Warwick and Clarke, 1991). In order to test the significance of differences between seasons, the computer program ANOSIM was used. The identification of the species responsible for the similarity and for the difference between groups was achieved with the computer program SIMPER (Clarke, 1993). All statistical programs are included in the PRIMER software package.

Species/Abundance/Biomass (SAB) curves were produced for visualising the community variations (Pearson and Rosenberg, 1978). In order to have a picture of the diversity change the most common indices (i.e. Shannon–Wiener diversity index (log e), Margalef's species richness and Pielou's evenness) were calculated on numerical abundances and plotted on the same graphs beside k dominance curves (Lambshed *et al.*, 1983).

RESULTS AND DISCUSSION

The role of *Ulva rigida*

In a stable environment a benthic community undergoes only small-scale quantitative and qualitative changes over time, the structure of the community is characterised by a high species number and biomass but by moderate abundance. Confined coastal water bodies, the so-called 'parhalic domain' (Guelorget and Perthuisot, 1983), where fluctuations of an environmental parameter are not caused by disturbed situations, harbour typical, well-adapted communities. In our case, the environmental conditions were subject to fluctuations far beyond the normal estuarine range (cf. other papers in this volume).

The peculiar topography of the Palude della Rosa allowed the growth of huge amounts of *Ulva rigida*. The joint action of wind and currents influenced the *Ulva rigida* distribution pattern, accumulating the macroalga at the northern reaches of the Palude, an area moreover subjected to river inflow.

During winter–spring 1991–92, *Ulva rigida* growth contributed to the water column oxygenation and became a suitable substratum for a rich community. The *Ulva rigida* bloom was followed, in summer, by a massive decline of the alga. The decaying thalli caused extreme alterations of the physico-chemical parameters (see Bendoricchio *et al.*, Chapter 26; Ravera *et al.*, Chapter 14). To begin with, oxygen depletion, especially at night, occurred so that from June to December the entire benthic community suffered a strong reduction in the number of species and individuals. In addition, the accumulation of a layer of slow decaying particles acted as a physical barrier at the water–sediment interface. During spring 1993 a negligible amount of *Ulva* was present.

Community features

In December 1991 structured communities were present at all stations. In winter–spring 1991–92 there was an initial high number of individuals and a moderately high biomass. From March 1992 to June, at all sampling stations, the abundance sharply declined and the community finally collapsed from June 1992 to January–February 1993. During this period, at all stations, only a few individuals belonging to a small number of resistant species could survive and their presence was characterised by an extremely high variability; whereas when the effects of anoxia became extreme, the macrofauna disappeared completely. Following this period of crisis in winter–spring 1993, a recolonization sequence took place by a few pioneer species represented by a high number of individuals. In summer 1993, in contrast with the previous year, there was no evidence of any dystrophic trend. On this basis, the study period can be divided into three parts: before the crisis, from December 1991 to May 1992; during the dystrophic crisis, from June 1992 to February 1993, and after the crisis from March to July 1993.

The highest number of individuals was recorded at station 31 in December 1991 (22,390 ind.m^{-2}, 14,700 of which were Chironomids larvae). The highest biomass was obtained at station 5 in March 1993 with 34.5 gm^{-2}. The mean abundance per surface unit was 3423 ind.m^{-2} (7964 ind.m^{-2} before the dystrophic crisis, 182 ind.m^{-2} during the crisis and 4716 ind.m^{-2} after the crisis). The mean biomass was 5.9 g m^{-2} (10 g m^{-2} before the dystrophic crisis, 2.2 g m^{-2} during the crisis and 8.2 g m^{-2} after the crisis). The stations with the highest mean biomass were stations 5 (8.7 g m^{-2}) and 38 (10.8 g m^{-2}), due to the presence of large bivalves. The minimal values of abundance and biomass were obviously found during the crisis when the community broke down.

The mean number of taxa, for the five stations pooled, was 19 (29 before the dystrophic crisis, 14 during the crisis, and 20 after the crisis). Station 38 showed the highest mean number of species (13) while station 31 and station 23 showed the lowest annual mean number of species (5). Considering all the samples examined (survey, seasonal and monthly campaigns), 65 taxa were identified (see Table 11.1)

The diagram in Figure 11.2 provides the perception of the seasonal evolution of the Palude della Rosa macrozoobenthic community.

Changes in species composition

The changes in community structures are outlined in Figures 11.3–11.7. The MDS ordination plots show quite clearly the cycle of events. SAB curves evidenced clearly that the study period can be divided into three phases:

(1)   The phase *before* the summer crisis, from December 1991 to May 1992,
(2)   The phase of summer crisis, from June 1992 to February 1993, when the number of individuals was extremely low, and
(3)   The phase *after* the summer crisis, from March to July 1993, characterised by recolonization.

The three distinct phases evidenced by SAB curves and the groups generated by the cluster analysis (not reported) were superimposed on the MDS plots. The ANOSIM test was then performed on groups resulting from SAB curves to test the difference between the three periods. The ANOSIM pairwise tests showed a significant difference between the three sets of groups ($0.002 < p < 0.029$). However, multiple comparisons had been considered carefully because of the risk of error accumulation. Taxa responsible for the similarity within each period, and for the dissimilarity between phases, were

Table 11.1 List of species found in the Palude della Rosa during the whole study

Anthozoa	*Cirriformia tentaculata* (Montagu)	Crustacea isopoda
	Capitella capitata (Fabricius)	*Idotea baltica* (Pallas)
Nemertea	Maldanidae	*Idoetea metallica* Bosc.
	Clymenura clypeata (Saint-Joseph)	*Cyathura carinata* Kroyer
Plathelminthes	*Ficopomatus enigmaticus* (Fauvel)	*Sphaeroma serratum* Fabricius
	Sabellidae	*Lekanesphaera hookeri* Leach
Nematoda		
	Polychaeta errantia	Crustacea amphipoda
Placophora	*Eumida sanguinea* (Oersted)	*Carophium insidiosum* Crawford
	Phyllodoce sp.	*Corophium orientale* Schellenberg
Mollusca gastropoda	*Syllis gracilis* Grube	*Melita palmata* (Montagu)
Bittium reticulatum (Da Costa)	*Hediste diversicolor* (O.F. Müller)	*Microdeutopus gryllotalpa* Costa
Hydrobia spp.	*Neanthes succinea* (Frey & Leukart)	*Gammarus aequicauda* (Martynov)
Cyclope neritea (Linné)	*Perinereis cultrifera* (Grube)	*Gammarus subtypicus* Stock
Haminoea navicula Da Costa	*Platynereis dumerlii* (Aud. & M.-Edw.)	
	Marphysa bellii (Aud & M.-Edw.)	Crustacea decapoda
Mollusca bivalvia	*Marphysa fallax* Marion & Bobretzky	*Carcinus aestuarii* Nardo
Mytilaser lineatus (Gmelin)	*Marphysa sanguinea* (Montagu)	*Upogebia pusilla* Petagna
Mytilaster minimus (Poh)	*Nematonereis unicornis* Schmarda	*Palaemon elegans* Rathke
Cerastoderma glaucum (Poiret)		
Scrobicularia plana (Da Costa)	Oligochaeta	Ascidiacea
Abra spp.	*Heterochaeta costata* Claparède	*Molgula manhattensis* (de Kay)
Tapes decussatus (Linné)	*Limnodriloides agnes* Hrabè	
Tapes philippinarum (Ad. & Reeve)	*Tubificoides swirenkowi* (Jarosenko)	Pisces
Paphia aurea (Gmelin)	*Parandis* Cf *grandis* Harman	*Pomatoschistus minutus* (Pall.)
		Syngnatus taenionotus Can.
Polychaeta sedentaria	Crustacea cirripedia	
Malacoceros fuliginosum (Claparède)	*Balanus amphitrite* Darwin	Insecta
Polydora ligni Webster		*Chironomus salinarius* Kieffer
Streblospio shrubsolii (Buchanan)	Crustacea cumacea	
Aphelochaeta marioni (Saint-Joseph)		
Cirratulus cirratus (O.F. Müller)	Crustacea tanaidacea	
Cirriformia filigera (Della Chiaie)	*Tanais dulongii* (Audouin)	

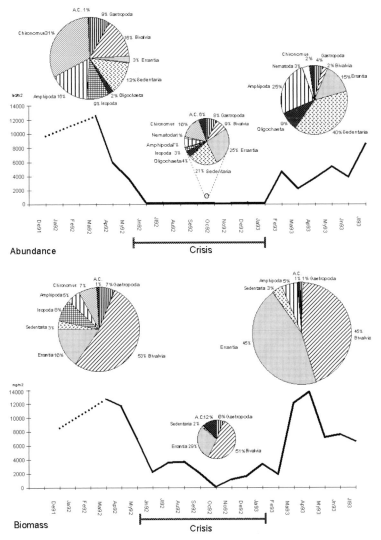

Figure 11.2 Mean values for abundance (ind.m^{-2}) and biomass (mg m^{-2}); given as average for the five stations. Pie diameters are proportional to mean values

explored with the computer program SIMPER. Species changes for each station are summarised below.

Station 5

The species characterising this station before the crisis were *Abra segmentum*, *Cyathura carinata*, *Cerastoderma glaucum* and *Haminaea navicula*, beside

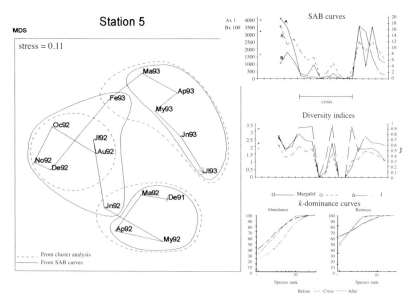

Figure 11.3   Main features of station 5

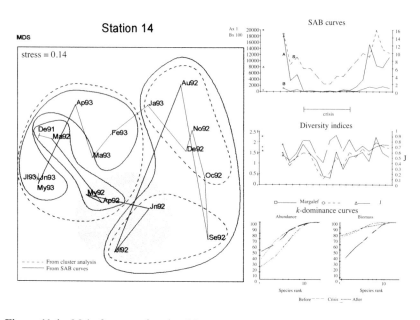

Figure 11.4   Main features of station 14

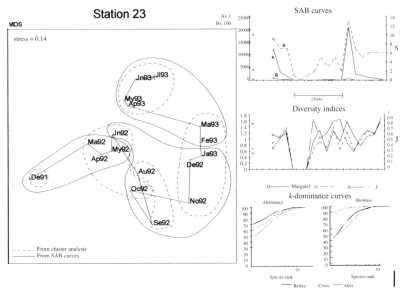

Figure 11.5   Main features of station 23

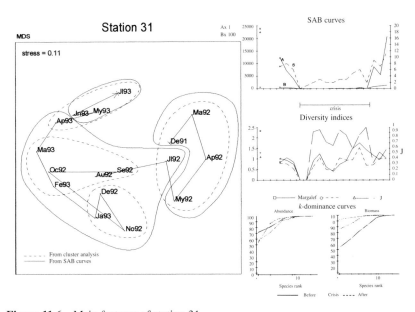

Figure 11.6   Main features of station 31

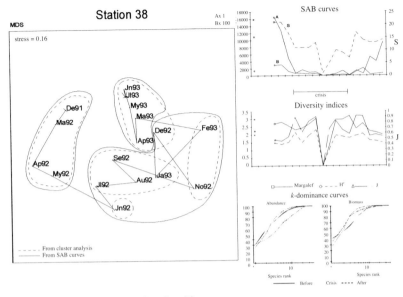

Figure 11.7 Main features of station 38

Hediste diversicolor, Chironomus salinarius and *Hydrobia ulvae* that were widely found in more confined environments. During the crisis, a few individuals of *Cyclope neritea, Cerastoderma glaucum* and *Chironomus salinarius* were present. After this period, recolonization took place with an increase of *Streblospio shrubsolii, Polidora ligni, Capitella capitata* and *Hediste diversicolor*, together with a general increment in Oligochaeta. *Nematoda* appeared at this time, followed by *Corophium* sp. and *Abra segmentum*. *Cyathura carinata* and *Haminaea navicula* continued to be absent still indicating a stress situation.

Station 14

Before the crisis the site was characterised by *Hediste diversicolor, Chironomus salinarius, Hydrobia ulvae, Polidora ligni* and *Corophium* sp., as expected considering its confined location. Isopods and amphipods living on *Ulva* thalli (i.e. *Idotea baltica, Lekanosphaera hookeri, Sphaeroma serratus* and *Gammarus* spp.) were abundant. During the crisis, although numerical abundance and biomass fell to very low values, no 'azoic' samples were collected. In this period *Chironomus salinarius* was the only species always present. The crisis seems to have affected the polychaets more than other taxa: this group disappeared almost completely except for *Hediste diversicolor*. Polychaets vanished suddenly in the first months of the critical period, while the other taxa were still present. The crisis was then more acute in its second half, from October

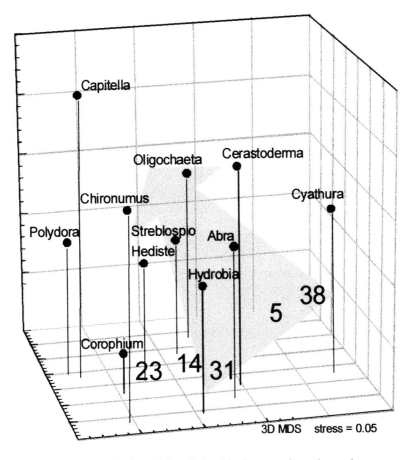

Figure 11.8   Schematic view of the relationships between the main species

1992 to January 1993. After the crisis *Hediste diversicolor* was constantly present, together with the classic colonisers *Capitella capitata, Polydora ligni* and *Streblospio shrubsolii. Chironomus salinarius* was still present and *Hydrobia ulvae* reappeared. The decomposition of the macroalga deprivated the crustaceans of a suitable substratum so that they disappeared almost completely.

Station 23

The trend at this station is less clear than at the other sites. *Corophium* sp., *Hediste diversicolor* and *Chironomus salinarius* typified this station before the summer breakdown. Beside this group of species, *Abra segmentum* and *Cyathura carinata* were less abundant, but constant. During the critical period *Streblospio shrubsolii, Hediste diversicolor* and Oligochaeta represented the characteristic taxa. This set of dominant species was unharmed and after the crisis their numbers multiply. In this last phase *Polydora ligni* was also present.

Station 31

This station at the beginning of our study presented a community dominated by *Chironomus salinarius*, *Hydrobia ulvae*, *Corophium* sp. and *Abra segmentum*. This early set of species was accompanied in December 1991 by another 15 species that were reduced to three in March 1992, underlining the rapidity at which this site reached dystrophic conditions. The general worst situation occurred at this station where, during the crisis, no animals were found. Only a few individuals of *Capitella capitata*, *Polidora ligni* and *Chironomus salinarius* survived from June 1992 to April 1993; the longest critical period detected in the Palude. The recolonization was performed by *Hediste diversicolor*, *Capitella capitata* and *Polidora ligni*, joined in the last months by *Corophium* sp., with numerical abundances up to 11,000 ind.m^{-2} in July 1993.

Station 38

The *zosteretum* present at station 38 in December 1991 gradually faded during the critical period, giving no sign of recovering in the following year. This station was still the most 'healthy' station throughout the study. Before the crisis the community was well structured and typified by *Abra segmentum*, *Cyathura carinata*, *Hediste diversicolor* and *Haminaea navicula* together with a good presence of *Streblospio shrubsolii* and *Polydora ligni*. Numbers of *Tapes decussatus* and *Cerastoderma glaucum* remained constant.

During the crisis, the community was much reduced and was characterised by *Hediste diversicolor* and *Streblospio shrubsolii*. The occurrence of *Abra segmentum* was strongly reduced and *Cyathura carinata* disappeared. *Hediste diversicolor*, *Polidora ligni*, *Streblospio shrubsolii*, Oligochaeta and *Corophium* sp. were the dominant taxa after the crisis. However, numbers of *Hediste diversicolor* and *Corophium* sp. were augmented in comparison with the previous year, while *Streblospio shrubsolii* and Oligochaeta became less abundant. The gammarid *Melita palmata* was well represented in 1993.

Recolonization patterns

After the dystrophic crisis in winter–spring 1993, a partial recovery of the community was evidenced. The rise in SAB curves was due to the succession of opportunistic species. A small amount of Polychaeta sedentaria, with *Capitella capitata*, *Streblospio shrubsolii* and *Polydora ligni*, played an important role as pioneer species. These are opportunistic species belonging to the trophic group of deposit feeders, well adapted to live in dystrophic environments. These characteristics have made many authors consider them to be indicator species for organically enriched environments. Pearson and Rosenberg (1978) consider *Capitella* and *Streblospio* as 'enrichment opportunists', that is, taxa able to colonise organically rich sediments in contrast with the 'general' opportunists, such as *Polydora ligni*, which are the initial colonisers of any available substratum. The trend observed in the Palude della Rosa confirm this statement: *Capitella capitata* and *Streblospio shrubsolii* showed numerical peaks from February to April 1993, while *Polydora ligni* exhibited its maximum from May to July 1993. The appearance of Nematoda and an increase in number of Oligochaeta, parallel to those of the 'enriched opportunists' was noticed. *Polydora ligni* was accompanied in the last months by the massive presence of *Corophium* spp.; these two taxa displayed abundances one order of magnitude higher than the other colonisers (several thousands ind.m^{-2}).

A similarity matrix was produced for the main species that typified or discriminated the three periods and was subsequently subjected to MDS in order to visualise their sociability. Figure 11.8 show the 3D configuration thus obtained; the grey arrows represent, in a schematic way, the direction of the shift in species composition during the period of study. The position of the stations in the graph are indicative of their situation at the beginning of this study.

Diversity

For each station, the diversity was evaluated by graphically visualising the temporal series of the most common indices (i.e. Shannon–Wiener diversity index (log e), Margalef's species richness, and Pielou's evenness). In addition, the k dominance curves, for numerical abundance and biomass before, during and after the crisis, were generated.

When the community is extremely poor, as in our case, during a dystrophic crisis, the majority of indices and statistics become meaningless or cannot be computed. Thus, in such a situation, the Shannon index and evenness are not suitable for the evaluation of diversity, while k dominance curves seems to be more appropriate. The interpretation of the statistics related to the crisis was problematic so the comparison between the periods before and after the crisis was preferred.

The highest diversity were those of the 'non-*Ulva* stations': station 5 (mean = 1.22, max = 2.07 in March 1992) and station 38 (mean = 1.64, max = 2.31 in September 1992), while the lowest diversity was recorded at station 31 (mean = 0.77, max = 1.53 in March 1993). Station 14 (mean = 1.30, max = 1.98 in May 1993) and station 23 (mean = 0.83, max = 1.58 in July 1993) fluctuated between these extremes. Pielou's evenness mean values ranged, for all stations, between 0.57 and 1.00 both values found at station 31.

The highest Margalef's species richness values were those again of station 38 (mean = 2.35, max = 3.21 in March 1993) and station 5 (mean = 1.61, max = 3.14 in December 1991), while the lowest values were recorded at station 31 (mean = 1.01, max = 2.33 in December 1991), station 14 (mean = 1.34, max = 2.24 in December 1991) and station 23 (mean = 1.34, max = 2.24 in December 1991).

As with the diversity indices, the diversity profiles (Figs 11.3–7) showed the higher diversity at station 5 and station 38, especially with reference to biomass. As a common trait, the dominance in abundance diminished, while for the biomass, the dominance was augmented. Comparing the k dominance curves of abundance and biomass, the tendency from 1991 to 1993 is a shift from a configuration A/B to a configuration B/A (mean W statistic: before = –0.056, during = 0.256 and after the crisis = 0.144 (Clarke, 1990)). However, in confined environments the application of the ABC method (Warwick, 1986) can be misleading because of the role played by the different taxonomic groups as exposed in detail by Beukema (1988) and Warwick and Clarke (1994).

An overview of the diversity and community structure showed some relevant differences between the stations located in the northern part of the Palude della Rosa (stations 14 and 31), subject to the direct influence of the *Ulva rigida* ('*Ulva* stations'), and the stations located in the southern part of the Palude della Rosa, along the incoming canals (stations 5 and 38, 'Non-*Ulva* stations'). In this context, station 23 located, remember, in a peripheral canal influenced by freshwater, should be considered as a separate case.

Trophic groups

The environmental conditions in the Palude della Rosa appear to most favour small deposit-feeding species; our results agree with the principle that trophic relationships are simplified as the gradient of organic input to an environment increases (Lindeman, 1942). As part of an estuary, the Palude della Rosa is already a natural habitat with high organic input and an ecosystem dominated by deposit-feeders. In such a situation the complex food web relationships, typical of healthy environments, gradually turn into simplified ones, with communities entirely composed of detritus feeders as eutrophic conditions are reached. At very high input levels, a simple trophic system composed of only non-selective deposit feeders and carnivores is established. The broad principles of this change in feeding habits were also described by other authors considering the effects of organic enrichment on benthic communities (O'Connor, 1972; Rosenberg, 1973; Grassle and Grassle, 1974; Pearson, 1975).

CONCLUSIONS

In the light of the results of this study we can conclude that:

- Considering that the observed fluctuations in benthic communities closely followed the *Ulva rigida* cycle, is clear that the Palude della Rosa benthic community was ruled by the overgrowth of this green macroalga.
- Significant differences in community structure were found between the stations more influenced by *Ulva rigida* (stations 14 and 31) presence and those less influenced (station 5 and station 38). Station 23 can be considered as a case apart, being located outside the factual Palude boundaries.
- Recolonization by 'enriched opportunist' species took place according to a known pattern, which has been recorded in other similar situations. In any case, the ecosystem showed a capacity for recovery, probably due to planktonic larvae brought into the Palude della Rosa by the tides or to the migration of benthic juveniles. There could also be some less affected areas within the Palude where a part of the population has survived.
- During the period of extreme crisis, when all communities were annihilated, macrobenthic communities showed a higher degree of vulnerability with respect to meiobenthic ones (Villano and Warwick, 1995 and Chapter 12). The results of meiobenthic studies (Villano and Warwick, 1995 and Chapter 12) and the enzymatic activity of the sediments (Sabil *et al.*, 1993, 1994 and Chapter 10), carried out in the same project, showed a strong dependency of the investigated subjects on the seasonal presence and distribution of *Ulva rigida*, and are in good agreement with our results.
- The pattern of SAB diagrams and the presence of *Zostera noltii* suggest that the conditions in the Palude in summer 1991 were not as bad as in summer 1992. SAB (species/Abundance/Biomass) suggest that even in normal situations a community adapted to organically enriched environments is characteristic of the Palude.

Since few detailed studies have been made on the benthic community of the Venice Lagoon, the present study (1993) provides a new starting point for future studies.

ACKNOWLEDGEMENTS

We are very grateful to Prof. O. Ravera (Università Ca' Foscari di Venezia, Dip. Di Sci. Ambientali, Venice), Dr R. M. Warwick (Plymouth Marine Laboratory, Plymouth), and Prof. A. Marzollo (UNESCO, Paris) for their kind support. The identification of the Oligochaeta specimens was kindly performed by Prof. Christer Erséus (Swedish Museum of Natural History, Stockholm), valuable support in Polychaeta sedentaria identification was given by Dr M. Kendall (Plymouth Marine Laboratory, Plymouth). We take the opportunity here to thank them very much. Finally we are grateful to Dr Allison Bannister for her friendly assistance.

REFERENCES

Beukema, J. J. (1988). An evaluation of the ABC method (Abundance/Biomass Comparison) as applied to macrozoobenthic communities living on tidal flats in the Dutch Wadden Sea. *Mar. Biol.*, **99**: 425–33

Clarke, K. R. (1990). Comparison of dominance curves. *J. Exp. Mar. Biol. Ecol.*, **138**: 143–57

Clarke, K. R. (1993). Non-parametric multivariate analysis of changes in community structure. *Australian Journal of Ecology*, **18**: 17–143

Grassle, J. F. and Grassle, J. P. (1974). Opportunistic life histories and genetic systems in Marine benthic polychaetes. *J. Mar. Res.*, **32**: 253–84

Guelorget, O. and Perthuisot, J. P. (1983). Le domain paralique – expressions géologiques, biologiques et économiques du confinment. *Trav. Lab. Gèol.*, École Norm. Sup. Paris, **16**: 136 pp.

Lambshed, P. J. D., Platt, H. M. and Shaw, K. M. (1983). The detection of differences among assemblages of benthic species based on an assessment of dominance and diversity. *Journal of Natural History,* **17**: 859–74

Lindeman, R. L. (1942). *Ecology*, **23**: 399–418

O'Connor, J. S. (1972). The benthic macrofauna of Moriches Bay. *New York, Biol. Bull.*, **142**: 84–102

Pearson, T. H. and Rosenberg, R. (1978). Macrobenthic succession in relation to organic enrichment and pollution of the marine environment. *Oceanogr. Mar. Biol. Ann. Rev.*, **16**: 229–311

Pearson, T. H. (1975). The benthic ecology of Loch Linnhe and Loch Eil, a sea-loch system on the west coast of Scotland. IV. Changes in the benthic fauna attributable to organic enrichment. *J. Exp. Mar. Biol. Ecol.*, **20**: 1–41

Rosenberg, R. (1973). Succession in benthic macrofauna in a Swedish fjord subsequent to the closure of a sulphite pulp mill. *Oikos*, **24**, 244–58

Sabil, N., Cherqui, A., Tagliapietra, D. and Coletti-Previero, M. A. (1994). Immobilized enzymatic activity in the Venice Lagoon sediments. *Water Research*, **28**: 77–84

Sabil, N., Tagliapietra, D. and Coletti-Previero, M. A. (1993). The insoluble degradative potential of the Venice Lagoon sediment. *Environmental Technology*, **14**: 1089–95

Tagliapietra, D., Pavan, M., Targa, C. and Wagner, C. (1977). La fauna macrobentonica della Palude della Rosa, Laguna di Venezia – Dati tabulati. *Lavori-Soc. Ven. Sc. Nat.*, **22**: 43–9

Villano, N. and Warwick, R. M. (1995). Meiobenthic communities associated with the seasonal cycle growth of *Ulva rigida* Agardh in the Palude della Rosa, Lagoon of Venice. *Estuarine, Coastal and Shelf Science*, **41**: 181–94

Wagner, C. (1995). Jahreszeitliche Veränderungen der makrozoobentischen Lebengemeinschaften im Palude della Rosa (Lagune von Venedig) im Zusammenhang mit Massenentwicklung der Makroalge *Ulva rigida* Agardt. Ph.D. Thesis, Marburg/Lahn, 176 pp.

Warwick, R. M. and Clarke, K. R. (1991). A comparison of some methods for analysing changes in benthic community structure. *J. Mar. Biol. Assoc. U.K.,* **71**: 225–44

Warwick, R. M. and Clarke, K. R. (1994). Relearning the ABC taxonomic changes and abundance/biomass relationships in disturbed benthic communities. *Mar. Biol.,* **118**: 739–44

Warwick, R. M. (1986). A new method for detecting pollution effects on marine benthic communities. *Mar. Biol.,* **92**: 557–62

CHAPTER 12

THE MEIOBENTHIC COMPONENT OF THE PALUDE DELLA ROSA, LAGOON OF VENICE

R.M. Warwick and N. Villano

SUMMARY

A survey of the meiofauna at 42 stations in the Palude della Rosa in 1991–1992 showed that there were two distinct associations of nematodes and total meiofauna (nematodes plus copepods), one associated with that part of the Palude where *Ulva* attains a high biomass, and another where *Ulva* biomass is low or absent. Species characteristic of the *Ulva* region include nematodes of the genus *Diplolaimella* and copepods of the genus *Tisbe*, both of which are known from elsewhere to be associated with decaying plant material and implicated in the decomposition process. In the *Ulva* region both diversity (as evidenced from *k*-dominance curves) and species composition (using non-metric MDS) varied considerably from season to season, associated with the seasonal cycle of growth and decay of *Ulva*, but diversity and species composition were seasonally more stable in the non-*Ulva* region. For the copepods alone, differences between the *Ulva* and non-*Ulva* regions were not so evident and species composition in both regions changed seasonally in the same way, although diversity was only reduced markedly in the summer at the *Ulva* stations. We conclude that it is not only likely that the *Ulva* cycle controls seasonal changes in meiobenthic community composition, but in turn, the latter has some control over the *Ulva* cycle.

INTRODUCTION

In shallow coastal waters and estuaries meiobenthos may be more important than macrobenthos in terms of energy flow and the role it plays in biogeochemical processes. There is a relatively large literature on the impacts of organic enrichment of various kinds on the meiofauna (Coull and Chandler, 1992), the majority of which concerns the effects of sewage rather than decaying macrophytes such as *Ulva*. The meiofaunal response in terms of changes in abundance is very variable. In field investigations, where organically polluted sites have been compared with control sites, about half report increases in meiofaunal abundance (Gowing and Hulings, 1976; Raffaelli, 1982; Vidakovic, 1983; Hennig *et al.*, 1983; Van Es *et al.*, 1980; Bouwman *et al.*, 1984; Moore and Pearson, 1986; Nichols, 1977; Arthington *et al.*, 1986; Khera and Randhawa, 1985; Lorenzen *et al.*, 1987). Other studies found decreased abundance at the polluted sites (Olsson *et al.*, 1973; Anger and

Scheibel, 1976; Hennig *et al.*, 1983; Aissa and Vitiello, 1984; Ansari *et al.*, 1984; Keller, 1984; 1985; Varshney, 1985; Vitiello and Aissa, 1985; Moore, 1987; Sandulli and Nicola-Guidici, 1990). In another study (Marcotte and Coull, 1974) the meiofaunal abundance increased in winter and decreased in summer, and in one study no effect on abundance was found (Austen *et al.*, 1989). This variability is most likely to depend on the condition of oxygenation of the water overlying the sediment, and also on the quantity of toxic contaminants (e.g. heavy metals) in the organic material, which in almost all the above cases was unknown. In a mesocosm system with exchange of water to the open sea, Gee *et al.* (1985) found that enrichment with macroalgal detritus (dried *Ascophyllum*) resulted in an increase in abundance of copepods but had no effect on nematodes in sublittoral mud sediments.

In field studies, where the meiobenthos have been identified to species, the response in terms of changes in diversity is more predictable: particulate organic enrichment invariably results in a reduction in diversity (Marcotte and Coull, 1974; Sandulli and Nicola-Guidici, 1990; Moore, 1987; Olsson *et al.*, 1974; Anger and Scheibel, 1976; Van Es *et al.*, 1980; Hennig *et al.*, 1983; Keller, 1986; Moore and Pearson, 1986; Vitiello and Aissa, 1985; Arthington *et al.*, 1986; Bouwman *et al.*, 1984; Khera and Randhawa, 1985; Lorenzen *et al.*, 1987; Hummon *et al.*, 1990). In the mesocosm study mentioned above, Gee *et al.* (1985) found that copepod diversity decreased, and dominance increased, with increased organic loading, but the diversity of nematodes was not affected.

The meiofaunal taxa that favour habitats which are highly enriched with organic matter are remarkable in two respects; first they are virtually confined to a few groups of nematodes and copepods, and second these same groups are of ubiquitous occurrence in such situations, at least in temperate latitudes where they have been most studied. The suite of species is small, but differs rather consistently between intertidal or shallow estuarine sites and those sites which are subtidal and more or less fully marine. Intertidally and in shallow water, nematodes are found in very high densities in decaying plant material derived from both macroalgae and terrestrial phanerogams. Odum and Heald (1972) found nematodes in 'extremely high numbers' in decaying mangrove leaves and suggested that they 'play an important role in the decomposition process' (see below). Koop *et al.* (1982) similarly reported 'large populations of nematodes' associated with beds of decomposing kelp in Southern Africa. These beds are dominated by a single species, *Rhabditis marina*. This species has a cosmopolitan distribution (Inglis, 1966) and is commonly associated with stranded decomposing algae (Inglis and Coles, 1961). Decaying marsh vegetation and mangrove leaves are usually dominated by two closely-related genera of the nematode family Monysteridae, namely *Diplolaimella* and *Diplolaimelloides* (Lorenzen, 1969; Hopper, 1970; Hopper *et al.*, 1973). These two genera are associated with brackish-water environments and may not be truly marine. A wide variety of environments enriched by particulate organic material are also characterised by the predominance of a limited number of copepod species. Notable among these are members of the genus *Tisbe* (Fava and Volkmann, 1975). This genus comprises a number of very closely related and morphologically similar species (Volkman, 1979) which are often found in multi-species guilds in organically enriched habitats (Bergmans, 1979). For example, Gee *et al.* (1985) found that sediments enriched with *Ascophyllum* detritus became dominated by a guild of five *Tisbe* species. These copepods are of more ubiquitous occurrence in both brackish and marine situations.

A common feature of Rhabditid and Monhysterid nematodes, and Tisbid copepods, is their ability to utilise a wide variety of organic particles as a food source. They also have a high reproductive potential and rapid rates of population growth, as shown by Warwick (1981) for nematodes and Battaglia (1970) and Bergmans (1981) for *Tisbe*. These characteristics enable them to exploit the erratic high inputs of organic matter such as those associated with blooms of *Ulva*. Rhabditid and monhysterid nematodes are selective bacterial feeders, as can be deduced from the structure of their buccal cavities (Wieser, 1953). It is now known that bacterial-feeding nematodes play an important role in the decomposition and mineralisation of organic detritus. Gerlach (1978) first speculated that the grazing of nematodes maintained exponential growth, whereas the bacterial populations would become senescent in the absence of grazing. The stimulation of bacterial metabolism, he postulated, is important for the breakdown of organic matter. Subsequently, Milton (1981) found that *Diplolaimella shiewoodi* enhanced bacterial numbers in sediments, and Findlay and Tenore (1982), in experiments with ^{14}C-labelled plant detritus, found that *Diplolaimella chitwoodi* enhanced mineralization rates (organic ^{14}C mineralized to $^{14}CO_2$). More recently Alkemade *et al.* (1992a,b) have found that both *Diplolaimelloides bruciei* and *Diplolaimella dievengatensis* also enhance decomposition and mineralization rates of detritus. The trophic position and role of benthic copepods in organically-enriched sediments is problematical and probably complex. Harpacticoids may graze bacterial cells off detrital particles or ingest some detrital fragments whole. The mucilage released from macroalgal and other detritus, with its associated microbiota, may also be an important resource (Hicks and Coull, 1983). Tisbiids can be maintained in laboratory culture on a wide variety of food resources (Hicks and Coull, 1983), but the exact method of utilisation of the resource is not known.

In addition to their potentially important role in increasing the rate and efficiency of decomposition and mineralization of organic detritus on the seabed (see above), the meiobenthos by their 'bioturbating' activity (burrow formation and sediment mixing) play an important role in maintaining oxic conditions in the surface layers of sediment (Cullen, 1972). Since aerobic respiration of bacteria is always of major importance in the mineralization of organic matter (Jorgensen, 1980; Fry, 1987), bioturbation by the meiobenthos will also contribute in an indirect way to the process of detrital decomposition. This has been shown experimentally by Alkemade *et al.* (1992b). In situations of gross enrichment, no truly macrobenthic species remain, so that this role is fulfilled by the meiobenthos only.

In view of these important ecological roles played by the meiobenthos, the 'health' of the ecosystem depends on the extent to which meiobenthic communities are affected by the stressful environmental conditions induced by the *Ulva* cycle. The objectives of this study were therefore to determine:

(1) The community composition of the meiobenthos in the Palude della Rosa in areas of high and low seasonal biomass of *Ulva*.
(2) The seasonal variation of the meiobenthos in response to environmental changes induced by the '*Ulva* cycle'.
(3) Whether the meiobenthic species present in the regions of high *Ulva* biomass are those likely to be involved in controlling its decomposition rate.

MATERIAL AND METHODS

Research on the meiobenthic component of the benthos of the Palude was carried out at the Plymouth Marine Laboratory, Plymouth, UK, where the samples from the four field surveys (December 1991, March, July and October 1992) were processed and analysed. In order to map the broad distribution of meiobenthic assemblages in the Palude, a first survey of 42 stations (Figure 12.1), in the Palude itself and along the canal around it, was carried out in December 1991. On the basis of the analysis of these data, sites typical of the major assemblages present were selected for a study of seasonal changes (Group 1: stations 9, 10, 29, 30, 31. Group 2: 5, 11, 17, 23, 38).

Field techniques

Sediment samples were taken using a hand-operated core, of 5 cm diameter; the top 5 cm, with the supernatant, was retained and preserved in 4% formalin. Additional sediment samples from each station were collected for analysis of sediment grain-size parameters and organic content.

Laboratory techniques

Preserved meiofauna were extracted from the sediment using the LUDOX TM technique, according to the methods described by McIntyre and Warwick (1984). After removing the silt and clay fraction of the sediment by sieving through a 63 μm mesh, the remaining sediment and meiofauna was suspended in 40% LUDOX in 150 ml glass beakers by stirring. Samples were left for 45 minutes and then the supernatant containing the meiofauna was poured off through a 63 μm sieve. The meiofauna was washed thoroughly with tap water. This procedure was repeated three times and then the meiofauna samples were

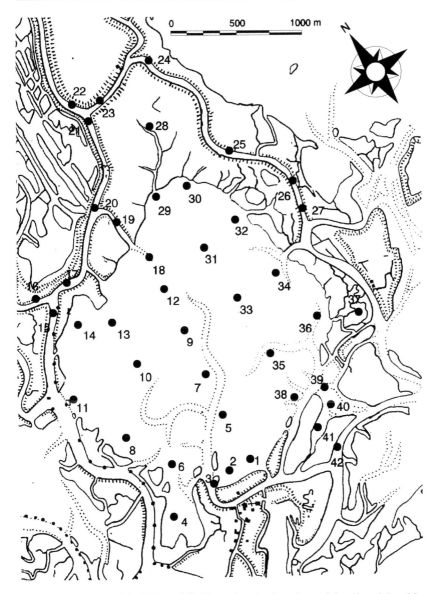

Figure 12.1 Map of the Palude della Rosa showing locations of the 42 meiobenthic sampling stations

preserved in 4% formalin. One hundred harpacticoid copepods were picked out of each sample (all of them if there were fewer) under a stereomicroscope and preserved in 4% formalin for further identification.

Subsamples for identification of nematodes were transferred to 10% glycerol solution in cavity blocks, placed on a hot plate and allowed to evaporate overnight to 100% glycerol. They were then mounted on microscope slides so that the nematodes could be identified and counted to the lowest possible taxonomic level using a compound microscope. To subsample the nematodes, the extracted samples were uniformly suspended in 2000 ml of tap water and subsamples of varying sizes (1/10, 1/5 or 1/2) were removed using a ladle such that approximately 100 nematode specimens were identified from each sample. Where the specimens were more than 100, these were identified and the remainder were counted. The final proportion of the core sample for which the nematodes were identified was calculated. The total abundance of each species within the core was then estimated after the appropriate correction by multiplication.

The percentage of silt/clay in the sediment samples was determined by wet-sieving the sediment through a 63 μm sieve, and the organic content determined by loss in weight on ignition at 550 °C in a muffle furnace.

Data analysis

All data on the abundance of meiobenthic taxa were analysed using univariate and multivariate statistical methods. Diversity profiles were visualised by plotting k-dominance curves (Lambshead *et al.*, 1983) for each sampling station (elevated curves indicate low diversity and increased dominance by one or few species). For the multivariate analysis, similarity matrices between stations were produced using the Bray–Curtis similarity measure on root-transformed species abundance data, followed by non-metric multidimensional scaling ordination (MDS) to identify groups of sites with a similar community composition and to monitor seasonal changes in community structure (Warwick and Clarke, 1991). Essentially, this technique produces a 2-dimensional 'map' in which the distances between samples reflect their similarity in species composition rather than their geographical distance apart. The significance of differences between sites and times was determined using the computer program ANOSIM, and the species which both characterised and distinguished between sites or times were determined using the computer program SIMPER (Clarke, 1993).

RESULTS AND DISCUSSION

The two main taxonomic groups characterising the meiobenthic community are the nematodes and the Harpacticoid copepods, and very few other taxa were encountered in this study. Thus all the results of this study refer to these two taxonomic groups. From the initial broad survey of 42 stations (Figure 12.1), 72 species of nematodes and 36 species of copepods were found (Table 12.1).

Table 12.1 Meiobenthic species list (December, 1991)

Nematodes		Copepods
Anoplostoma viviparum	*Setosabatieria* sp.	*Longipedia minor*
Crenopharynx, sp.	*Desmodora* sp.	*Brianola stebleri*
Halalaimus sp.	*Metachromadora* sp.	*Canuella* sp.
Nemanema sp.	*Molgolaimus* sp.	*Ectinosoma* sp.
Oxystomina sp.	*Aponema* sp.	*Halectinosoma* sp.
Paroxystomina sp.	*Antomicron* sp.	*Microarthridion fallax*
Thalassoalaimus sp.	*Camacolaimus* sp.	*Harpacticus* sp. 1
Adoncholaimus sp.	*Deontolaimus* sp.	*Harpacticus* sp. 2
Meyersia sp.	*Leptolaimus* sp.	*Stenhelia elizabethae*
Oncholaimellus sp.	*Onchium* sp.	*Stenhelia normani*
Oncholaimus sp.	*Procamacolaimus* sp.	*Stenhelia palustris bispinosa*
Viscosia sp.	*Paramicrolaimus* sp.	*Robertsonia knoxi*
Calyptronema sp.	*Diplolaimella occellata*	*Bulbamphiascus aff. inermis*
Ditlevsenella sp.	*Monhystera* sp.	*Amonardia normani*
Eurystomina sp.	*Ammotheristus* sp.	*Amphiascopsis* sp.
Symplocostoma sp.	*Daptonema* sp.	*Robertgurneya similis*
Chromadoridae	*Paramonhystera* sp.	*Typhamphiascus* sp.
Chromadora sp.	*Theristus* sp.	*Amphiascoides aff. sterilis*
Chromadorella sp.	*Sphaerolaimus* sp.	*Schizopera compacta*
Chromadorina germanica	Linhomoeidae	*Ameira parvula*
Chromadorita sp.	*Desmolaimus* sp.	*Nitoka aff. minor*
Chromadorita/Innocuonema	*Linhomoeus* sp.	*Dactylopusia tisboides*
Dichromadora sp.	*Metalinhomoeus* sp.	*Parastenhelia* sp.
Hypodontolaimus sp.	*Paralinhomoeus* sp.	*Metis ignea*
Neochromadora sp.	*Terschellingia communis*	*Mesochra pontica*
Ptycholaimellus sp.	*Terschellingia gourbaultae*	*Mesochra* sp.
Spilophorella sp.	*Terschellingia longicaudata*	*Cletocamptus confluens*
Steineridora sp.	Axonolaimidae	*Enhydrosoma propinquum*
Cyatholaimus sp.	*Ascolaimus* sp.	*Enhydrosoma gariene*
Paracanthonchus sp.	*Axonolaimus* sp.	*Nannopus palustris*
Halichoanolaimus sp.	*Odontophora* sp.	*Asellopsis salmatica*
Dorylaimopsis sp.	*Odontophoroides* sp.	*Paralaophonte congenera*
Paracomesoma dubium	*Pseudolella* sp.	*Tisbe holothuriae*
Paramesonchium sp.	*Synodontium* sp.	*Tisbe histriana*
Sabatieria juv.	Rhabditidae	*Cyclopoida* sp. 1
Sabatieria praedatrix		*Cyclopoida* sp. 2
Sabatieria pulchra		Calanoida

The MDS ordination (Figure 12.2) and the hierarchical agglomerative clustering (not shown) for the total meiofauna show a quite clear distinction between two main groups of stations for the nematodes and total meiofauna (nematodes + copepods), but not for the copepods alone.

Two distinct assemblages of meiobenthos occupy specific areas of the Palude:

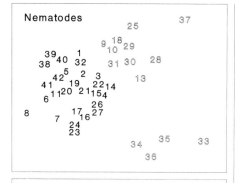

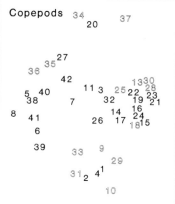

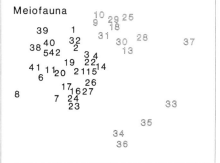

Figure 12.2   MDS ordination by stations for the meiobenthos in December 1991

GROUP 1   The northern and central part of the Palude; sites identified as having a high biomass of *Ulva* (UNESCO, 1993) (stations 9, 10, 13, 18, 25, 28–31, 33–37).

GROUP 2   The southern part of the Palude and canali around it; sites with little or no *Ulva* present (all remaining stations).

These two meiobenthic associations correspond to two ecologically different situations in relation to the presence or absence of *Ulva*. Average diversity profiles as indicated by *k*-dominance curves (Figure 12.3) are rather similar in each site group, but for nematodes and total meiobenthos, diversity is slightly higher at the *Ulva* stations, whereas for the copepods diversity cannot be compared adequately because the curves cross. It is clear from Figure 12.4 that the two site groupings for nematodes do not relate to the grain size of the sediment. This Figure depicts the *same* MDS configuration as in Figure 12.2, but with the station numbers replaced by symbols scaled in size to represent the percentage silt/clay in the sediment. There is no distinction in sediment grain size between the right- and left-hand groups (*Ulva* and non-*Ulva* groups). However, the organic content of the sediment does appear to correlate

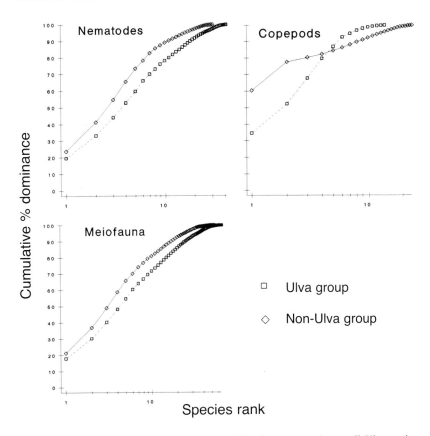

Figure 12.3 k-dominance curves for the meiobenthos averaged over all *Ulva* stations (squares and dashed lines) and non-*Ulva* stations (diamonds and solid lines) in December 1991

with the MDS configuration. In Figure 12.4 the organic content (loss in weight on ignition) is superimposed as before, and it is clear that all the stations with the higher organic content are found in the right-hand (*Ulva*) group.

The main species of nematodes and copepods characterising the *Ulva* and non-*Ulva* stations in December (i.e. the species contributing 90% of the total Bray–Curtis similarity among stations within each group) are given in Table 12.2. The nematode, *Diplolaimella ocellata,* was the top-ranked characterising species in the *Ulva* group, but did not appear at all in the rankings for the non-*Ulva* group, and the harpacticoid copepods, *Tisbe histriana* and *Tisbe holothuriae,* were the third and fourth ranked characterising species in the *Ulva* group but again were not characterising species in the non-*Ulva* group.

On the basis of these results, and considering the time needed to analyse the samples, the number of sampling stations for the seasonal study was reduced to ten, five from each assemblage group. The MDS ordination for the nema-

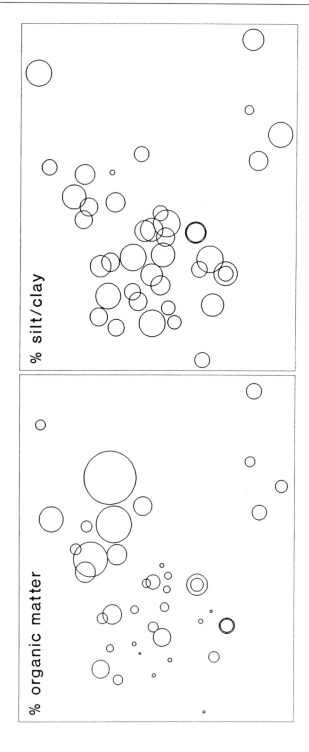

Figure 12.4 MDS ordination as in Figure 12.2, with station numbers replaced by circles scaled in size to represent measured values of organic content and % silt/clay (< 63 μm particle diameter)

Table 12.2 Species contributions to Bray–Curtis similarity among stations (December, 1991)

Species	Average no. per core	Average contribution	Cumulative %
Stations with high biomass of *Ulva*			
Nematodes (average similarity among stations 57.2%)			
Diplolaimella ocellata	246.1	14.1	24.6
Terschellingia longicaudata	100.0	9.7	41.7
Daptonema sp.	289.9	9.6	58.4
Sabatieria juv.	70.72	6.9	70.4
Neochromadora sp.	56.2	5.8	80.6
Theristus sp.	32.6	3.8	87.3
Chromadorina germanica	56.5	2.8	92.3
Copepods (average similarity among stations 46.9%)			
Canuella sp.	104.32	18.3	39
Cyclopoida sp. 2	53.4	9.4	59
Tisbe histriana	35.9	6.4	72.7
Tisbe holothuriae	46.37	4.9	83.2
Amonardia normani	21.7	3.9	91.6
Non-*Ulva* stations			
Nematodes (average similarity among stations 44.1%)			
Terschellingia longicaudata	1265.8	12.5	28.3
Sabatieria juv.	629.6	9.2	49.1
Daptonema sp.	565.4	8.0	67.2
Anoplostoma viviparum	256.1	3.2	74.4
Molgolaimus sp.	160.0	1.8	78.5
Terschellingia communis	75.12	1.5	81.9
Odontophora sp.	142.74	1.4	85.2
Paracomesoma dubium	27.24	0.9	87.3
Sabatieria pulchra	39.7	0.9	89.4
Sphaerolaiumus sp.	37.14	0.8	91.2
Copepods (average simiarity among stations 46.9%)			
Canuella sp.	137.7	19.2	47.1
Brianola stebleri	39.2	4.6	58.5
Stenhelia elizabethae	4.9	3.3	66.7
Asellopsis salmatica	4.1	3.0	73.9
Amphiascopsis sp.	4.3	1.7	78.1
Microarthridion fallax	6.2	1.4	81.4
Harpacticus sp. 1	2.9	1.4	84.7
Enhydrosoma gariene	3.8	1.4	88.1
Amonardia normani	1.85	1.2	91.0

todes and total meiobenthos shows a quite clear trend of change in community composition in the *Ulva* group through the four seasons, while in the non-*Ulva* group the similarity is closer between the same stations in different seasons than between seasons. The open symbols in Figure 12.5 denote the *Ulva* (right) stations which are clearly separated from the solid symbols denoting the non-*Ulva* (left) stations in December, March and July, but this separation is not evident in October. It is in fact clear that for the nematodes and total meiofauna, the *Ulva* and non-*Ulva* stations, were indistinguishable from each other in their community structure in October 1992 (at the end of the study). However, at all other times the *Ulva* and non-*Ulva* stations were significantly different from each other, this difference being greatest in July when the *Ulva* bloom crashes.

The ANOSIM test gives high values for the *R*-statistic and significant (< 5%) seasonal changes in community structure of the nematodes in the *Ulva* group, but not in the non-*Ulva* group (Table 12.3). This could be interpreted as a direct effect of the '*Ulva* cycle' on the nematode community: where there is a high seasonal biomass of *Ulva* the community is affected, but where *Ulva* is absent the species characterising the sites are always the same. This is

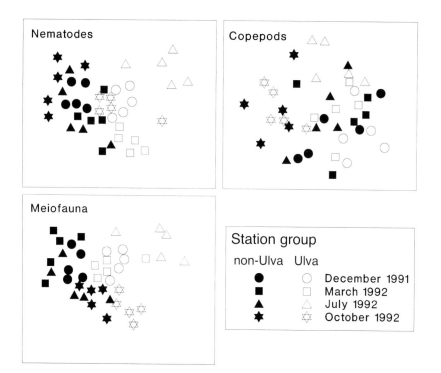

Figure 12.5 MDS ordinations for 5 *Ulva* stations and 5 non-*Ulva* stations in different seasons of the year (see text for further explanation)

Table 12.3 Results of the ANOSIM test for differences between consecutive sampling periods

Months compared	*Ulva* stations		Non-*Ulva* stations	
	R statistic value	% significance	R statistic value	% significance
Nematodes				
Dec–Mar	0.51	1.6	0.08	27
Mar–July	0.90	0.8	0.16	16
July–Oct	0.96	0.8	0.08	28
Copepods				
Dec–Mar	0.06	26	0.04	26
Mar–July	0.40	0.8	0.15	8.3
July–Oct	0.79	0.8	0.39	1.6

confirmed by the results of the data analysis with the computer program, SIMPER: the non-*Ulva* group is mainly characterised by *Terschellingia longicaudata*, *Sabatieria* and *Daptonema* in all four seasons, while the *Ulva* group is characterised by *Diplolaimella ocellata* and *Terschellingia longicaudata* in December; *Daptonema* and *Terschellingia longicaudata* in March; *Molgolaimus* and *Theristus* in July; and *Daptonema* and *Sabatieria* in October. The diversity profiles, as indicated in Figure 12.6, are very similar in the non-*Ulva* group, but they differ between seasons in the *Ulva* group, especially in July and October when the diversity is very low with few species present. The cyclic changes in community structure must clearly be related to the cycle of growth and decay of *Ulva*, and it is unfortunate that exactly parallel field data on *Ulva* are not available from these stations and sampling times, hence these conclusions remain speculative.

The results of the MDS ordination for the copepods show a pattern different from those of the nematodes, reflecting more a seasonal cycle, and there is no clear distinction between the two site groups. At all stations the copepod community seems to change quite uniformly through the four seasons. This is clearly shown in Figure 12.5 where the stations from the same season cluster together. The ANOSIM test for copepods (Table 12.3) indicates significant seasonal changes in the community structure in both groups, except that differences between the December and March samples are not significant, and neither are the March–July samples for the non-*Ulva* group of stations. The results of the data analysis with the computer program SIMPER show that the average similarity is low in all the months for both groups (*Ulva* and non-*Ulva*). The dominant species, responsible for the homogeneity through the months is *Canuella*, but it seems to be extremely reduced in October. The diversity profiles as shown in Figure 12.7, are rather similar in the non-*Ulva* group through the four seasons, while they change in the *Ulva* group with the lowest diversity shown in July.

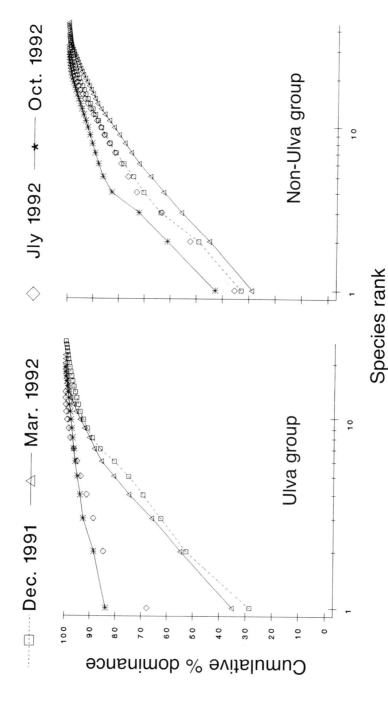

Figure 12.6 *k*-dominance curves for nematodes averaged over 5 *Ulva* and 5 non-*Ulva* stations in different seasons of the year

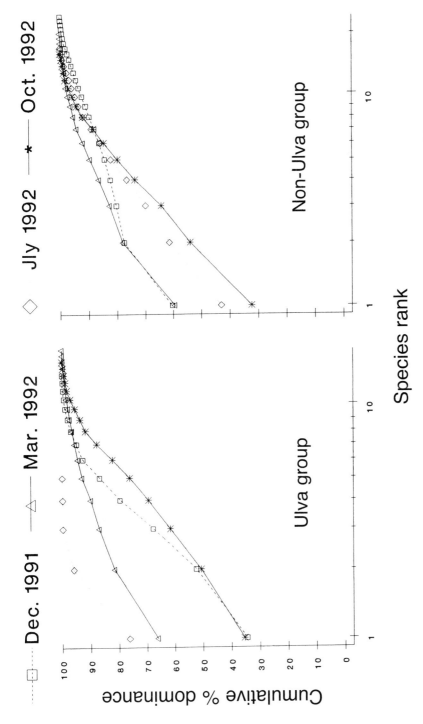

Figure 12.7 *k*-dominance curves for copepods averaged over 5 *Ulva* and 5 non-*Ulva* stations in different seasons of the year

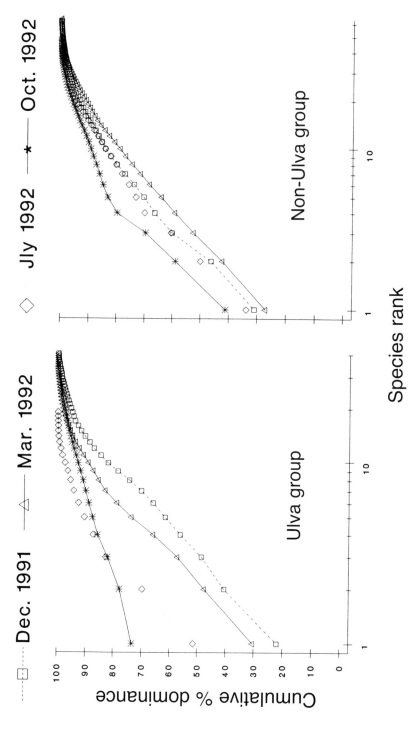

Figure 12.8 *k*-dominance curves for the total meiobenthos (nematodes + copepods) averaged over 5 *Ulva* and 5 non-*Ulva* stations in different seasons of the year

CONCLUSIONS

Since the macrobenthos completely disappeared during the decay-phase of *Ulva*, the survival of a taxonomically and functionally diverse community of meiobenthos may be crucial for the 'health' of the ecosystem, providing the link between primary producers and higher carnivores, and playing important roles in the decomposition of *Ulva* and the bioturbation of sediments. The results of this study on the meiobenthic component of the Palude della Rosa lead to the conclusion that there is a strong seasonal effect of the *Ulva* cycle on meiobenthic community composition. Several meiobenthic taxa which are prominent in the regions of the Palude with high *Ulva* biomass, but not in the regions where *Ulva* is not prominent, are known to be associated with the process of macrophyte decomposition, e.g. nematodes of the genus *Diplolaimella* and copepods of the genus *Tisbe*. Thus, it is not only likely that the *Ulva* cycle controls seasonal changes in meiobenthic community composition, but in turn the meiobenthic community has some control over the '*Ulva* cycle'. Experimental studies to evaluate the exact role played by these meiobenthic organisms in the decomposition of *Ulva* would be most valuable.

REFERENCES

Aissa, P. and Vitiello, P. (1984). Impact de la pollution et de la variabilité des conditions ambiantes sur la densité du meiobenthos de la Lagune de Tunis. *Rev. Fac. Sci. Tunis.*, **3**: 155–77

Alkemade, R., Wielemaker, A. and Hemminga, M. A. (1992a). Stimulation of decomposition of *Spartina anglica* leaves by the bactivorous marine nematode *Diplolaimelloides bruciei* (Monhysteridae). *J. Exp. Mar. Biol. Ecol.*, **159**: 267–78

Alkemade, R., Wielemaker, A., De Jong, S. A. and Sandee, A. J. J. (1992b). Experimental evidence for the role of bioturbation by the marine nematode *Diplolaimella dievengatensis* in stimulating the mineralization of *Spartina anglica* detritus. *Mar. Ecol. Prog. Ser.*, **90**: 149–55

Anger, K. and Scheibel, W. (1976). Die benthische Copepodenfauna in einem ufernahen Verschmutzungsgebiet der westlichen Ostsee. *Hegoländer Wiss. Meeresunters.*, **28**: 19–30

Ansari, Z. A., Chatterji, A. and Parulekar, A. H. (1984). Effect of domestic sewage on sand beach meiofauna at Goa, India. *Hydrobiologia*, **111**: 229–33

Arthington, A. H., Yeates, G. W. and Conrick, D. L. (1986). Nematodes, including a new record of *Tobrilus diversipapillatus* in Australia, as potential indictors of sewage efflent pollution. *Aust. J. Mar. Freshwater. Res.*, **37**: 159–66

Austen, M. C., Warwick, R. M. and Rosado, M. C. (1989). Meiobenthic and macrobenthic community structure along a putative pollution gradient in southern Portugal. *Mar. Poll. Bull.*, **20**: 398–405

Battaglia, B. (1970). Cultivation of marine copepods for genetic and evolutionary research. *Helgoländer Wiss. Meeresunters.*, **20**: 385–92

Bergmans, M. (1979). Taxonomic notes on species of *Tisbe* (Copepoda: Harpacticoida) from a Belgian sluice dock. *Zool. Scripta*, **8**: 211–20

Bergmans, M. (1981). A demographic study of the life cycle of *Tisbe furcata* (Baird 1837) (Copepoda: Harpacticoida). *J. Mar. Biol. Assoc., UK*, **61**: 691–705

Bouwman, L. A., Romeyn, K. and Admiraal, W. (1984). On the ecology of meiofauna in an organically polluted estuarine mudflat. *Est. Cstl. Shelf Sci.*, **19**: 633–53

Clarke, K. R. (1993). Non-parametric multivariate analysis of changes in community structure. *Aust. J. Ecol.*, **18**: 117–43

Coull, B. C. and Chandler, G. T. (1992). Pollution and meiofauna: field, laboratory, and mesocosm studies. *Oceanogr. Mar. Biol. Ann. Rev.*, **30**: 191–271

Cullen, D. J. (1972). Bioturbation of superficial marine sediments by interstitial meio-benthos. *Nature*, **242**: 323–4

Fava, G. and Volkmann, G. (1975). *Tisbe* (Copepoda: Harpacticoida) species from the lagoon of Venice 1. Seasonal fluctuations and ecology. *Mar. Biol.*, **30**: 151–66

Findlay, S. E. G. and Tenore, K. R. (1982). Effect of free-living marine nematode (*Diplolaimella chitwodi*) on detrital carbon mineralization. *Mar. Ecol. Prog. Ser.*, **8**: 161–6

Fry, J. C. (1987). Functional roles of the major groups of bacteria associated with detritus. In Moriarty, D.J.W. and Pullin, R.S.V. (eds), *Detritus and Microbial Ecology in Aquaculture*. ICLARM, Manila, 83–121

Gee, J. M., Warwick, R. M., Schaanning, M., Berge, J. A. and Ambrose, Jr W. G. (1985). Effects of organic enrichment on meiofaunal abundance and community structure in sublittoral soft sediments. *J. Exp. Mar. Biol. Ecol.*, **91**: 247–62

Gerlach, S. A. (1978). Food-chain relationships in subtidal silty marine sediments and the role of meiofauna in stimulating bacterial productivity. *Oecologia, Berl.*, **33**: 55–69

Gowing, M. M. and Hulings, N. C. (1976). A spatial study of the meiofauna on a sewage-polluted Lebanese beach. *Acta Adriat.*, **18**: 339–63

Hennig, H.F.-H.O., Eagle, G. A., Fielder, L., Fricke, A. H., Gledhill, W. J., Greenwood, P. J. and Orren, M. J. (1983). Ratio and population density of psammolittoral meiofauna as a perturbation indicator of sandy beaches in South Africa. *Envir. Monitor. Assess.*, **3**: 45–60

Hicks, G. R. F. and Coull, B. C. (1983). The ecology of marine meiobenthic harpacti-coid copepods. *Oceanogr. Mar. Biol. Ann. Rev.*, **21**: 67–175

Hopper, B. E. (1970). *Diplolaimelloides bruciei*: n. sp. (Monhysteridae: Nematoda), prevalent in marsh grass, *Spartina alterniflora* Loisel. *Can. J. Zool.*, **48**: 573–5

Hopper, B. E., Fell, J. W. and Cefalu, R. C. (1973). Effect of temperature on the life cycles of nematodes associated with the mangrove (*Rhizophora mangle*) detrital system. *Mar. Biol.*, **23**: 293–6

Hummon, W. D., Todaro, M. A., Balsamo, M. and Tongiorgi, P. (1990). Effects of pollution on marine Gastrotricha in the northwestern Adriatic Sea. *Mar. Poll. Bull.*, **21**: 242–3

Inglis, W. G. (1966). The occurrence of *Rhabdits marina* on Western Australian beaches. *Nematologica*, **12**: 643

Inglis, W. G. and Coles, J. W. (1961). The species of *Rhabditis* (Nematoda) found in rotting seaweed on British beaches. *Bull. Brit. Mus. Nat. Hist. (Zoology)*, **7**: 320–33

Jorgensen, B. B. (1980). Mineralization and bacterial cycling of carbon, nitrogen and sulphur in marine sediments. In Ellwood, D.C., Hedger, J.N., Latham, M.J., Lynch, J.M. and Slater, J.H. (eds) *Contemporary Microbial Ecology*. Academic Press, London, 239–51

Keller, M. (1984). Effects du deversement en mer du grand collecteur de l'agglomeration marseillaise sur les populations meiobenthiques. *C.R. Acad. Sci., Ser III.*, **229**: 756–68

Keller, M. (1985). Distribution quantitative de la meiofaune dans l'aire d'épandage de l'égoût de Marseille. *Mar. Biol.*, **89**: 293–302

Keller, M. (1986). Structure des peuplements meiobenthiques dans le secteur pollué par le reject en mer de l'égoût de Marseille. *Ann. Inst. Oceanogr., Paris*, **62**: 13–36

Khera, S. and Randhawa, N. (1985). Benthic nematodes as indicators of water pollution. *Res. Bull. Panjab Univ. Sci.*, **36**: 401–3

Koop, K., Newell, R. C. and Lucas, M. I. (1982). Biodegradation and carbon flow based on kelp (*Ecklonia maxima*) debris in a sandy beach microcosm. *Mar. Ecol. Prog. Ser.*, **7**: 315–26

Lambshead, P. J. D., Platt, H. M. and Shaw, K. M. (1983). The detection of differences among assemblages of benthic species based on an assessment of dominance and diversity. *J. Nat. Hist.*, **17**: 859–74

Lorenzen, S. (1969). Freilebende meeresnematoden aus dem Schlickwatt und den Salzwiesen der Nordseekuste. *Veroff. Inst. Merresforsch. Bremerh.*, **11**: 195–238

Lorenzen, S., Prein, M. and Valentin, C. (1987). Mass aggregations of the free-living marine nematode *Potonema vulgare* (Oncholaimidae) in organically polluted fjords. *Mar. Ecol. Prog. Ser.*, **37**: 27–34

Marcotte, B. M. and Coull, B. C. (1974). Pollution, diversity and meiobenthic communities in the North Adriatic (Bay of Piran, Yugoslavia). *Vie Milieu.*, **24**B: 281–330

McIntyre, A. D. and Warwick, R. M. (1984). Meiofauna techniques. In Holme, N.A. and McIntyre, A.D. (eds), *Methods for the Study of Marine Benthos*. IBP Handbook no. 16 (2nd edition). Blackwell, Oxford, 217–44

Milton, R. (1981). The effect of nematode (*Diplolaimella shiewoodi*) presence on density of marine sediment bacteria. *Am. Zool.*, **21**: 972

Moore, C. G. (1987). Meiofauna of the industrialised estuary and Firth of Forth, Scotland. *Proc. R. Soc. Edinburgh, Sect. B.*, **93**: 415–30

Moore, C. G. and Pearson, T. H. (1986). Response of a marine benthic copepod assemblage to organic enrichment. In Schriever, G. *et al.* (eds), *Proceedings of the Second International Conference on Copepoda, Ottowa, Canada, 13–17 August 1984*. National Museum of Canada, Ottawa, 369–73

Nichols, J. A. (1977). Benthic community structure near the Woods Hole sewage outfall. *Int. Rev. Ges. Hydrobiol.*, **62**: 235–44

Odum, W. E. and Heald E. J. (1972). Trophic analysis of an estuarine mangrove community. *Bull. Mar. Sci.*, **22**: 671–738

Olsson, I., Rosenberg, R. and Olundh, E. (1973). Benthic fauna and zooplankton in some polluted Swedish estuaries. *Ambio*, **2**: 158–63

Raffaelli, D. (1982). An assessment of the potential of major meiofauna groups for monitoring organic pollution. *Mar Env. Res.*, **7**: 151–64

Sandulli, R. and Nicola-Guidici, M. de (1990). Pollution effects on the structure of meiofaunal communities in the Bay of Naples. *Mar. Poll. Bull.*, **21**: 144–53

UNESCO (1993). The Venetian lagoon. *UNESCO Sources*, **44**: 12–13

Van Es, F. B., Van Arkel, M. A., Bouwman, L. A. and Schroder, H. G. J. (1980). Influence of organic pollution on bacterial, macrobenthic and meiobenthic populations in intertidal flats of the Dollard. *Neth. J. Sea Res.*, **14**: 288–304

Varshney, P. K. (1985). Meiobenthic study of Mahim (Bombay) in relation to prevailing organic pollution. *Mahasagar-Bull. Nat. Inst. Oceanogr., India*, **18**: 27–35

Vidakovic, J. (1983). The influence of raw domestic sewage on density and distribution of meiofauna. *Mar. Poll. Bull.*, **14**: 84–8

Vitiello, P. and Aissa, P. (1985). Structure des peuplements de nematodes en milieu lagunaire pollue. *100 Congr. Nat. Soc. Savantes. Montpellier, Sci.*, **2**: 115–26

Volkmann, B. (1979). A revision of the genus *Tisbe* (Copepoda: Harpacticoida). Part 1. *Archo. Oceanogr. Limnol. Suppl.*, **19**: 121–284

Warwick, R. M. (1981). The influence of temperature and salinity on energy partitioning in the marine nematode *Diplolaimelloides bruciei. Oecologia, Berl.*, **51**: 318–25

Warwick, R. M. and Clarke, K. R. (1991). A comparison of some methods for analysing changes in benthic community structure. *J. Mar. Biol. Assoc. UK*, **71**: 225–44

Wieser, W. (1953). Die Beziehung zwischen Mundhohlengestalt, Ernahrungsweise und Vorkommen bei freilebenden marinen Nematoden. *Ark. Zool.*, **4**(2): 439–84

CHAPTER 13

ULVA RIGIDA AGARDH. STUDIES

N. Riccardi and S. Foltran

SUMMARY

Over the last two decades increases in nutrient load along with poor water renewal caused an abnormal proliferation of *Ulva* in the Venice Lagoon. Experiments under controlled and semi-natural conditions were carried out to assess the influence of environmental variables on *Ulva* growth. Net production, growth and nutrient uptake rates increased with temperature from 10 to 20 °C and decreased slightly at 30 °C. The net production rate increased with light intensity up to about 24,000 lux and decreased at higher light intensities. Net production occurred even at light intensities lower than 1% of the surface incident radiation and only a slight inhibition of photosynthesis was caused by light intensities as high as 120,000 lux in the field. The results demonstrate the vast capacity of adaptation to environmental variations and the high growth potential of this macroalga.

INTRODUCTION

Field studies alone are inadequate to quantify the influence of the most important factors (e.g. temperature, light, nutrients) on the production and decomposition of *Ulva* and the influence, in turn, exerted by *Ulva* on water quality. Experimental studies under controlled conditions are needed to substantiate the results obtained from field studies. With this objective, experiments on *Ulva* in thermostat chambers and under semi-natural conditions were carried out.

MATERIAL AND METHODS

The experiments were carried out in temperature-controlled rooms with provisions for programming light intensity and dark–light cycles. Illumination was provided by a bank of daylight fluorescent tubes which produced a maximum light intensity of 11,500 lux in the laboratory. In addition, some experiments under semi-natural conditions (see below) were carried out.

Productivity measurements

Productivity was measured both in the field and in the laboratory by the 'light and dark bottle method'. *Ulva* thalli of about 0 2 g wet weight were incubated in 300 ml BOD bottles for 1 h. Production and respiration rates were estimated by the increase and decrease of oxygen concentration, respectively, in the bottles exposed to light and kept in the dark.

Measurements in the field were done in summer at noon when the incident light on the surface was as high as 100,000 lux. Lower light intensities were obtained by reducing solar radiation by white and black sheets while an intensity of 120,000 lux was obtained by using reflecting surfaces. Ambient temperature was maintained by a continuous flow of Lagoon water around the incubation vessels. Measurements in the laboratory at the same temperature as Lagoon waters were carried out concurrently.

Nutrient uptake and growth rate measurements

Nutrient uptake and growth rates were measured both in batch and under continuous flow conditions.

Experiments in batch were done in 1000 ml beakers, each one containing about 0.2 g wet weight of *Ulva* and 1 l of filtered Lagoon water. Continuous flow experiments were carried out in an apparatus where Lagoon water was circulated at a constant rate by a peristaltic pump. Algae were grown in filtered (0.45 μm) Lagoon water enriched with nutrients (nitrate, ammonium and phosphate). Since *Ulva* growth was reported to be less dependent on phosphorus than on nitrogen availability (Lavery and McCoomb, 1991) Lagoon water with and without phosphate addition was used in one experiment.

Uptake rates were calculated from the variations in dissolved nutrient concentrations. Specific growth rates (μ, as percentage increase in wet weight per day) were calculated as:

$$\mu = \ln \frac{Wt/Wi}{t} \times 100$$

where Wt = final weight,
Wi = initial weight, and
t = time (De Boer *et al.*, 1978).

Chemical analyses

Dissolved N and P compounds were analysed following the methods given by Parsons *et al.* (1984). Dissolved oxygen was measured by Winkler titration. Chlorophyll *a* and phaeopigments were analysed following the methods given by Golterman (1969).

^{15}N experiments

Ulva thalli (about 0.2 g wet weight) were incubated at 25 °C and 11,500 lux in 300 ml BOD bottles filled with filtered (0.45 μm) Lagoon water to which ^{15}N-NO$_3$ and ^{15}N-NH$_4$ were added at about 10% of the ambient (nitrate or ammonium) concentration. Algal samples were removed after 1, 2, 4 and 8 h, dried at 50 °C for 24 h and ground to fine particles. An algal sample without ^{15}N was used to measure the total N content and the natural abundance of ^{15}N in the tissue. Isotopic enrichment (atom % excess of ^{15}N) was measured by emission spectrometry and total N content, by CHN analyser.

Specific transport rate (V) was calculated as:

$$V = \frac{\Delta \,(\text{atom } \% \,^{15}N \text{ of N pool})}{t \times R}$$

where t = incubation time, and
R = atom % enrichment of the medium.

Absolute uptake rate (ρ) was calculated as:

$$\rho = V \times PN$$

where PN = amount of N in the algal tissue.

Oxygen consumption during *Ulva* decomposition

Oxygen consumption during the decomposition of dead *Ulva* was measured in the laboratory at 10, 20 and 25 °C. *Ulva* thalli, dried at 50 °C, were incubated in 1000 ml bottles filled with Lagoon water filtered through ordinary filter papers.

RESULTS

Light effects

The relationship between light intensity and net production is shown in Figure 13.1. Net production increased until about 24,000 lux but decreased at higher light intensities. The data obtained in the laboratory experiments agreed well with those from semi-natural conditions.

Temperature effects

The relationship between light intensity and net production rate is affected by temperature (Figure 13.2). Production increased with temperature from 10 to 20 °C and decreased slightly at 30 °C. The effect of temperature is pronounced at light intensities higher than 1000 lux. Net production was measured even at very low light intensities, such as 100 lux at 10 °C and 20 °C, while at 30 °C oxygen consumption was observed. This suggests that the compensation point (the light intensity at which photosynthetic oxygen production equals respiration) is higher at a temperature of 30 °C.

Nutrient uptake and growth rates of *Ulva* thalli varied with temperature in the same way as did production; both nutrient uptake and growth rates increased from 10 to 20 °C and slightly decreased at 30 °C (Table 13.1).

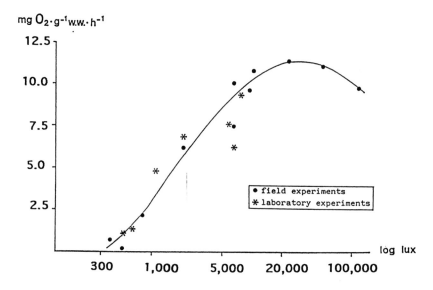

Figure 13.1 Net production (mg O$_2$ / g wet weight / h) of *Ulva rigida* as a function of light intensity (log lux) measured in seminatural * and controlled • conditions at 27 °C

Nutrient uptake

Ulva proliferation depends on nutrient availability and in turn, this affects chemical characteristics of the water by increasing oxygen and decreasing nutrient concentrations. The influence of algal uptake on nutrient concentrations and their ratios in Lagoon waters can be seen from Tables 13.2 and 13.3.

DIN/PO$_4$-P ratios decrease over time since algae require much more N than P. Through this change of the N/P ratios, *Ulva* proliferation can be expected to affect phytoplankton growth, since N/P ratios lower than 15 have been reported to limit such growth (Redfield, 1958). This limitation becomes evident in summer when a massive proliferation of macroalgae results in a decrease of nutrient concentrations in water (cf. Ravera *et al.*, Chapter 14). The increase of chl *a* at station 31 soon after the decline of *Ulva* supports this hypothesis (cf. Ravera *et al.*, Chapter 14).

The increase in NO$_3$-N/NH$_4$-N ratios (Figures 13.3 and 13.4) indicates that uptake of ammonium becomes predominant over that of nitrate with time. In conformity with this result, lowest concentrations of ammonium in the Palude della Rosa were measured in summer at stations 14 and 31 where *Ulva* biomass was at its highest (cf. Ravera *et al.*, Chapter 14).

The data in Table 13.4 show that the N uptake rates of *Ulva* are not influenced by the concentration of P in water. On the other hand, the growth rate of *Ulva* on enrichment with P was twice that without (Table 13.5). These results can be explained by the ability of *Ulva* to accumulate N reserves which

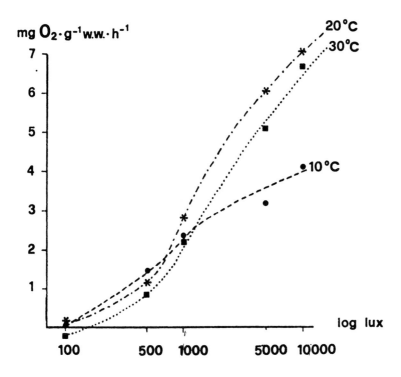

Figure 13.2 Net production (mg O_2 / g wet weight / h) of *Ulva rigida* as a function of light intensity (log lux) measured in the laboratory at 10 °, 20 ° and 30 °C

Table 13.1 Nutrient uptake rates (μg/g wet weight/h) and specific growth rates (μ; %/day) at 10 °, 20 ° and 30 °C. Light intensity 8500 lux

$t°C$	PO_4-P	NH_4-N	NO_3-N	μ
10	0.56 ± 0.05	7.39 ± 0.61	12.19 ± 1.41	10.6
20	0.53 ± 0.27	10.52 ± 0.65	18.54 ± 1.26	37.3
30	0.63 ± 0.21	8.88 ± 0.91	15.93 ± 1.86	30.0

are then used for growth when favourable conditions occur (Rosenberg and Ramus, 1982). This causes an uncoupling of N uptake and growth when limitation by other factors occurs. In our experiment, phosphate limitation caused a reduction of growth rate without affecting the uptake of nitrogen which was probably stored in the thallus.

[15]N uptake (Figure 13.5) showed that uptake of nitrate was higher than that of ammonium. The relative importance of nitrate and ammonium uptake in total N uptake appears to be related to their concentration ratios in the ambient waters. In fact, ambient NH_4-N/NO_3-N ratio at the beginning of the experi-

Table 13.2 Nutrient concentrations (μg/l) in water and their ratios 7 h, 24 h 30 min, and 28 h after introduction of *Ulva* in the continuous flow system. Water volume: 13.7 l; initial wet weight of *Ulva* 8.66 g; t: 20 ± 1 °C; light intensity: 8100 lux

Nutrient	t = 0	t = 7 h	t = 24 h 30 min	t = 28 h
PO$_4$-P	92	47	22	21
NO$_2$-N	14	12	7	5
NO$_3$-N	748	634	86	41
NH$_4$-N	298	23	8	8
DIN	1060	669	101	54
DIN/PO$_4$-P	11.5	14.2	4.5	2.5
NO$_3$-N/NO$_2$-N	53.4	2.8	12.3	8.2
NO$_3$-N/NH$_4$-N	2.5	27.6	10.7	5.1

Table 13.3 Nutrient concentrations (μg/l) and their ratios 2 h 30 min and 7 h 30 min after introduction of *Ulva* in the continuous flow system. Lagoon waters with (A) phosphate addition and (B) without. Water volume in each system: 11.4 l; t: 2 ± 1 °C; light intensity: 8300 lux

Nutrient	A			B		
	t = 0	t = 2 h 30 min	t = 7 h 30 min	t = 0	t = 2 h 30 min	t = 7 h 30 min
PO$_4$-P	90.3	54.9	23.6	3.0	3.0	3.0
NO$_2$-N	6.9	4.8	3.0	7.2	5.5	4.0
NO$_3$-N	572.2	331.8	133.5	635.3	429.7	164.9
NH$_4$-N	298.9	78.1	11.3	263.8	56.6	13.1
DIN	878.0	414.6	147.8	906.3	491.7	181.9
DIN/PO$_4$-P	9.7	7.6	6.3	302.1	163.9	60.6
NO$_3$-N/NO$_2$-N	82.9	69.1	44.5	88.2	78.1	41.2
NO$_3$-N/NH$_4$-N	1.9	4.2	11.8	2.4	7.6	12.6

ment was about 0.32. The NH$_4$-N/NO$_3$-N ratio in the algal tissue ranged from 0.81 (after 1 h incubation) to 0.56 (after 8 h incubation), becoming progressively similar to that in the water.

Oxygen consumption due to *Ulva* decomposition

During the decomposition of *Ulva*, anoxic conditions and a consequent decline of the macrobenthic community occurred at some stations in the Palude della Rosa (cf. Ravera *et al.*, Chapter 14; Tagliapietra *et al.*, Chapter 11). Oxygen consumption rates during decomposition of *Ulva* measured as a function of temperature in the laboratory showed an increase from 1.04 mg/g wet weight/h at 10 °C to 1.17 and 2.14 mg/g wet weight/h at 20 and 25 °C, respectively.

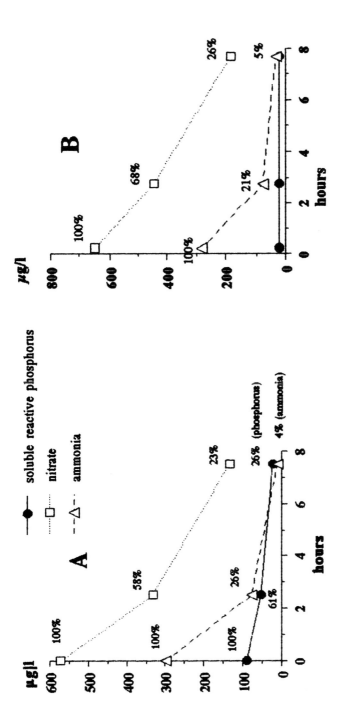

Figure 13.3 Decrease of nutrient concentrations (μg / l) in water and increase of *Ulva* wet weight (g) over time in continuous flow experiments at 20 °C and 8300 lux. A = Lagoon water with phosphate addition; B = Lagoon water without phosphate addition

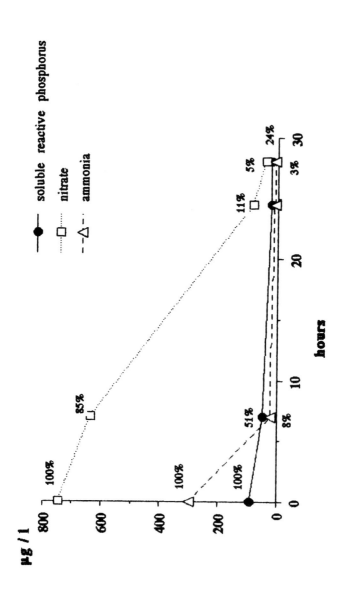

Figure 13.4 Decrease of nutrient concentrations (μg / l) in water and increase of *Ulva* wet weight (g) over time in continuous flow conditions at 20 °C and 8100 lux

Table 13.4 Nutrient uptake rates (μg / g wet weight / h) during the first 2 h 30 min (1) and the subsequent 5 h (2) in Lagoon water with (A) and without (B) phosphate addition. Water volume in each system: 11.4 l. Temperature: 20 ± 1 °C; light intensity: 8300 lux

Nutrient	A (1)	B (2)	A (1)	B (2)
PO_4-P	6.0	2.6	0.0	0.0
NO_2-N	0.4	0.2	0.3	0.1
NO_3-N	40.7	16.8	36.2	22.8
NH_4-N	37.5	5.6	36.1	3.7

Table 13.5 Initial and final wet weight (g) and specific growth rate (μ, % / day) of *Ulva* in Lagoon water with (A) and without (B) phosphate addition. Temperature: 20 ± 1 °C. Light intensity: 8300 lux

	wet wt (t = 0)	wet wt (t = 31 h)	μ
A	25.95	36.75	26.9
B	25.32	30.21	13.7

Assuming a mean *Ulva* biomass of 1 kg wet weight/m^2, the daily oxygen consumption in the Palude della Rosa during decomposition of *Ulva* at summer temperatures (about 25 °C) is about 2 g/m^2. With a mean water depth of 0.5 m and a mean oxygen concentration of 10 g/m^3, the mass balance of oxygen in the water column would become strongly negative within about 3 days if there are no fresh inputs from the atmosphere and tidal waters. Therefore, anoxic conditions can be expected to occur as soon as the whole macroalgal biomass begins to decompose.

DISCUSSION AND CONCLUSIONS

Our results underscore the great capacity of *Ulva* to adapt to environmental variations and its high growth potential.

Production and growth of *Ulva* were optimal at 20 °C, decreased slightly at 30 °C and were not severely limited at 10 °C. *Ulva* is well adapted to a wide range of light intensities, as is typical of eulittoral species. Light saturation of photosynthesis occurs at about 24,000 lux (Figure 13.1) and a slight inhibition of photosynthetic activity occurs at higher light intensities. In fact, only a 15% decrease of oxygen production (with respect to the maximum value at 24,000 lux) was observed even in the algae exposed to 120,000 lux. Similar responses to light by various *Ulva* species have been reported (Fortes and Luning, 1980; Arnold and Murray, 1980). *Ulva rigida* is able to photosynthesize at very low irradiances typical of waters with high turbidity, such as

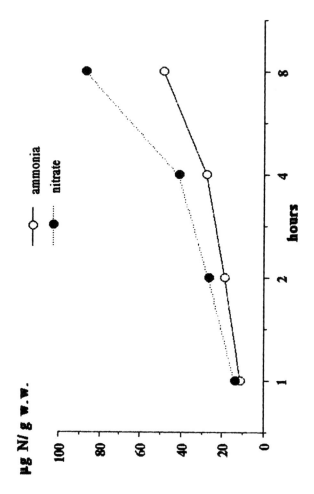

Figure 13.5 Uptake of nitrate and ammonium as a function of time measured by ^{15}N

the Venice Lagoon. In fact, net production was measured at light intensities as low as 100 lux, far below the 1% of surface light considered to limit phytoplankton growth. The high photosynthetic efficiency of *Ulva* species, also observed in other studies (Vermaat and Sand-Jensen, 1987; Sand-Jensen, 1988a,b) accounts for its ability to develop dense mats, without being limited by self-shading.

The success of *Ulva* in colonization of eutrophic coastal waters, enabling it to dominate other macroalgae, has been attributed to its high nutrient uptake capability, high photosynthetic rate and fast growth during favourable periods (Steffensen, 1976; Lapointe and Tenore, 1981; Rosenberg and Ramus, 1984). Our results confirm this and show that, under favourable conditions, *Ulva rigida* is characterized by high nitrogen uptake rates and an explosive growth capacity. For instance, the mean growth rate at 20 °C was as high as 37%/day, which is far higher than those reported for other species of *Ulva* (e.g. Davies, 1981; Lapointe and Tenore, 1981; Fujita, 1985; Duke *et al.*, 1986; Smith and Horne, 1988; Fujita *et al.*, 1989).

Phosphorus availability does not affect N uptake rate while growth rate is consistently reduced under P limitation. The ability of *Ulva* to store nitrogen reserves under unfavorable conditions (e.g. Rosenberg and Ramus, 1982; Fujita, 1985; Fujita *et al.*, 1989) is the most probable reason for uncoupling observed in N uptake by *Ulva* and growth.

Due to high nutrient uptake rates of *Ulva,* nutrient depletion occurs in water when its proliferation is at its highest. Our experiments show a consistent decrease of the N/P ratios in water during *Ulva* growth, suggesting that resource competition for inorganic N could be an important cause of the absence of phytoplankton development during the growth period of *Ulva*. A similar conclusion, that *Ulva* may outcompete phytoplankton by reducing N to levels below those necessary to support phytoplankton growth, was arrived at by Smith and Horne (1988), based on mesocosm experiments.

A preference for ammonium as a N source can be hypothesized on the basis of our results. In fact, the ratio of ammonium taken up to its concentration in water was higher than that of nitrate. But this preference was masked by the high availability of nitrate in the Lagoon waters, which accounted for the higher amount of nitrate taken up by the algae.

The high oxygen consumption during *Ulva* decomposition causes anoxic conditions in areas characterized by abundant macroalgal biomass and slow water renewal. Macrobenthic communities are, consequently, modified by the *Ulva* cycle, as demonstrated by the study in the Palude della Rosa (cf. Tagliapietra *et al.*, Chapter 11).

In conclusion, its ability to tolerate wide environmental variations and its high growth potential make *Ulva* a successful competitor in the Venice Lagoon. Owing to the high levels of biomass it attains, this macroalga strongly influences the physicochemical characteristics of the environment and, consequently, the community structure.

ACKNOWLEDGEMENTS

We are especially grateful to Dr P. Morin, Prof. P. LeCorre and Dr C. Madec for their collaboration in ^{15}N experiments and the analyses of ^{15}N samples. We wish to thank Prof. E. de Fraja Frangipane, Prof. O. Ravera and Prof. A. Marzollo for many helpful suggestions during experimental work and discussion of the results.

REFERENCES

Arnold, K. E. and Murray, S. N. (1980). Relationships between irradiance and photosynthesis for marine benthic green algae (Chlorophyta) of differing morphologies. *J. Exp. Mar. Biol. Ecol.*, **43**: 183–92

Davies, M. (1981). Production dynamics of sediment-associated algae in two Oregon estuaries. Ph.D thesis. Oregon State University

De Boer, J. A., Guigli, H. J., Israel, T. L. and D'Elia, C. F. (1978). Nutritional studies of two red algae. Growth rate as a function of nitrogen source and concentration. *J. Phycol.*, **14**: 261–6

Duke, C. S., Lapointe, B. E. and Ramus, J. (1986). Effects of light on growth, RuBPCase activity and chemical composition of *Ulva* species (Chlorophyta). *J. Phycol.*, **22**: 362–70

Fortes, M. D. and Luning, K. (1980). Growth rates of North Sea macroalgae in relation to temperature, irradiance and photoperiod. *Helgoländer Meeresunters*, **34**: 15–29

Fujita, R. M. (1985). The role of nitrogen status transient in ammonium uptake and nitrogen storage by macroalgae. *J. Exp. Mar. Biol. Ecol.*, **92**: 283–301

Fujita, R. M., Wheeler, P. A. and Edwards, R. L. (1989). Assessment of macroalgal nitrogen limitation in a seasonal upwelling region. *Mar. Ecol. Progr. Ser.*, **53**: 293–303

Golterman, H. L. (1969). Methods for chemical analysis of freshwater. *IBP Handbook No. 8*. Blackwell Sci. Publ. Oxford. UK

Lapointe, B. E. and Tenore, K. R. (1981). Experimental outdoor studies with *Ulva fasciata* Delile. I. Interactions of light and nitrogen on nutrient uptake, growth and biochemical composition. *J. Exp. Mar. Biol. Ecol.*, **53**: 135–52

Lavery, P. S. and McCoomb, A. J. (1991). The nutritional ecophysiology of *Chaetomorfa linum* and *Ulva rigida*. In *Peel Inlet, Western Australia. Botanica Marina*, **34**: 251–60

Parson, T. R., Maita, Y. and Lalli, C. M., (1984). *A manual of chemical and biological methods for seawater analysis*. Pergamon Press, Oxford, UK

Redfield, A. C. (1958). The biological control of chemical factors in the environment. *Am. Sci.*, **46**: 205–21

Rosenberg, G. and Ramus, J. (1982). Ecological growth strategies in the seaweeds *Gracilaria foliifera* (Rhodophyceae) and *Ulva* sp. (Chlorophyceae): soluble nitrogen and reserve carbohydrates. *Mar. Biol.*, **66**: 251–9

Rosenberg, G. and Ramus, J. (1984). Uptake of inorganic nitrogen and seaweed surface area: volume ratios. *Aquat. Bot.*, **19**: 65–72

Sand-Jensen, K. (1988a). Minimum light requirements for growth in *Ulva lactuca*. *Mar. Ecol. Progr. Ser.*, **50**: 187–93

Sand-Jensen, K. (1988b). Photosynthetic responses of *Ulva lactuca* at very low light.

Mar. Ecol. Progr. Ser., **50**: 195–201

Smith, D. W. and Horne, A. J. (1988). Experimental measurement of resource competition between planktonic microalgae and macroalgae (seaweeds) in mesocosms simulating the San Francisco Bay-Estuary, California. *Hydrobiologia*, **159**: 259–68

Soeder, C. J. and Talling, J. F. (1969). Measurements (*in situ*) on isolated samples of natural communities. In Vollenweider, R. A. (ed.). *A manual on methods for measuring primary production in aquatic environments*. IBP Handbook No. 12, Blackwell Sci. Publ., Oxford, 80–9

Steffensen, D. A. (1976). The effect of nutrient enrichment and temperature on the growth in culture of *Ulva lactuca* L. *Aquatic Botany*, **2**: 337–51

Vermaat, J. E. and Sand-Jensen, K. (1987). Survival, metabolism and growth of *Ulva lactuca* L. under winter conditions: a laboratory study of bottlenecks in the life cycle. *Mar. Biol.*, **95**: 55–61

CHAPTER 14

CHEMICAL CHARACTERISTICS OF THE PALUDE DELLA ROSA

O. Ravera, A. Piva and S. Foltran

SUMMARY

From a study carried out on the chemical aspects of the Palude della Rosa between December 1991 and April 1993, the following conclusions were drawn:

(1) The physical and chemical gradients from north to south of the marsh control the spatial distribution, abundance and structure of the macrobenthic community, whereas the cycle of *Ulva rigida* deeply influences the seasonal variations of the physical, chemical and biological characteristics of the marsh.

(2) The importance of this macroalga is evident from the ratio between the biomass of *Ulva* and that of the macrobenthos; this ratio is equal to 22 in terms of carbon biomass and 11 in terms of nitrogen biomass.

(3) The anoxic conditions due to the decomposition of *Ulva* produce a dramatic decrease in biomass and number of species of the macrobenthos.

INTRODUCTION

Since an ecosystem is an integration of the physical environment and its community, both compartments must be studied simultaneously to draw ecologically reliable conclusions. With this approach, the structure and seasonal variations of the community can be explained in relation to physical and chemical characteristics and the influence of the community, in turn, on the physical environment. Also, the nutrient contents, usually in terms of nitrogen and carbon, in the most important taxonomic groups are useful indices of the trophic status. Unfortunately, this information is very scarce. These considerations led to the present study on the chemical characteristics of the environment and nutrient content of macrobenthic organisms of the Palude della Rosa. Such data, when obtained along with those on macrobenthic populations, can prove very useful in environmental interpretation. With this aim, samples for benthic studies and chemical analyses were collected simultaneously and from the same stations, with a sampling frequency adequate to cover the seasonal variations of the parameters taken.

MATERIAL AND METHODS

Samples of water, sediment and macrobenthos were collected at monthly intervals from five stations (5, 14, 23, 31, 38) in the Palude della Rosa (northeastern part of the Venice Lagoon) from May 1992 to April 1993.

Temperature, salinity and pH were measured *in situ*. Dissolved oxygen, soluble reactive phosphorus (SRP), ammonium, nitrate, nitrite and sulfate were analysed following standard methods (Parsons *et al.*, 1984). Chlorophyll *a* was measured following the method of Golterman (1969).

Soon after collection, sediment and macrobenthos samples were dried at 105 °C for 24 h, and a fraction ashed at 450 °C. Both the dried and ashed samples were analysed for C and N in a Carlo Erba model EA1108 CHNS-O analyser. Total phosphorus in sediment samples was measured after digestion with hydrochloric acid (Aspila *et al.*, 1976). Biological samples were autoclaved and treated with potassium persulfate and sulfuric acid (Donazzolo, pers. comm.) for total phosphorus estimations.

RESULTS

Water chemistry

Figure 14.1 shows the seasonal patterns of water temperature, salinity, pH and dissolved oxygen at the five stations. The annual mean concentrations of dissolved oxygen at each station are given in Table 14.1.

Differences in water temperature among the stations were small. Salinity at station 23 ranged from 7 to 19 PSU whereas at other stations, it varied from 16 to 35 PSU. The low salinity at station 23 was due to the freshwater flow from the River Sile.

Differences in pH values among the stations were also small. Summer pH values (8.6–9.0) were higher than winter (7.0–7.4) values. The high summer pH values were due to active photosynthesis; in fact, during the greatest proliferation of *Ulva**, the highest values were measured at stations 14 and 31. During June 1992, when the biomass of *Ulva* at station 31 decreased from 3.5 to 0.2 kg wet weight/m², the pH decreased from 8.7 to 7.3. Soon after the decline in *Ulva*, the chl *a* concentration increased from 17 to 39 μg/l and pH values increased again from 7.3 to 8.1. Evidently, the pH increase during this period was due to phytoplankton production.

Seasonal variations in dissolved oxygen concentrations were observed only at two stations (14 and 31) which had a rich *Ulva* biomass (Table 14.2). At other stations, the coefficients of variation were small (about 0.2). For example, dissolved oxygen was as high as 27 mg/l at station 14 during the proliferation of *Ulva* but decreased to 1 mg/l at station 31 during the decomposition of *Ulva*. Average percentage saturation of oxygen at stations where

---

*The species referred to as *Ulva rigida* in the Venice Lagoon is in fact *U. laetevirens*

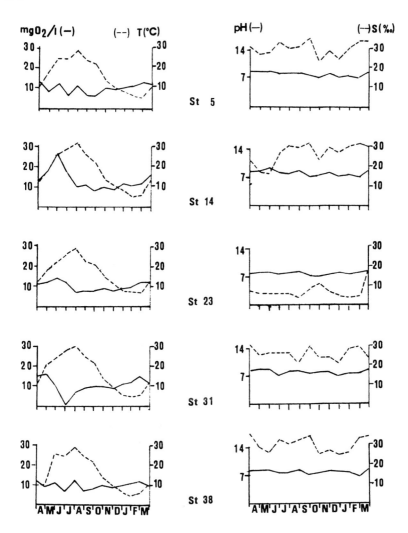

Figure 14.1 Seasonal variations in dissolved oxygen concentration (mg O_2/l), temperature, pH and salinity at the five sampling stations in the Palude della Rosa

Ulva was abundant was about 100 but rose to 350 when *Ulva* growth peaked.

From the concentrations of nutrients (Table 14.1) and their seasonal changes (Figure 14.2), the following conclusions can be drawn:

(1) The annual average soluble reactive phosphorus (SRP) was similar at stations 5, 14, 31 and 38, whereas it was very high at station 23. Seasonal variations of SRP were small when compared with those of total inorganic nitrogen (TIN);

Table 14.1 Mean concentration ($\bar{x}$), standard deviation (sd) and coefficient of variation (vc) of dissolved oxygen, soluble reactive phosphorus (SRP), total inorganic nitrogen (TIN), ammonia, nitrate and sulfate in the water at the five stations

Station	mgO_2/l		$\mu g\ SRP/l$		$\mu g\ TIN/l$		$\mu g\ N\text{-}NH_3/l$		$\mu g\ N\text{-}NO_3/l$		$mg\ S\text{-}SO_4/l$	
	$\bar{x} \pm sd$	vc	$\bar{x} \pm sd$	vc	$\bar{x} \pm sd$	vc	$\bar{x} \pm sd$	vc	$\bar{x} \pm sd$	vc	$\bar{x} \pm sd$	vc
5	9.72 ± 2.27	0.23	10 ± 8	0.8	957 ± 722	0.8	172 ± 221	1.3	767 ± 621	0.8	0.87 ± 0.60	0.7
14	13.36 ± 4.99	0.37	12 ± 7	0.6	903 ± 850	0.9	104 ± 67	0.6	780 ± 805	1.0	0.74 ± 0.35	0.5
23	10.06 ± 2.09	0.21	64 ± 24	0.4	2004 ± 1135	0.6	307 ± 464	1.5	1653 ± 827	0.5	0.44 ± 0.48	1.1
31	10.48 ± 3.73	0.36	10 ± 7	0.7	788 ± 361	0.5	99 ± 109	1.1	657 ± 302	0.5	0.67 ± 0.27	0.4
38	9.65 ± 1.87	0.19	12 ± 9	0.8	951 ± 902	0.9	142 ± 117	0.8	789 ± 871	1.1	0.76 ± 0.30	0.4

Table 14.2 Seasonal variations of *Ulva* biomass (kg wet weight/m^2) at stations 14 and 31

	Station	
Date	*14*	*31*
April 1992	1.3	1.6
May	0.6	3.5
June	2.3	0.2
July	2.2	0
August	0	0
September	0	0
October	0	0
November	0	0.2
December	0	0.4
January 1993	0	0.3
February	0	0.2
March	0.2	0.5

(2) Since TIN consisted mainly of nitrate (75%), its seasonal changes were similar to those of nitrate;

(3) Nitrate concentrations were always high at all stations, and especially at station 23. Highest concentrations were measured during late winter and early spring;

(4) Ammonium concentrations showed large seasonal variations. The highest concentrations were measured at station 23 and the lowest, at stations 14 and 31. The relatively low concentrations at the latter stations can be explained by an uptake by *Ulva* when its biomass reached peak values at these stations;

(5) Nitrite accounted for 2–3% of TIN concentration. Only at station 31 when anoxic conditions set in, did this percentage increase to 10.

Differences in the annual average sulfate concentrations (Table 14.1) between the stations were due to the degree of water renewal by the tides; it was very rapid at stations 5 and 38, slow at stations 31 and 14 and very slow at station 23. In fact, the highest concentrations of sulfate were measured at stations 5 and 38 and the lowest at stations 23, where seawater was diluted by freshwater. Sulfate concentrations showed larger seasonal variations at this station than at others, as evident from the relatively high coefficient of variation (1.1).

Sediment chemistry

Table 14.3 presents average concentrations of elements in the sediments. Seasonal variations in their concentrations were not evident. The differences between the stations were generally rather small. The high percentage of ash

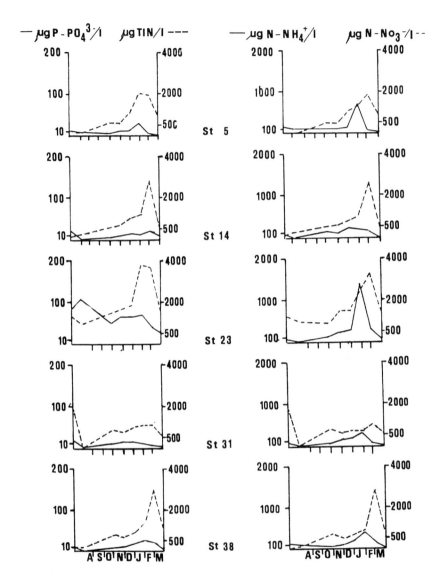

Figure 14.2 Seasonal variations in soluble reactive phosphorus, total inorganic nitrogen, ammonia and nitrates at the five sampling stations in the Palude della Rosa

indicates a high mineral content in sediments. Assuming a C/N ratio of 5 for the organic matter in sediments, the N content of sediments was multiplied by 5 to obtain the organic C content. Inorganic C content was obtained as the difference between total C and organic C. The inorganic C/total C ratios were

Table 14.3 Concentrations of some important elements in the sediments from the five stations. The total C, N and S are expressed in mg/g dry weight in C = inorganic carbon

Station	5	14	23	31	38
C	78.9	80.8	78.0	80.1	80.6
N	1.4	2.5	2.7	3.1	2.0
S	5.4	6.0	7.6	7.2	5.6
org. C	7.0	12.5	13.5	15.5	10.6
in. C	71.9	68.3	64.5	64.6	70.0
in. C / C (%)	91.1	84.5	82.7	80.7	86.9
ash / dry weight (%)	93.8	93.2	92.4	91.5	94.7

always lower than the ash/dry weight ratios, indicating that the greatest part of the C in the sediments is inorganic and that the ashes are composed of carbonates and other minerals, probably clay minerals.

Chemical composition of the biota

Variations in N and P contents of *Ulva* between the seasons and the stations were similar to those of ambient N and P nutrients. For example, N and P contents of *Ulva* samples collected from station 23 on May 28th 1992 were 35.9 mg N/g dry weight and 3.14 mg P/g dry weight whereas at the other 4 stations they were in the range of 16–17 mg N/g dry weight and 0.77–1.24 mg P/g dry weight. This is due to high concentrations of dissolved nutrients at station 23 compared with other stations. Similarly, N and P contents in *Ulva* were also higher during winter when nutrient concentrations were at their highest. For example, *Ulva* samples collected at station 14 in May 1992 had 16.8 mg N/g dry weight and 1.18 mg P/g dry weight whereas samples from the same station collected in December of the same year had 41.1 mg N/g dry weight and 3.31 mg P/g dry weight.

Zostera noltii Hornem is a phanerogam which was widely distributed in the Lagoon in the past but is now restricted to relatively small areas. During the period of this study, *Zostera noltii* was found only at stations 23 and 38. It is very difficult to identify the causes of this decline but it is possible that changes in the environment have directly or indirectly influenced distribution of this plant. Unlike *Ulva*, the N and P concentrations in *Zostera* did not vary between stations and seasons. For example, *Zostera* samples collected in May 1992 from stations 23 and 38 had, respectively, 32.7 and 32.2 mg N/g dry weight, and a similar concentration of P (3.5 mg P/g dry weight). It is of interest to note that this constancy was maintained in spite of the large differences in dissolved nutrient concentrations between the two stations. Consequently, the N/P ratio was fairly constant (about 8) whereas that calculated for *Ulva* varied from 12 to 41. A comparison of these ratios also shows that *Ulva* requires a larger supply of N than *Zostera*. This could be one of the reasons for the partial replacement of *Zostera* by *Ulva*, which was favoured by the increased nutrient load in the Lagoon. The S concentration in *Ulva* was very high (mean = 36 mg S/g dry weight) compared with that in *Zostera* (mean = 6 mg S/g dry weight) and with those measured in other benthic species (7–16 mg S/g dry weight).

C, N and P concentrations in the macrozoobenthos at different seasons and at all stations were rather constant and did not differ significantly among species. Table 14.4 gives the mean concentrations of C, N and P in the soft tissues of the most important benthic taxa.

Table 14.5 presents the seasonal variations of total macrobenthic biomass in terms of C and N, calculated by multiplying the dry weight by the average C and N content (Table 14.4), at the 5 stations. The dry weight biomass (g dry weight/0.1 m^2) and the density (numbers/0.1m^2) of the macrozoobenthos are reported elsewhere (see Tagliapietra *et al.*, Chapter 11). The data listed in Tables 14.5 and 14.6 demonstrate seasonal variations of the total C and N content in the macrobenthos. Since the C and N concentrations did not vary significantly with species, these seasonal variations were due to biomass variations and not to those of structure of the benthic community. The benthos at the stations under tidal influence (stations 5 and 38) had higher C and N contents.

The mean biomass of *Ulva* colonies in the Palude della Rosa, at their peak growth, was 20 g dry weight/0.1 m^2, equivalent to 1 kg wet weight/m^2. The C and N contents of the *Ulva* biomass were 6542 mg/0.1 m^2 and 720 mg/0.1 m^2, respectively. These values are very high compared with the average C and N contents in macrozoobenthos biomass which were 295 mg/0.1 m^2 and 68 mg/0.1 m^2, respectively. The mean C/N ratio (4.37) of the benthos was also very low compared with that of *Ulva* (9.09). This difference was due to the higher concentration of N in benthic animals than in *Ulva*.

Table 14.4 C, N and P concentrations (mg/g dry weight ash-free) in the most important taxa of macrozoobenthos (nm = not measured)

Taxa	C	N	P	C/N	N/P
Nereis sp.	525.3	134.6	n.m.	3.90	–
Cerastoderma glaucum	503.9	115.9	7.2	4.35	16.10
Gammaridae	495.3	122.1	7.6	4.06	16.07
Idotea baltica	498.4	118.9	5.3	4.19	22.43
Cyatura carinata	520.7	113.7	n.m.	4.58	–
Sphaeroma serratum	544.1	119.0	5.1	4.57	23.33
Chironomus salinarius	535.3	108.4	6.2	4.94	17.48
$\bar{x}$	517.6	118.9	6.3	4.37	19.08

Table 14.5 Seasonal variations of macrozoobenthos biomass (in mg C/0.1 m² calculated on ash-free dry weight basis) at the five stations

Station	Dec	March	April	May	June	July	Aug	Sept	Oct	Nov	Dec	Jan	Feb	Mar	April	$\bar{x}$
5	804	509	885	592	167	88	474	0	0	182	1	0	8	1787	1584	472.1
14	237	549	89	328	60	3	10	2	13	3	160	148	87	442	839	198.0
23	191	447	221	131	1	0	1	0	1	2	9	13	371	4	54	96.4
31	293	254	233	135	0	0	0	0	0	0	5	0	5	233	0	77.2
38	696	1530	1597	627	333	837	485	509	52	102	261	716	11	628	1070	630.3
$\bar{x}$	444.2	657.8	605.2	362.6	112.2	185.6	194.0	102.2	13.2	57.8	87.2	175.4	96.4	618.8	709.4	294.8

Table 14.6 Seasonal variations of macrozoobenthos biomass (in mg N/0.1 m² calculated on ash-free dry weight basis) at the five stations

Station	Dec	March	April	May	June	July	Aug	Sept	Oct	Nov	Dec	Jan	Feb	Mar	April	$\bar{x}$
5	185	117	203	136	38	20	109	0	0	42	0	0	2	411	364	108.5
14	54	126	20	75	14	0	2	0	3	1	37	34	20	105	193	45.6
23	44	103	51	30	0	0	0	0	0	0	2	3	85	1	12	22.1
31	67	58	54	31	0	0	0	0	0	0	1	0	1	54	0	17.7
38	160	351	367	144	77	192	111	117	12	23	60	165	3	144	246	144.8
$\bar{x}$	102.0	151.0	139.0	83.2	25.8	42.4	44.4	23.4	3.0	13.2	20.0	40.4	22.2	143.0	163.0	67.7

DISCUSSION AND CONCLUSIONS

The chemical characteristics of the water did not show significant differences among the stations, except at station 23 which received a high influx of freshwater and nutrients from the watershed. In fact, the total P concentration in the sediments (Table 14.7) decreased gradually from the station more influenced by the P load from the watershed (station 23) to those more under tidal influence (stations 5 and 38). On the other hand, differences in SRP concentrations (10–12 μg/l) were not significant, except at station 23 (64 μg/l).

The TIN/SRP ratio decreased with the decrease in tidal influence from station 5 to station 23 (Table 14.8). At station 23, due to the low influence of the sea, salinity showed important seasonal variations and its values, as well as those of sulfate concentration, were always lower than those measured at other stations.

Oxygen concentrations and pH were controlled by *Ulva* during its growth and by phytoplankton when *Ulva* was scarce. When the proliferation of *Ulva* was maximum, oxygen concentrations supersaturated. The consumption of oxygen during the decomposition of the *Ulva* bloom was up to the point of producing anoxic conditions with a consequent decline of the macrobenthic community (see Tagliapietra *et al.*, Chapter 11). Our results show anoxic conditions at station 31 in June 1992. The full impact of oxygen depletion can best be understood only with measurements at the sediment–water interface and during the night.

Table 14.7   Total phosphorus (TP) concentrations in the sediments of the five stations

Station	mgTP/g dry weight
5	0.324
38	0.393
31	0.427
14	0.443
23	0.540

Table 14.8   TIN/SRP ratios at the five stations

Station	TIN/SRP
5	95.70
38	79.25
31	78.80
14	75.25
23	46.94

As evident from the lower salinity and sulfate concentrations, water renewal by tidal exchange at stations 14 and 31 was comparatively less rapid than at stations 5 and 38. Consequently, these two stations offered the best conditions for proliferation of *Ulva*, which attained the greatest biomass at these sites. It is probable that this macroalga had a negative impact on the macrobenthos, which had the lowest C and N biomass at these stations (excluding station 23). The relatively low ammonium concentrations at stations 14 and 31 were probably a consequence of its uptake by the huge biomass of *Ulva*.

The N and P concentrations in *Zostera* and macrobenthic animals were fairly constant, whereas in *Ulva* they varied in relation to their concentrations in water. This suggests that *Ulva* does not possess regulatory mechanisms for these elements. The N content was always higher in macrobenthic animals than in *Zostera* and *Ulva*. The amount of C biomass of *Ulva* per unit surface was 22 times greater than that of the macrobenthos biomass, and the N biomass was 11 times higher.

According to Show (1985) the sulfur content of the macroalgae varies from 0.5 to 1% of dry weight. Brault and Briand (1985) calculated a value of 4% of dry weight for *Ulva*. The mean value calculated for *Ulva* in this study is of the same order of magnitude as that reported by Brault and Briand (1985). The high concentration of sulfur in *Ulva* must have a great influence on the cycle of this element in the Lagoon.

It is probable that a high percentage of the sulfur present in *Ulva* is tied to polysaccharides. During the decomposition of *Ulva*, sulfur is oxidised to sulfate. Under anaerobic conditions sulfate is reduced to sulfide and the sulfur released as sulfuric acid. The toxicity of this gas is related to temperature and pH (Khan and Trottier, 1978). Sulfuric acid is partly transferred to the atmosphere and in some years, such as 1989–1990, its concentration in the air of Venice was so high as to disturb the inhabitants. If the amount of soluble metal compounds is proportional to sulfuric acid concentration, they would then precipitate as sulfides. This precipitation would eliminate the sulfuric acid from the water column, decrease its toxicity to plankton and nekton and increase the concentration of metal sulfides in the surface sediments. In conclusion, the importance of *Ulva* in the economy of the Palude della Rosa is fundamental.

REFERENCES

Aspila, K. J., Agemian, H. and Chau, A. S. J. (1976). A semiautomated method for the determination of inorganic, organic and total phosphate in sediments. *Analyst*, **101**: 187–97

Brault, D. and Briand, X. (1985). Les marées vertes: mise au point d'une technique de stockage de l'algue *Ulva* sp. faisant office de prétraitement pour sa méthanisation. *Rapport de contract no. 3-320-1847. Pleubian, France: Centre d'Experimentation et de Recherche Appliquée en Algologie*, pp. 106

Golterman, H. L. (1969). *Methods for chemical analysis of freshwater*. IBP Handbook no. 8. Blackwell Sc. Publ., Oxford, UK

Khan, A. W. and Trottier, T. M. (1978). Effect of sulphur-containing compounds on anaerobic degradation of cellulose to methane by mixed cultures obtained from sewage sludge. *Appl. Env. Microbiol.*, **35**: 1027–34

Parsons, T. R., Maita, Y. and Lalli, C. M. (1984). *A manual of chemical and biological methods for seawater analysis*. Pergamon Press, Oxford, UK

Show, I. T, Jr. (1985). Marine Biomass. In Sofer, S. S. and Zaborsky, O. R. (eds). *Biomass Conversion Processes for Energy and Fuel*. Plenum Press, New York, 57–77

CHAPTER 15

BENTHIC EUTROPHICATION STUDIES (BEST)

M.C.Th. Scholten, R.G. Jak, B. Pavoni, A. Sfriso,
C.J.M. Philippart and H. de Heij

SUMMARY

The Benthic Eutrophication Studies (BEST) project was conceived to elucidate the responses of coastal marine benthic systems to progressive fertilisation, with special emphasis on the Wadden Sea and the Venice Lagoon. This paper summarises the results obtained in the Venice Lagoon.

The average nutrient loading in the Venice Lagoon was approximately 6 μmol N and 0.5 μmol P per litre per day. Release of ammonium from the sediments constitutes an additional equivalent source of nitrogen, resulting in enhanced macroalgal growth. The absence of gammarid grazers may contribute to the build-up of high macroalgal biomass over vast areas of the Lagoon. Phytoplankton growth was inhibited during periods of macroalgal growth. Motile species, like nereid worms and crabs, benefit from the growth and decay of macroalgae. The absence of macroalgae in parts of the Venice Lagoon seems to be related to hydrodynamic conditions

INTRODUCTION

The aim of the BEST project was to elucidate the response of coastal marine benthic systems to eutrophication, with special emphasis on the comparison of the situations in the Venice Lagoon (northern Adriatic Sea) and in the Wadden Sea (southern North Sea); in both areas increased nutrient load has led to eutrophication. The multidisciplinary study examined four key processes: sediment–water exchange of nutrients (N, P), competition between benthic macroalgae and planktonic microalgae, control of algal density by grazing, and the effects of eutrophication on benthic macrofauna.

This chapter summarises the results obtained from field studies in the Venice Lagoon. The central part of the Venice Lagoon has undergone major environmental changes during the twentieth century (Pavoni *et al.*, 1992; Sfriso *et al.*, 1992a; Marcomini *et al.*, 1993) mainly because lack of a suitable sewage treatment plant increased the anthropogenic impacts from the Porto Marghera industrial area, the town of Mestre and the surrounding villages, and this has been exacerbated by intensive agriculture practices. The central Lagoon has also suffered marked hydrodynamic changes following digging of wide (80–150 m) and deep (12–20 m) canals (Vittorio Emanuele, 1919–1930 and Malamocco Marghera, 1961–1969) which have led to a rapid expansion of tides into the innermost parts of the basin, thereby preventing efficient water renewal in large areas near the watersheds. As a result, from the 1970s onwards, the flora and phytobenthic associations of the central Lagoon were severely affected (Sfriso, 1987; Sfriso and Pavoni, 1994; Sfriso and Marcomini, 1994) and monospecific populations of *Ulva rigida*, C.Ag. have progressively replaced the autochthonal rhizophyte (*viz. Zostera marina*) and smaller macroalgal populations (*viz.* the red algae *Gracilaria verrucosa, Porphyra leucosticta, Ceramium rubrum* and *Polysiphonia sanguinea*, and the brown alga *Dictyota dichotoma*) and these have outcompeted the phytoplankton. Within a few years *Ulva* became the dominant species, attaining biomasses of over 20 kg wet weight m^{-2} over vast areas of the central Lagoon, mostly where water renewal and turbulence were negligible (Sfriso *et al.*, 1993).

The uncontrolled spread and growth of *Ulva*, which has filled the entire water column up to the water surface for tens of square kilometres (Sfriso *et al.*, 1989), caused frequent and sudden biomass decays, leading to anoxia and mortality of all benthic populations (Sfriso *et al.*, 1987, 1988). A marked proliferation of mosquitoes of the genus *Chironomus* has also occurred (Ceretti *et al.*, 1985).

MATERIAL AND METHODS

Stations

From a preliminary survey in 1990–1991, four stations (Alberoni, Sacca Sessola, San Giuliano and Fusina – Figure 15.1), which differed in their eutrophication characteristics and macroalgal abundance, were retained for detailed investigations. During 1992 regular field studies were undertaken at these stations.

Station A (Alberoni) was located close to the Malamocco–Marghera canal entrance and, therefore, experienced very effective tidal water exchange. It was also far from sources of anthropogenic inputs. This station featured a high variety of macroalgal species (*Enteromorpha, Gracilaria, Dictyota, Codium, Ulva*).

Station B was near Sacca Sessola on the watershed of the central Lagoon and, therefore, experienced a longer residence time of water. The flora was characterised by a single species of macroalgae, *Ulva rigida*. Phytoplankton blooms occurred when the macroalgae were decomposing. Anoxic conditions have been observed at this site in the past. In 1992, an additional station (B1), which had a maximum *Ulva* abundance, was selected near station B, where *Ulva* had disappeared.

Station C (San Giuliano) was close to the inner border of the Lagoon and, therefore, experienced very poor water exchange. It was located close to both urban and agricultural water inputs. The flora of this station consisted mainly of two species of macroalgae, *Ulva rigida* and *Gracilaria verrucosa*.

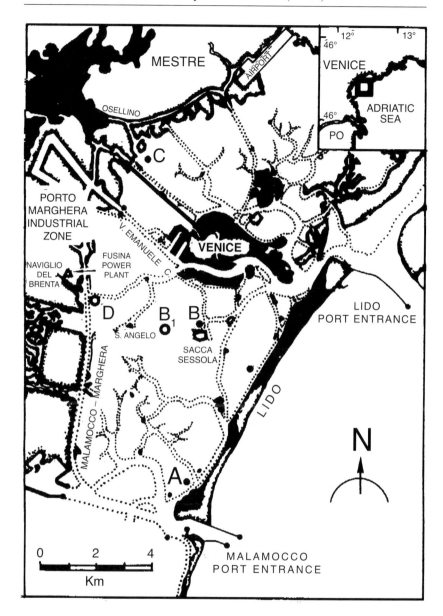

Figure 15.1 The central Venice Lagoon and the sampling stations

Phytoplankton blooms occurred frequently, again when the macroalgae were decomposing. Anoxic situations had been observed earlier.

Station D (Fusina) was located close to the inner border of the Lagoon, at the mouth of the River Brenta and the inner extremity of the Malamocco–Marghera canal. This canal connects the Marghera industrial area with the

Adriatic Sea, resulting in intense water circulation. This station was close to direct urban and agricultural water inputs. Macroalgae were absent but phytoplankton blooms occurred frequently. Anoxic conditions had been observed earlier.

FIELD SURVEYS

Preliminary monitoring (at 5–7-day intervals in spring–summer and 15-19-day intervals in winter) was carried out at stations A, B and C from March 1989 to April 1990 in order to identify the main phytobenthic associations and estimate macroalgal production. This was extended to June 1991 to study trends in macroalgal growth and major physicochemical parameters. Subsequently, between April and December 1992, the study was shifted to two other areas of the Lagoon, station D and an additional station (B1) at Sacca Sessola, in order to differentiate between the presence and absence of macro-algae and/or phytoplankton and to obtain better estimates of the production and decomposition of algal biomass.

In the period June 25–27 1991, a more extensive survey was carried out at stations A, B and C and data on sediment, algal characteristics, species composition, and standing stocks of zooplankton and zoobenthos were obtained. In 1992, an intensive study was carried out at stations A, B1, C and D with the aim of monitoring the processes at the sediment–water interface (nutrient and oxygen dynamics, composition of the macrobenthos) and in the water column (dynamics of nutrients, plankton and macroalgae). This was carried out in four specific periods: in April, at the time of spring phyto-plankton bloom and the beginning of growth of macroalgae; in May, when macroalgae blooms were intense; in July, when macroalgal blooms collapsed and situations of anoxia began to occur; and in September, when mineralisa-tion of organic material at the sediment surface took place.

RESULTS

Physical conditions

The physicochemical and biological parameters measured in the regular monitoring programmes during April 1990–June 1991 (stations A, B, C) and April–November 1992 (stations B1, D) are summarized in Table 15.1. Water temperature varied between 5°C (January–February) and 25–30°C (July–August) and was higher by ~1°C at station D than at other stations because of the discharge of cooling waters from the nearby electric power plant. Salinity was almost constant, with only a small difference between summer (high) and winter (low) values. At stations A and B, salinity varied between 22 PSU (winter) and 31 PSU (summer) with a median of 28 PSU. Stations C and D had a somewhat lower salinity with median values of 24 and 20 PSU,

Table 15.1 Synoptic listing of physicochemical and biological parameters measured at the various sampling stations in April 1990–June 1991 and April–November 1992. Pore water parameters were taken at 5 cm in 1990–91 and at 2 cm in 1992

Station		Biomass (wet weight) g/m²/y	Temp. water °C	pH water	pH sed.	Eh (mV) water	Eh (mV) sed.	Chlor. ×1000	$O_{2,sat}$ %	Chl a mg/m²	Phaeo a mg/m²	RP	Surface NH₄	NO₂	NO₃	DIN	Pore RP	Pore NH₄
1990–91																		
Alberoni (A)	Mean	580	18.6	8.42	7.78	379	55	17.2	150	1.3	0.9	0.59	6.66	0.58	6.80	14.04	4.0	79
	max	2855	28.2	9.26	8.20	445	228	18.9	281	11.4	9.9	1.50	42.81	2.47	48.51	92.25	9.5	261
	min	17	3.6	7.98	7.43	312	−114	15.8	96	0.1	0.01	0.10	0.78	0.07	0.12	1.88	1.0	18
Sacca Sessola (B)	Mean	1452	18.3	8.39	7.71	360	−2	16.7	139	1.7	1.4	0.84	9.46	1.06	8.27	18.79	4.7	169
	max	10652	27.0	9.56	8.15	432	142	18.6	254	7.4	5.6	3.00	31.67	3.38	44.74	61.35	17.7	618
	min	0	2.0	7.60	7.23	250	−203	13.0	60	0.04	0.07	0.05	2.94	0.26	0.69	5.09	1.4	34
San Giuliano (C)	Mean	1589	18.5	8.36	7.70	360	−85	14.7	130	8.0	9.4	2.72	11.93	1.76	11.40	25.09	11.3	208
	max	8228	27.6	9.35	8.21	419	62	18.3	243	86.0	109.8	16.25	44.73	5.34	47.58	94.20	31.8	581
	min	0	4.0	7.28	7.05	−110	−229	9.1	0	0.19	0.28	0.28	1.70	0.21	0.37	5.10	2.3	16
1992																		
Sacca Sessola (B1)	Mean	4207	20.6	8.34	7.44	268	−132	17.0	117	3.2	0.7	0.86	20.83	1.29	9.31	31.43	53.5	1114
	max	19785	28.6	9.42	7.90	400	49	18.0	237	12.1	15.8	6.25	107.75	3.62	48.53	143.91	209	2508
	min	0	9.5	7.42	7.04	−18	−243	14.7	0	0.28	0.12	0.05	3.20	0.10	0.13	4.42	2.5	222
Fusina (D)	Mean	0	22.4	8.28	7.60	293	12	16.0	139	9.1	16.5	1.61	20.48	2.19	25.89	48.56	21.4	552
	max	0	30.2	8.95	8.12	397	84	17.8	274	57.8	47.5	7.05	98.80	4.93	61.15	144.81	153.5	1633
	min	0	9.8	7.48	7.06	31	−119	14.5	86	0.43	2.00	0.10	1.89	0.31	0.65	3.39	7.1	73

respectively. The pH was generally 8.3–8.4, but increased to higher values (up to 9.6) during macroalgal blooms and decreased to lower levels (down to 7.3) during their decay. More details on these parameters are given by Sfriso *et al.*, Chapter 18.

Macroalgae

Two major macroalgal associations were identified in the central Venice Lagoon: one characterised by the almost monospecific dominance (~ 95% of the biomass and ~ 99% of the total macroalgal production) of the nitrophilic species *Ulva rigida*, with excessive blooms in the April–June period (stations B, B1 and C – Figure 15.2); the other by growth of a few species at different times of the year with a nearly equivalent annual production (station A – Figure 15.2). Macroalgae were absent from station D.

Since 1989, a regression of *Ulva rigida* patches and a general increase in those species considered to be more sensitive to eutrophication has been observed. In 1990–1991, in the infralittoral zone (i.e. 0.8 m below mean tide level), many specimens of a new Rhodophycea, *viz. Grateloupia doryphora* (Mont.) Howe and a remarkable recolonization by *Zostera marina* L., which had disappeared in the 1970s, were noticed.

Biomass production differed substantially between stations, seasons and consecutive years. It was highest at stations B and B1 (15–20 kg wet weight

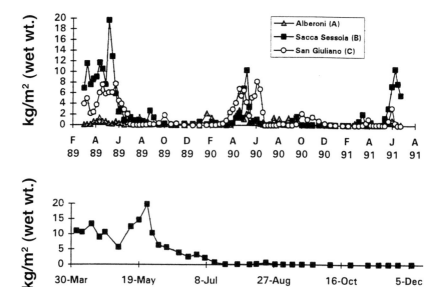

Figure 15.2 Trends in macroalgal biomass at various stations in 1989–1991 (above) and 1992 (below)

m^{-2} y^{-1}), whereas at station C it decreased from ~16 to ~7 kg wet weight m^{-2} y^{-1} between 1990 and 1991. Production at station A was lower and decreased, as at Station C, from ~9 to ~6 kg wet weight m^{-2} y^{-1} in the same period. Peak production periods, in May–June, usually lasted ~10 days with production rates as high as ~1500 g wet weight m^{-2} d^{-1}, equivalent to ~33 g C m^{-2} d^{-1}. The highest decomposition rates (~2400 g wet weight m^{-2} d^{-1}, ~51 g C m^{-2} d^{-1}) were usually observed soon after the period of highest production.

At stations B, B1 and D, *Ulva* fronds of ~150 g wet weight were regularly placed in net cages at various depths. The growth rates measured in these experimental fronds were remarkably higher than those of naturally occurring macroalgal fronds under optimal growth conditions, even during periods of decay and anoxia. In general, growth was higher at the surface, than in the bottom, cages. In April–May, macroalgal growth in surface cages was higher at station D than at Sacca Sessola, but in August–December this situation was reversed. During these experiments many crustaceans, such as shrimps and amphipods, fed on the exposed macroalgae. The effect of their grazing pressure, estimated from the difference in growth rates of *Ulva* in cages with a smaller (1 mm) and a larger (1 cm) mesh, was ~ 4–15% day^{-1}.

Growth in the bottom cages at Fusina was always lower than at Sacca Sessola. This was due to turbulence caused by tidal currents in the Malamocco–Marghera channel and wave motion due to shipping. This resulted in a resuspension of surface sediments and a reduction in transparency, as well as deposition of a thin film of particulate matter over algal fronds, spores and gametes. These effects might explain the absence of *in situ* macroalgal growth at Fusina. The average photosynthetically active radiation (PAR) at the water surface was approximately 2000 μE m^{-2} s^{-1}, which was reduced near the sediments to ~700 μE m^{-2} s^{-1} at Sacca Sessola and to <100 μE m^{-2} s^{-1} at Fusina. The latter PAR is suboptimal for growth of *Ulva* and *Gracilaria* (Keith and Marray, 1980, Hanisak, 1987). Sedimentation of particulate matter at station D was ~850 kg m^{-2} y^{-1}; about 3 times higher than at Sacca Sessola (B1).

Plankton

Phytoplankton was of secondary importance in the abundance and production of phytobiomass. Its production exceeded that of macroalgae only in certain areas, such as station D (Figure. 15.3). Elsewhere, phytoplankton production was significant only in summer after macroalgal decay. The largest phytoplankton blooms (up to ~150 μg chl *a* l^{-1}) were observed outside the sampling sites in the present study (i.e. between Sacca Sessola and station D, near the island of S. Angelo della Polvere), at the edges of the areas covered by decomposing macroalgal biomass. The influence of tidal flow coming through the Malamocco–Marghera channel was negligible at this spot.

Zooplankton population consisted of both holoplankton and meroplankton. The former was dominated by calanoid (*viz. Acartia*) and harpacticoid copepods and the latter included larvae of polychaetes, bivalves, echinoderms and barnacles. Highest plankton densities were observed at station D, followed by stations B and C and the lowest, at station A. The density of *Acartia* correlated negatively with chlorophyll at station D in April–May, indicating control of algal density by grazing.

Sediment

The organic matter content of the sediments was relatively high, ranging from 2% of dry weight at station A, 3–4% at station D, 4–5% at station B and 9% at station C. The latter two stations were dominated by macroalgae. Highest organic content was measured in July, after the peak in macroalgal biomass changes. The percentage organic matter content in the upper 2 cm of the sediment was generally about the same as in the 2–10 cm depth layer, except at station C where it was higher in the upper layer. The sediment surface under the macroalgal cover at Sacca Sessola was richer in organic matter than in areas without macroalgae.

Pigment concentrations in the sediments were also highest at station C, followed by stations B, D and A. Surface sediments had higher pigment concentrations than deeper sediments. Phaeopigment concentrations, compared to chlorophyll, were very high and this may have been due to sedimentation of macroalgal debris.

Nutrients dynamics

Highest dissolved inorganic nitrogen (DIN) concentrations (50–150 μmol N l^{-1}) in surface waters were measured in winter, during stormy weather and/or after periods of anoxia. The annual average DIN concentration was lower at station A (~14 μmol N l^{-1}) than at station B (~20 μmol N l^{-1}) and station C (~25 μmol N l^{-1}). The average concentrations measured in 1992 were ~30 μmol N l^{-1} at station B and ~50 μmol N l^{-1} at station D. As nitrate was taken up preferentially by the macroalgae, its concentration was high only in autumn–winter when the latter were absent. At stations A and D, which were characterised by a low abundance or a total absence of macroalgal biomass

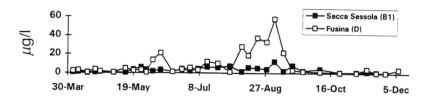

Figure 15.3 Trends of chlorophyll *a* concentrations at Sacca Sessola and Fusina in 1992

and were influenced by the sea- and the fresh-waters of the Naviglio del Brenta, nitrate concentrations were higher than those of ammonium. In contrast, at station B, which was influenced by marked macroalgal and phytoplankton blooms and subsequent decay, concentrations of ammonium were equal to, or even higher than, those of nitrate. Nitrite concentrations were generally about one tenth of those of nitrate and ammonium.

The annual average phosphate concentration was also lower at stations A (~ 0.6 μmol P l^{-1}) and B (~ 0.8 μmol P l^{-1}) than at stations D (~1.6 μmol P l^{-1}) and C (~2.7 μmol P l^{-1}). Phosphate concentrations, in general, did not show large temporal variations. Silicon was depleted during the phytoplankton bloom (July–August) but not during the period of macroalgal growth (May–June).

Ammonium and phosphate concentrations in the sediment pore water (Figure 15.4) increased markedly with depth, especially at stations B and C. At station A concentrations in the sediments were only slightly higher than in the overlying waters and did not show a depth profile. An intermediate nutrient profile was observed at station D. Nitrate and nitrite concentrations (not shown) did not show any clear depth profile.

A substantial flux of ammonium and phosphate from the sediments (Figure 15.5) was observed during summer (April–July) of 1991 at station B, and to a lesser extent at stations D and C (not shown on Figure 15.5). At station A the flux was much smaller and increased gradually during summer, with the highest rates in September. The ammonium flux in the central Venice Lagoon was, on average, ~5 mmol N m^{-2} day^{-1} and the phosphate flux was ~ 0.1 mmol P m^{-2} day^{-1}. The fluxes of nitrate and nitrite were negligible.

Oxygen dynamics

Oxygen concentrations were highly variable during the period of algal blooms (up to 280%) and decay (down to 0%). Hypoxic and anoxic conditions occurred occasionally at the Sacca Sessola stations and at station C between May and July. From September to December, dissolved oxygen concentrations were always at ~100% saturation.

The Eh of top sediment layers (2–5 cm) ranged between –50 and +150 mV at stations A and D. At stations B, B1 and C, the redox potential of top sediment layers fluctuated between –150 and –250 mV when macroalgal blooms were at their peak. With the disappearance of macroalgae, Eh values increased progressively at these stations and, in October–December, ranged between –100 and +100 mV.

Benthic community respiration was generally 200–400 μmol O$_2$ m^{-2} d^{-1}, and increased by 2–4 times in summer (July–August) at the macroalgae-dominated stations B and C, compared to stations A and D.

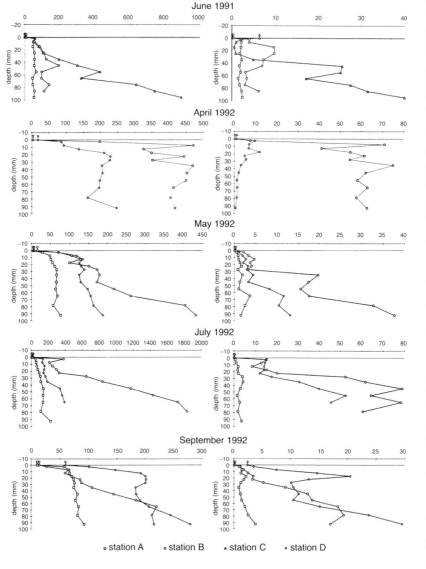

Figure 15.4 Profiles of ammonium and phosphate in pore waters at various stations during 1991 (June) and 1992 (April, May, July, September)

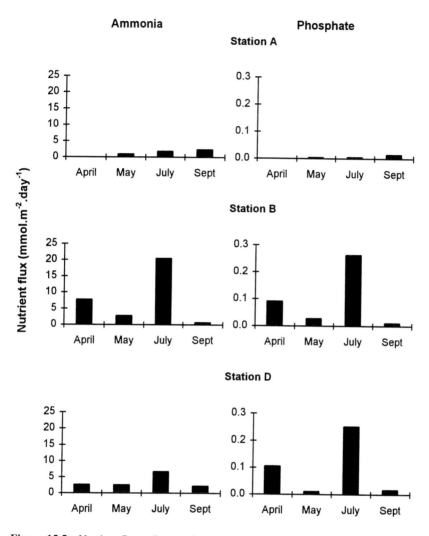

Figure 15.5 Nutrient fluxes from sediments into the overlying water at various stations in 1992

Benthic fauna

Biomass (Figure 15.6) and species composition (Figure 15.7) of the benthic macrofauna varied considerably in space and time. In general, biomass was highest at station A and lowest at station B. Some resident species (e.g. *Cerastoderma* sp.) had a negative relation with eutrophic parameters, whereas some motile species (e.g. *Upogebia deltaura* and *Carcinus mediterraneus*) had a positive relation with them. More detailed information on macrofauna is given by Phillipart and Dankers in Chapter 17.

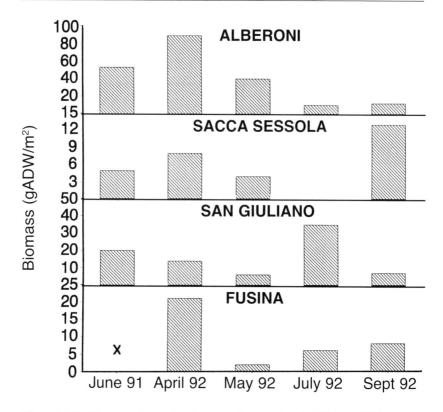

Figure 15.6   Biomass of macrobenthos at various stations in 1991 and 1992

DISCUSSION

The Venice Lagoon is heavily loaded with pollutants from domestic and industrial waste waters and agricultural runoff. The calculated N loading from these sources is approximately 6 μmol N l^{-1} d^{-1}, with another 5 μmol N l^{-1} d^{-1} coming from sediments. Phosphorus enrichment from these two sources are about 0.4 and 0.1 μmol P l^{-1} d^{-1} respectively. The overall nutrient loading in the Venice Lagoon is thus approximately 5 times higher than in the Wadden Sea, in particular with reference to nitrogen. However, the highest (winter) nutrient concentrations in the Venice Lagoon were only 50–150 μmol N l^{-1} and 1–6 μmol P l^{-1}, which are similar to the highest concentrations in the Wadden Sea. This might be due to a larger allocation of N and P to the sediments; inorganic N and P concentrations in pore waters were approximately 5 times higher, and the organic matter content of sediments was 5–10 times higher in the Venice Lagoon than in the Wadden Sea (Scholten *et al.*, 1995).

The organic matter content of the sediments was quite high at locations with large accumulation and decay of macroalgal debris. Mineralisation of this

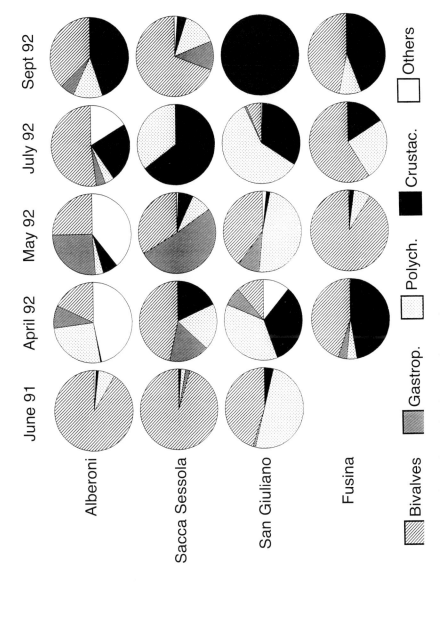

Figure 15.7 Composition of macrobenthos at various stations

material increased oxygen consumption and nutrient efflux rates, with a consequent reduction of the redox potential to negative values, even in surface layers, and of the oxygen concentrations in the overlying waters to hypoxic or even anoxic levels. The N flux, as ammonium, from the sediments was almost equal to the N input from anthropogenic and riverine sources. The phosphate flux was of minor importance and it appears to be a limiting factor during the growing season, but it is nitrogen that controls any macroalgal bloom sizes.

Primary production was stimulated by nutrient enrichment. However, the prolific macroalgal biomass increase observed in the Venice Lagoon occurred in a Wadden Sea mesocosm experiment only at a 4–8 times higher nutrient loading (Scholten *et al.*, 1995). This can be partly explained by the higher temperature and early onset of the growing season for the fast-growing *Ulva rigida* in the Venice Lagoon, compared with the lower temperature and late onset of the growing season for *Ulva lactuca* and *Enteromorpha compressa* in the Wadden Sea mesocosm study. In the Venice Lagoon, macroalgae success-fully outcompeted phytoplankton; the blooms of the latter were restricted to periods following the decay of macroalgae and to those areas where macro-algae cannot grow due to enhanced resuspension and sedimentation of fine materials (e.g. Fusina) (Sfriso *et al.*, 1992b). In the Wadden Sea, however, macroalgal growth was preceded by a spring phytoplankton bloom and nutrient supply was at its highest during the first phase of the growing season.

Macroalgae are generally thought to be less subject to grazing but, in this study, *Gammarus aequicauda* grazed 4–15% of the *Ulva* biomass day^{-1}, which is substantial compared to a daily growth rate of $\approx 20\%$. This indicates that grazing can be a significant factor in controlling macroalgal densities and that the absence of gammarids may thus play a role in the build-up of high macroalgal densities over vast areas of the Venice Lagoon.

In heavily polluted coastal areas, high concentrations of nutrients generally coincide with high concentrations of toxic compounds. It has been shown that such toxic compounds were responsible for inhibition of the development of zooplankton which subsequently resulted in increased phytoplankton biomass sustained by increased nutrient enrichment (Jak, 1997). The absence of gammarids in the most eutrophic stations of the Venice Lagoon may thus be due not only to the avoidance of periodically anoxic conditions but also to an inhibition by high concentration of metals, pesticides and other organochlo-rines (Marcomini *et al.*, 1993; Maroli *et al.*, 1993). Further investigations are needed to find out which toxic substance caused a reduction in grazing and thus contributed to eutrophication.

The mean benthic macrofaunal biomass of 20 g ADW m^{-2} found in the central Venice Lagoon was high compared to other temperate estuaries (Knox, 1986) but similar to that in the subtidal areas of the Wadden Sea (Dekker, 1989). In general, the density and biomass of resident species (e.g. bivalves like *Cerastoderma*) were substantially reduced at those stations in the Venice Lagoon affected by high levels of eutrophication. Motile species (e.g.

crustaceans like *Upogebia* and *Carcinus*) and polychaete worms (Nereidae), on the other hand, appear to benefit from such conditions. The benthic macrofaunal biomass at eutrophied stations showed seasonal dynamics, with highest levels in spring followed by a decrease during the collapse of the macroalgal bloom and a recovery a few months later.

The decrease in eutrophication observed in the Venice Lagoon over the last few years is likely to be due to a reduction in nutrient loadings and concentrations of toxic substances. This should be continued by installing more wastewater treatment plants. In addition, removal of excess algal material and protection of the infra- or sublittoral eelgrass and seaweed beds can enhance restoration of affected areas.

ACKNOWLEDGEMENTS

Thanks are due to all our colleagues for assistance in sampling and analyses. Henk van het Groenewoud organized the BEST project, Pim de Kock (TNO) and Peter de Wilde (NIOZ) provided scientific advice and Marie-Cécile Scholten (TNO) took care of text preparation. This paper is dedicated to our late professor, Angelo A. Orio, who was the initiator of the BEST project.

The BEST project was partly funded by the European Commission (DG XII) within the framework of the STEP environmental research programme (contract STEP-CT-91-0120).

REFERENCES

Ceretti, G., Ferrarese, U. and Scattolin, M. (1985). I Chironomidi nella laguna di Venezia. *Arsenale, Venezia*, pp. 59

Dekker, R. (1989). The macrozoobenthos of the subtidal western Dutch Wadden Sea. I. Biomass and species richness. *Neth. J. Sea Res.*, **23**: 57–68

Hanisak, M. D. (1987). Cultivation of *Gracilaria* and other macroalgae in Florida for energy production. In Bird, K. T. and Benson, P. H. (eds). *Seaweed Cultivation for Renewable Resources*, Elsevier, Amsterdam, 191–218

Jak, R. G., Scholten, M. C. Th. and Van Der Meer, M. (1993). SEDEX: Intertidal mesocosm studies on the ecological impact of the marine disposal of dredged materials, Part 4: Zooplankton. *TNO report IMW-R93/255c*

Jak, R. G. (1997). Toxicant-induced changes in zooplankton communities and consequences for phytoplankton development, Thesis, The Free University, Amsterdam

Keith, E. A. and Marray, N. M. (1980). Relationship between irradiance and photosynthesis for marine benthic algae (Chlorophyta) of differing morphologies. *J. Exp. Mar. Biol. Ecol.*, **43**: 183–92

Knox, G. A. (1986). *Estuarine Ecosystems: A Systems Approach*. CRC Press, Boca Raton

Marcomini, A., Sfriso, A., Pavoni, B. and Orio, A. A. (1993). Eutrophication of the lagoon of Venice: nutrient loads and exchanges. In McComb, A. J. (ed.) *Eutrophication in Shallow Estuaries and Lagoon*. CRC Press, Boca Raton, Fl., USA

Maroli, L., Pavoni, B., Sfriso, A. and Raccanelli, S. (1993). Concentrations of poly-chlorinated biphenyls and pesticides in different species of macroalgae from the lagoon of Venice. *Mar. Pollut. Bull.* **26**: 553–8

Pavoni, B., Marcomini, A., Sfriso, A., Donazzolo, R. and Orio, A. A. (1992). Changes in an estuarine ecosystem. The lagoon of Venice as a case study. In Dunnette, D. A. and O'Brien, R. J. (eds), *The Science of Global Change*, American Chemical Society, Washington, DC, USA, 287–305

Scholten, M. C. Th., Jak, R. G., Pavoni, B., Sfriso, A., Phillipart, C. J. M., Dankers, N., De Heij, H. and Helder, W. (1995). BEST, Benthic Eutrophication Studies (Synthesis report). *TNO Report MW R94/049*

Sfriso, A. (1987). Flora and vertical distribution of macroalgae in the lagoon of Venice: A comparison with previous studies. *Giornale Botanico Italiano*, **121**: 69–85

Sfriso, A. and Marcomini, A. (1994). Gross primary production and nutrient behaviours in shallow lagoon waters. *Biores. Technol.*, **47**: 59–66

Sfriso, A. and Pavoni, B. (1994). Macroalgae and phytoplankton competition in the central Venice lagoon. *Environ. Technol.*, **15**: 1–14

Sfriso, A., Marcomini, A. and Pavoni, B. (1987). Relationship between macroalgal biomass and nutrient concentrations in a hypertrophic area of the Venice lagoon. *Mar. Environ. Res.*, **22**: 297–312

Sfriso, A., Pavoni, B., Marcomini, A. and Orio, A. A. (1988). Annual variation of nutrients in the lagoon of Venice. *Mar. Pollut. Bull.*, **19**: 54–60

Sfriso, A., Pavoni, B. and Marcomini, A. (1989). Macroalgae and phytoplankton standing crops in the central Venice lagoon: primary production and nutrient balance. *Sci. Total Environ.*, **80**: 139–59

Sfriso, A., Pavoni, B., Marcomini, A. and Orio, A. A. (1992a). Macroalgae, nutrient cycles and pollutants in the lagoon of Venice. *Estuaries*, **15**: 517–28

Sfriso, A., Pavoni, B., Marcomini, A., Raccanelli, S. and Orio, A. A. (1992b). Particulate matter deposition and nutrient fluxes into the sediments of the Venice lagoon. *Environ. Technol.*, **13**: 473–83

Sfriso, A., Marcomini, A., Pavoni, B. and Orio, A. A. (1993). Species composition, biomass, and net primary production in shallow coastal waters: the Venice lagoon. *Biores. Technol.*, **44**: 235–50

CHAPTER 16

ENVIRONMENTAL MONITORING IN THE PALUDE DELLA ROSA, LAGOON OF VENICE

G.M. Carrer, G. Todesco and M. Bocci

SUMMARY

The growth cycle of *Ulva rigida* (Chlorophyceae) was monitored at three stations in the Palude della Rosa (Lagoon of Venice) from April 1992 to July 1993. Water samples and algae were collected for measurement of nutrients (dissolved and total nitrogen and phosphorus) in the water, the biomass of *Ulva rigida* and the concentrations of N and P in algal thalli. Physicochemical parameters were monitored using an electrochemical probe during different periods of the growth cycle. The pattern of biomass changes differed between the three stations. Notwithstanding a reduction of nutrients in the water column in spring–summer 1992, internal nutrient concentrations of the algae were high enough to maintain fast growth during almost the whole of the study period. Dissolved oxygen concentrations in the water column recorded over a day–night cycle at different periods of the growth season was a good indicator of the behaviour of the system.

INTRODUCTION

Increased pollution and eutrophication since the 1960s have affected several confined areas of the Venice Lagoon. As a result of these long-term changes in the physicochemical properties of the water and sediments, the species composition and abundance of biological communities, mostly in the central lagoon, have altered in recent years (Sfriso *et al.*, 1990). Proliferation of species of green algae, known for their high adaptability, is one of several such impacts observed (Sfriso, 1987), and a few of them – *Ulva rigida*, *Enteromorpha* spp. and *Cladophora* spp. – have shown massive growth (Sfriso *et al.*, 1993). Of these, *Ulva rigida* is the most abundant and its proliferation was one of the major ecological problems in the Venice Lagoon during the 1980s and early 1990s (Sfriso *et al.*, 1989). It generally occurs in areas with low water exchange, high light penetration, high influx of nutrients from watersheds and sediments with high organic matter and nutrient loads, and can attain a biomass of 20 kg wet wt m^{-2} (Sfriso *et al.*, 1993)

The growth cycle of this macrophyte can be described as follows. In spring, initial growth is supported by high nutrients and increasing temperature and light intensity. A luxury consumption at this time leads to a build-up of reserve nutrients in the algal cells. Several minor blooms and partial declines follow the first bloom. During the blooms, the macroalgal biomass becomes stratified and the lower layers that do not receive enough light die off, decompose and cause a continuous deposition of organic matter on the Lagoon bottom. Along with a reduction in photosynthetic oxygen production, this process increases the oxygen deficit, which, in turn, is accentuated by high temperatures in late spring to the extent of causing anoxia of the whole water column. The ensuing total collapse of the bloom is accompanied by release of H_2S from the sediments and a mass mortality of all aerobic organisms.

The aim of the present study was to collect data that could help in description and interpretation of the life cycle of *Ulva rigida*. Such a data base can be useful in validation of the mathematical model of *Ulva* growth presented by Coffaro and Sfriso (Chapter 28).

MATERIAL AND METHODS

The study was carried out in the Palude della Rosa from April 1992 to July 1993. Carrer *et al.* (1993) provide a detailed report of monitoring at this site.

The Palude della Rosa, in the northeastern part of the Venice Lagoon, is a confined shallow water area (1.2 m mean depth). *Ulva rigida* is the most important autotroph in the Palude in terms of biomass and production. The seawater flow from the Torcello canal and the freshwater flow from the Silone canal into the Palude creates a gradient in physicochemical properties. The freshwater flow also brings in a large amount of nutrients – N and P compounds – resulting from intense agricultural activities within the northern watershed of the Venice Lagoon. A mass balance of nutrients in the Palude della Rosa has been constructed by Fossato *et al.* (1994).

Three sampling stations (Figure 16.1), all of the same depth (1 m), in a representative area of the Palude were chosen on the basis of the salinity gradient, geomorphology and previous knowledge on *Ulva* growth. Samples at these stations were obtained once a week during the blooms and at larger intervals after their decline and during autumn and winter. At each station, macroalgae and water samples were collected and the physicochemical parameters of water were recorded.

Six random samples at each station were taken for measurements of *Ulva* biomass. Each sample was isolated by a cylindrical plastic tube (39 cm ID) laid from the water surface onto the Lagoon bottom. The macroalgal thalli trapped inside were cut around the walls, extracted and carefully washed on a 1 mm sieve to remove sediments and animals. After allowing excess water to drip off, the algae were centrifuged to remove remaining water, and their wet weight was recorded.

A plexiglas tube (150 cm ht, 4 cm ID) that can be closed at one end was used to collect water samples from the whole column. Part of each sample was kept aside in polyethylene bottles for total N and P measurements and the

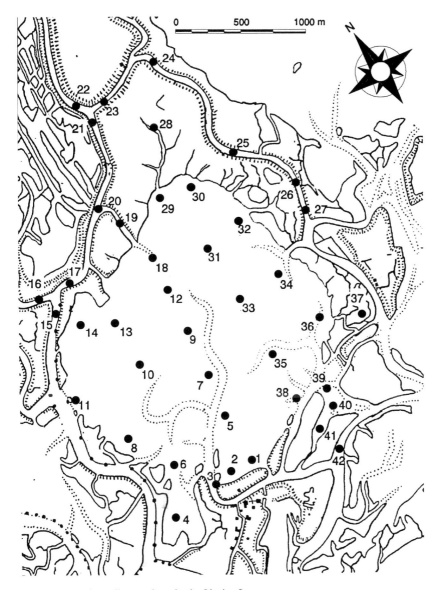

Figure 16.1 Sampling stations in the Venice Lagoon

remainder filtered immediately through cellulose acetate filters. Bacterial and phytoplankton activity in both aliquots were arrested by addition of dilute sulfuric acid (0.5 ml to 250 ml of sample). The filters were kept frozen in the dark until analysis of chlorophyll *a*.

An electrochemical probe was used for continuous recording of water temperature, conductivity, dissolved oxygen, pH and redox potential during the sampling period. At different periods of the *Ulva* growth cycle, the probe was left *in situ* for 42 h for continuous monitoring of these parameters.

Nutrient (ammonium, nitrite, nitrate, phosphate, total N and total P) concentrations were measured spectrophotometrically following standard methods (Strickland and Parsons, 1972; Grasshoff *et al.*, 1983). Total N and P concentrations in the algal tissue were determined by alkaline persulfate digestion. A small quantity of finely-ground algal matter was dried for 24 h at 100 °C, weighed (5 mg) and placed in 250 ml glass bottles with screw caps. Fifty ml of oxidising reagent (25 g $K_2S_2O_8$ and 15 g H_3BO_3 dissolved in 500 ml of 0.375 M NaOH) were added to each bottle. The samples were autoclaved at 120 °C for 45 min, cooled to room temperature, another 50 ml of oxidising reagent were added and the samples were autoclaved again. After cooling, they were filtered through 0.45 μm glass fibre filters and the nitrate and phosphate in the filtrate were measured colorimetrically (Strickland and Parsons, 1972). All samples were analysed in triplicate.

RESULTS

Macroalgal biomass

During the two growth seasons (spring–summer of 1992 and 1993) *Ulva rigida* showed different cycles at the three sampling stations (Figure 16.2). When the study began in 1992, *Ulva* had already started growing. Maximum biomass in this year was recorded at station 1 (6.6 kg wet weight m^{-2} in late April). A lower peak in biomass (4.8 kg wet weight m^{-2}) was recorded later (middle of May) at station 3. Station 2 had the highest biomass (3 kg wet weight m^{-2}) at the beginning of the study period (late April) and in the middle of June (2.4 kg

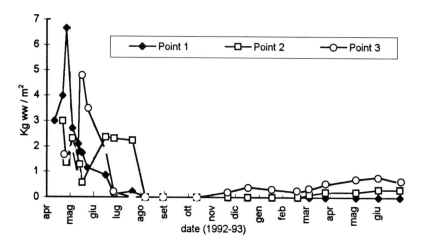

Figure 16.2 Seasonal changes in *Ulva* biomass

wet weight m^{-2}). These differences can be interpreted in terms of substantial water movement due to internal circulation in the Palude, since the macroalgae are easily transported. Strong winds were observed to have the same effect.

After every biomass peak there was a pronounced decomposition of *Ulva* at all stations. At this time oxygen concentrations in the water and in the sediments decreased sharply and many benthic organisms (*Cerastoderma glaucum, Carcinus mediterraneus, Gammarus* spp. and *Haminaea navicula*) were killed.

After the collapse of the bloom in summer, *Ulva* practically disappeared. At station 3, a small, second peak was observed in December, probably due to an accumulation of pieces of thallus from other areas of the Palude.

In the following year (1993) *Ulva* appeared in low quantities at stations 2 and 3 and not at all at station 1. At stations 2 and 3 the growth was very uniform and low compared with the previous year. At station 3, for example, the maximum biomass was attained at the beginning of June, but it was only 0.8 kg wet weight m^{-2} (20% of that of previous year). Benthic diatoms and small quantities of other macroalgal species were present in 1993.

Between the two years solar radiation (recorded by Ente Zona Industriale Porto Marghera) and water temperature did not show significant differences. Nutrient concentrations in the water column and in the algal thallus (see below) do not suggest nutrient limitation of growth. On the other hand, the fraction of macroalgae that survived the winter between the two years could have played an important role in causing this difference. The size of this fraction differed between 1992 and 1993. In fact, during an on-the-spot investigation before the commencement of the study, it was observed that the biomass of *Ulva* in the winter of 1992 was about 1 kg wet weight m^{-2} whereas in the winter of 1992–93 the algae were completely absent. Moreover, the persistence of anoxic conditions during the winter of 1992–93 also probably inhibited settlement and growth of young thalli.

Data on chlorophyll, nutrients in the water column and the algae, and the dissolved oxygen obtained from station 3 only are presented below, since they show typical variations in these parameters better than those of the other two stations.

Chlorophyll *a*

Changes of chl *a* concentrations showed an inverse trend compared with those of macroalgal biomass (Figure 16.3). Several hypotheses can be advanced to explain this. The first is that phytoplankton blooms appear after the decline of *Ulva* blooms, when nutrients once again become available. The second is that there could be a release of spores by the macroalgae immediately before the collapse. The third possibility is a combination of the previous two. The two peaks (in 1992 and in 1993) might owe their origin to two different processes: the first to sporulation and the second to phytoplankton bloom in the absence of macroalgal growth.

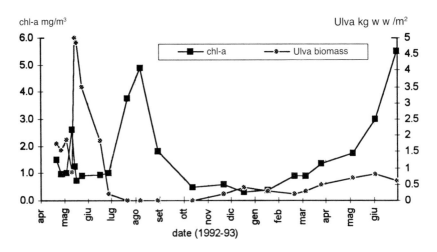

Figure 16.3 Seasonal changes in *Ulva* biomass and chl *a* in the water column

A microscopic analysis of the sample composition would have helped in identifying the origin of these chl *a* peaks. Such an analysis should be included in future studies.

Dissolved nutrients

There is a clear difference in dissolved N concentrations between the two years (Figure 16.4). The presence of large quantities of algae in the first year of observations caused a substantial reduction of total N and oxidised forms of N. Concentrations of ammonium were, however, more variable. Total and reactive phosphorus concentrations also showed a depletion during April and May 1992 (Figure 16.5).

Nutrients in the macroalgae

The cell quota of N and P in the algal thalli are more useful in interpreting the growth cycle of *Ulva* than the concentrations of nutrients in water. It has been shown experimentally (Fujita and Rodney, 1985; Lundberg *et al*., 1989; Lavery and McComb, 1991) that past history and nutritional state of *Ulva rigida* can affect its nutrient uptake velocity. Lavery and McComb (1991) have shown that the net growth rate of *Ulva rigida* is affected by variations in internal concentrations of nutrients (N and P). These observations support the idea of a variable internal quota of nutrients. According to Lavery and McComb (1991), *Ulva* experiences nutrient-limited growth when the internal quota of N is below 2% and P below 0.025%. These levels can be defined as critical quota.

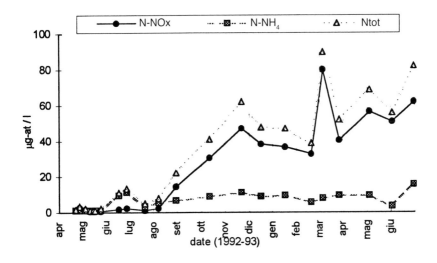

Figure 16.4 Changes in nitrogen concentrations

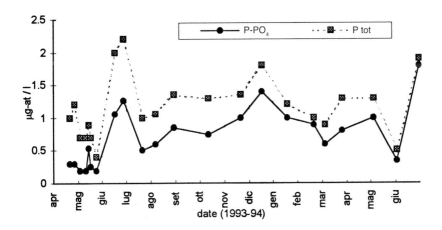

Figure 16.5 Changes in total and reactive phosphorus concentrations

The data (Figure 16.6) show that, in the macroalgae collected at station 3, the internal quota of P is almost an order of magnitude higher than the critical quota. The N concentrations fall below the critical value only during the period of maximum growth in 1992. From August 1992 to January 1993 data for the internal nutrient quota are not reported because the algae were almost

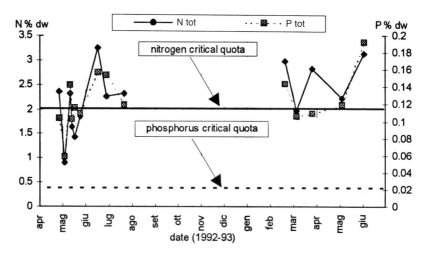

Figure 16.6   Concentrations of N and P in the macroalgae

completely absent at the sampling stations. Only small pieces of thallus were found at station 3 but their quantity was too small and their condition not sufficiently good to carry out measurements of internal nutrient concentrations.

Dissolved oxygen

The dissolved oxygen data are part of a larger data set obtained by continuous recording at station 3. Their concentrations (Figure 16.7) show the short-term changes in water quality better than other parameters and a comparison of concentrations at different periods of the *Ulva* cycle (Figure 16.2) shows clearly the progression of anoxic conditions in the water. Changes in oxygen concentrations monitored for 42 h from noon of the first day to early morning of the next day are shown in Figure 16.7 (The incomplete recording for 17–19 May 1992 was due to an instrument failure). The four curves on Figure 16.7 correspond to the periods when the standing stock of macroalgae was decreasing (Figure 16.2); concomitant with this decrease, organic detritus quantity and oxygen demand increased. Figure 16.7 also shows that nighttime oxygen minimum becomes progressively more pronounced while daytime maximum becomes more and more shallow. The lowermost curve was obtained a few days before total collapse of the bloom and onset of anaerobic conditions in the water column.

CONCLUSIONS

The environmental variables (light, temperature and nutrients) that are normally known to influence the population dynamics of the macroalgae were not adequate to explain the differences observed between 1992 and 1993. Other

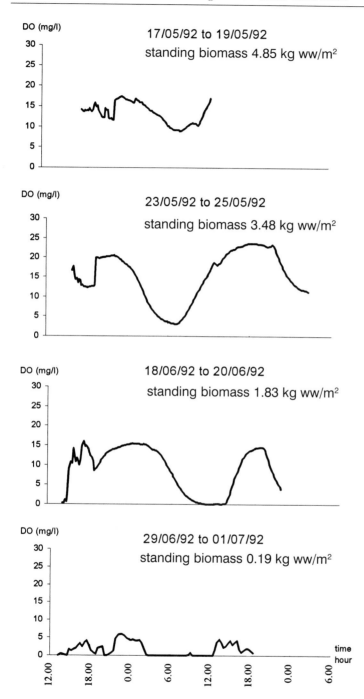

Figure 16.7 Dissolved oxygen concentration monitored continuously at different periods of the *Ulva* growth cycle

factors, such as the 'past history' of the standing stock (i.e. macroalgae that survived the winter) at the study area, also appear to be important.

The Palude della Rosa is a heterogeneous system where different areas show different situations of macroalgae growth: water circulation in the Palude and strong winds can play an important role in their transport and, as a consequence, in biomass distribution.

The internal nutrient quota can be successfully used to identify nutrient-limited growth and these data can, therefore, be more useful than *in situ* nutrient concentrations.

The chl *a* measurement alone is probably insufficient to describe phytoplankton–macroalgae competition when periods with high biomass are followed by a rapid decline in it and chl *a* peaks; a qualitative microscopic analysis is also required in such cases.

Continuous recordings with electrochemical probes are very useful to follow short-term changes of the *Ulva* cycle.

Finally, the choice of a few key parameters and their monitoring during the year is important not only for a quantitative description, but also for interpretation of the observed variations and to provide an opportunity to construct a clear and simple conceptual model of the growth dynamics of the macroalgae.

ACKNOWLEDGEMENT

This work was partially supported by the EU – MAS2-CT – 92-0036 contract.

REFERENCES

Carrer, G. M., Coffaro, G. and Bendoricchio, G. (1993). *Environmental Monitoring in the Venice Lagoon*. Final report to UNESCO

Fossato, V. U., Bergamasco, A. and Malvasi, G. (1994). Bilanci di massa per acqua, sale, solidi sospesi ed azoto totale in soluzione per la Palude della Rosa (Laguna veneta). *S.It.E. Atti VI Congresso Nazionale. Venezia 26-29/09/94*

Fujita, D. and Rodney, M. (1985). The role of nitrogen status in regulating transient ammonium uptake and nitrogen storage by macrolagae. *J. Exp. Mar. Biol. Ecol.*, **92**: 283–301

Grasshoff, K., Ehrhardt, M. and Kremling, K. (1983). *Methods of Seawater Analysis*. Verlag Chemie Weinheim, pp. 419

Lavery, P. S. and McComb, A. J. (1991). The nutritional ecophysiology of *Chaetomorpha linum* and *Ulva rigida* in Peel Inlet, Western Australia. *Botanica Marina*, **34**: 251–60

Lundberg, P., Weich, R. G., Jensen, P. and Vogel, H. J. (1989). Phosphorus-31 and Nitrogen-14 NMR studies of the uptake of phosphorus and nitrogen compounds in the marine macroalgae. *Ulva lactuca. Plant Physiol*, **89**: 1380–7

Sfriso, A. (1987). Flora and vertical distribution of macroalgae in the Lagoon of Venice: a comparison with previous studies. *Giornale Botanico Italiano*, **121**: 69–85

Sfriso, A., Marcomini, A. and Pavoni, B. (1989). Macroalgae and phytoplankton standing crops in the central Venice Lagoon: primary production and nutrient balance. *Science of Total Environ.*, **80**: 139–59

Sfriso, A., Marcomini, A., Pavoni, B. and Orio, A. A. (1990). Eutrofizzazione e macroalghe: la Laguna di Venezia come caso esemplare. *Inquinamento*, **4**: 62–78

Sfriso, A., Marcomini, A., Pavoni, B. and Orio, A. A. (1993). Species composition, biomass, and net primary production in shallow coastal waters: the Venice Lagoon. *Bioresource Technology*, **44**: 235–50

Strickland, J. D. H. and Parsons, T. R. (1972). *A Practical Handbook of Seawater Analyses*. Fisheries Research Board of Canada, Ottawa, Canada, pp. 310

CHAPTER 17

DYNAMICS OF BENTHIC FAUNA IN THE VENICE LAGOON

C.J.M. Philippart and N. Dankers

SUMMARY

The Venice Lagoon is a hypertrophic shallow embayment characterized by massive macroalgal blooms, followed by anoxic conditions with mass mortality of aerobic organisms. A pilot study showed that the benthic fauna seemed to have responded to eutrophication by favouring mobile species like crabs and Gammaridae rather than resident species such as bivalves. However, many other environmental factors such as sediment composition, temperature, predation and pollutants may play a role in the regulation of biomass and species composition. It is therefore difficult to distinguish between the effects of eutrophication and other environmental factors.

INTRODUCTION

The Venice Lagoon is a shallow marine embayment connected to the Adriatic Sea by means of three entrances. It receives large amounts of nutrients from domestic, industrial and agricultural sources and, as a result, conditions in parts of the lagoon are hypertrophic with massive *Ulva* spp blooms (Sfriso *et al.*, 1989). The EC project on Benthic Eutrophication Studies (BEST) was intended to elucidate the responses of coastal marine benthic systems to nutrient addition, with special emphasis on the Venice Lagoon. This chapter deals with the effects of eutrophication on benthic fauna, i.e. meiofauna and macrofauna.

Nutrient addition may result in an increase in biomass and primary production of phytoplankton (Oviatt *et al.*, 1986; Doering *et al.*, 1989; Keller *et al.*, 1990; Sullivan and Banzon, 1990; Sampou and Oviatt, 1991; Carlsson and Granéli, 1993; Hofmann and Höfle, 1993), microphytobenthos and macroalgae (Nilsson *et al.*, 1991). This may subsequently result in an increase of consumers, such as zooplankton (Doering *et al.*, 1989; Keller *et al.*, 1990; France *et al.*, 1992; Hofmann and Höfle, 1993), meiofauna (Widbom and Elmgren, 1988; Nilsson *et al.*, 1991), macrofauna (Beukema and Cadée, 1991; Kamermans, 1992) and fish (Keller *et al.*, 1990). However, under field conditions, as demonstrated for meiofauna in mesocosm experiments (Widbom and Elmgren, 1988), eutrophication does not generally result in an overall increase of species already present but rather in a shift of species composition and biomass. In hypertrophic areas, such as the Venice Lagoon, macroalgal blooms are generally followed by anoxic conditions with mass mortalities of aerobic organisms (Sfriso *et al.*, 1989). It can, therefore, be hypothesised that mobile animals such as crustaceans, polychaetes and gastropods will move away from areas of low oxygen concentrations whereas sedentary animals die off due to oxygen starvation (Figure 17.1).

A pilot study on benthic biomass and species composition in the Venice Lagoon was carried out in 1991–92 by sampling at four locations which were expected to show a gradient in nutrient loads. This chapter reports on the results from this study and discusses them in the light of the hypothesis that nutrient increase causes, through hypoxic and anoxic events, a change in the faunal composition.

MATERIAL AND METHODS

Sampling stations

The sampling was done at four stations in the Venice Lagoon, i.e. Alberoni, Sacca Sessola, San Giuliano and Fusina (Figure 17.2).

Station A (Alberoni) was located close to the Malamocco–Marghera channel entrance and therefore experienced very effective tidal water exchange. It was also relatively far from sources of anthropogenic inputs. This station was characterised by a high variety of macroalgal species (*Enteromorpha, Gracilaria, Dictyota, Codium, Ulva*) and very few phytoplankton blooms. Incidents of anoxia are unknown from this area.

Station B was near Sacca Sessola on the watershed of the central lagoon and therefore experienced a longer residence time of water. It was located halfway between station A and station C (see below), at some distance from urban water inflows. The flora was characterised by a single species of macroalgae, *Ulva rigida*. Phytoplankton blooms occurred when the macroalgae were decomposing. Anoxic conditions have been observed at this site in the past.

Station C (San Giuliano) was close to the inner border of the lagoon and therefore experienced very poor water exchange. It was located close to both urban and agricultural water inputs. The flora of this station consisted mainly of two species of macroalgae, *Ulva rigida* and *Gracilaria verrucosa*. Phytoplankton blooms occurred frequently, again when the macroalgae were decomposing. Anoxic situations had been observed earlier.

Station D (Fusina) was located close to the inner border of the lagoon, at the mouth of the river Brenta and the inner extremity of the Malamocco–Marghera channel. This channel connects the Marghera industrial area with the Adriatic Sea, resulting in intense water circulation. This station was close to direct urban and agricultural water inputs. Macroalgae were absent but phytoplankton blooms occurred frequently. Anoxic conditions had been observed earlier.

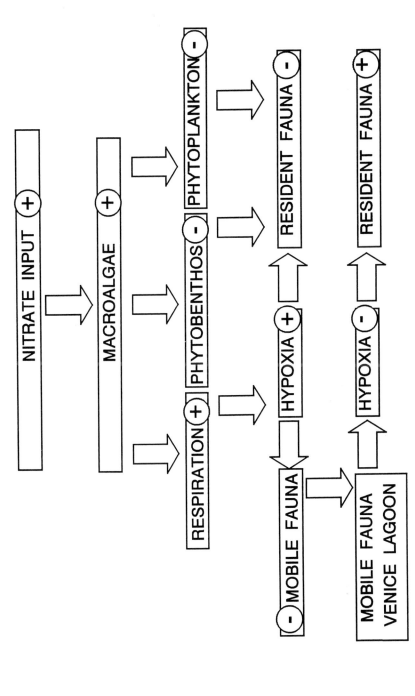

Figure 17.1 Hypothesis on the effect of eutrophication on the benthic fauna in the Venice Lagoon

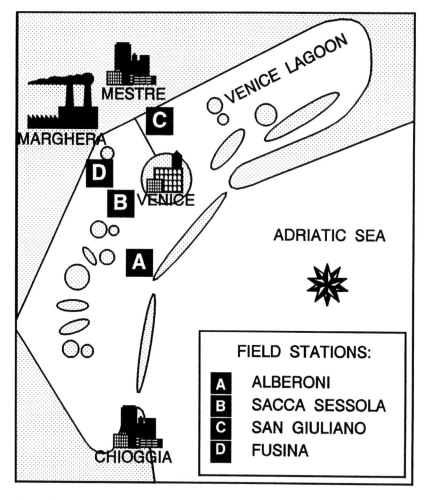

Figure 17.2 Location of the sampling stations for benthic fauna in the Venice Lagoon

Meiofauna

Meiofauna at stations B, A and C were sampled, respectively, on 25, 26 and 27 June 1991. Samples, to a depth of 5 cm, were taken in triplicate with a 32 mm ID core at stations A and C and with a 36 mm ID core at station B; preserved in 4% formalin and analysed later in the laboratory at Texel. The meiofauna was sorted into four main groups – Foraminifera, Nematoda, Annelida and Crustacea – and the number of animals in each group was recorded. Samples were placed in a petri dish with a mm grid painted on the bottom and counted under ×25 magnification with a stereomicroscope. When the samples had a high density of meiofauna, counting was confined to five grids of 1 cm² area chosen at random. When the density was low, animals from all the grids were counted.

Macrofauna

Macrofauna were sampled with sediment cores of 19 cm ID in 1991 and 30.4 cm ID in 1992. The sampled material was washed in a 1 mm mesh sieve and preserved in 80% alcohol. In 1991 no attempt was made to quantify the fauna present on the macroalgae, but in 1992 the macrofauna present on the macroalgae were also sampled. A preliminary sorting of the fauna into those with shells (e.g. bivalves and crustaceans) and without shells (e.g. poly-chaetes) was done at the University of Venice laboratory, in order to prevent damage to the soft material during transport to The Netherlands.

Samples were obtained once in 1991 and on four occasions in 1992 (Table 17.1). All samples were taken from the sites where the bell jars were placed, and immediately after bell jar measurements. Due to unfavourable weather conditions, samples in April 1992 were obtained using a smaller sediment core of 8 cm ID; in this instance, 10 samples were taken per station. All organisms were identified down to species level, except when their abundance was low and/or when precise identification was not possible. The ash-free dry weight (ADW) was estimated individually for all animals when their sizes were measured, and otherwise as the total biomass of animals of one species in a sample. ADW was taken as the difference between dry weight (DW; 24 h at 60°C) and ash weight (AW; 2 h at 560°C).

Environmental factors

Concentrations of dissolved nitrogen (ammonium and nitrate), biomass of primary producers (phytoplankton, microphytobenthos and macroalgae) and oxygen saturation were measured by the University of Venice team (Sfriso *et al.*, 1994) just before sampling for benthic fauna. Relationships between environmental factors and benthic biomass were tested by means of linear regressions.

Table 17.1 Dates of sampling for macrofauna in the Venice Lagoon

Year	Alberoni	Sacca Sessola	San Giuliano	Fusina
1991	26 June	25 June	27 June	–
1992	9 April	7 April	9 April	8 April
1992	14 May	13 May	12 May	12 May
1992	9 July	8 July	6 July	7 July
1992	15 Sep	17 Sep	15 Sep	14 Sep

RESULTS

Meiofauna

Densities of major meiofauna groups varied significantly at, and between, the stations (Table 17.2). Density of foraminifers was highest at station C and lowest at station B. The ratio of their mean densities among stations C, A and B was roughly 100:10:1. Density of nematodes was more or less similar at stations A and C but lower at station B, with a mean density ratio of 10:10:1 at these (C:A:B) stations. Density of annelids was minimum at station B and maximum at station A, and the ratio between their mean densities among stations A, C and B was roughly 20:10:1. Density of crustaceans was minimum at station B and maximum at station A. The ratio between their mean densities from station A through station C to station B was roughly 3:2:1.

Macrofauna

Total biomass of macrofauna varied considerably in space and time. In general, biomass was highest at station A and lowest at station B. The proportions of total benthic biomass from station A through stations C and D to station B was roughly 20:4:2:1 (Figure 17.3).

Station A

Biomass was highest in April 1992 and lowest in July 1992 (Figure 17.3). In June 1991, bivalves contributed to more than 90% of the total biomass (Figure 17.4), with *Tapes aureus* alone accounting for more than 40% of it. The bivalve, *Tellina donacina*, and an unidentified small polychaete were the numerically abundant species, with densities exceeding 600 animals m^{-2}. In April 1992, two species of the families Actiniidae (40%) and Nereidae (10%) accounted for more than half of the total biomass. Cirratudilidae (Polychaeta) species were most abundant in numbers, with almost 900 animals m^{-2}. The major bivalves contributing to biomass were *T. aureus* (10%), *Mytilus edulis* (10%) and *Loripes lacteus* (5%). In terms of abundance, *L. lacteus* was the

Table 17.2 Densities (mean SD, in m^{-2}) of the major meiofauna groups at stations A, B and C in June 1991

	Station A		Station B		Station C	
	n = 3		n = 3		n = 2	
Foraminifera	5388	±5436	655	±1310	86416	±81767
Nematoda	1199050	±667355	181424	±103881	1414986	±513462
Annelida	127241	±101952	982	±1134	6839	±7913
Crustacea	157912	±93125	7532	±9239	199565	±180239

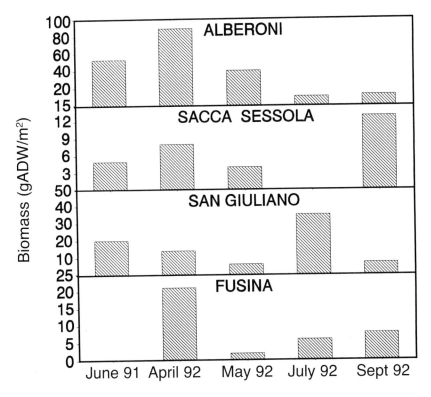

Figure 17.3   Total biomass of macrofauna at the 4 stations in 1991–92. Note differences in scales

most important, with a density of about 750 animals m^{-2}. In May 1992 representatives of the family Actiniidae were still dominant in biomass, contributing 40% of it. The gastropods *Cerithium vulgata* and *Gibbula albida* together contributed about 15% of the total biomass. The major bivalves contributing to biomass were still *T. aureus* (5%) and *L. lacteus* (5%), accompanied by *Tapes decussata* (5%) and *T. donacina* (5%). In terms of abundance, *L. lacteus* was the most important, with a density of almost 1000 animals m^{-2}. In July 1992 85% of the total biomass was represented by at least five species: Acteniidae (15%), the crustacean *Upogebia deltaura* (20%), and the bivalves *T. aureus* (20%), *T. decussata* (15%), and *L. lacteus* (10%). Most important in abundance were the Gammaridae with a density of more than 125 animals m^{-2}. In September 1992 the crustaceans *Anthura gracilis* (35%) and *U. deltaura* (10%) were the most important contributors to biomass. *L. lacterus*, *T. aureus* and *T. donacina* each accounted for approximately 10% of the total biomass. In terms of abundance, *L. lacterus* was still the most important species, although its density had decreased to approximately 275 animals m^{-2}.

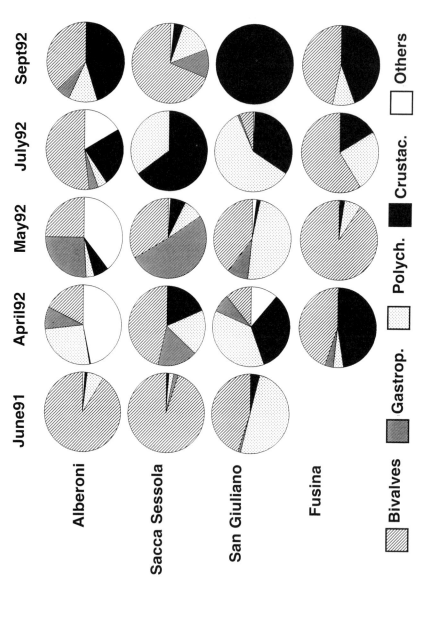

Figure 17.4 Relative mean biomass of major macrofauna groups at the 4 stations in 1991–92

Station B

The biomass was highest in September 1992 and lowest in July 1992 (Figure 17.3). In June 1991 bivalves contributed almost 95% of the total biomass (Figure 17.4); *T. tenuis* was the most dominant (40%), followed by *T. decussata* (30%) and *Tapes aureus* (20%). The gastropod *Hydrobia ulvae* was the most abundant species, with a density exceeding 600 animals m^{-2}. In April 1992 *T. donacina* was the most important species in biomass (35%), accompanied by Gammaridae (20%) and the gastropod *G. albida* (15%). *T. donacina* was also the most abundant, with a density of almost 300 animals m^{-2}. In May 1992, *T. donacina* was still the most important species in biomass (35%) and in density (over 200 animals m^{-2}). The contribution of the Gammaridae to biomass increased to almost 30%, and the gastropod *Cyclope neritae* contributed approximately 20%. In July 1992 only the Gammaridae and Polychaeta were still present, representing respectively 65% and 35% of total biomass. The polychaetes were numerically dominant. In September 1992, *T. aureus* was the most important species in biomass (40%), followed by *T. decussata* (20%) and *Cerastoderma* sp. (15%). Gammaridae and Polychaeta were still present, together accounting for approximately 10% of the total biomass. *T. tenuis* and the polychaetes *Nereis* sp. were the most abundant species, with densities greater than 50 and 45 animals m^{-2}, respectively.

Station C

The biomass was highest in July 1992 and lowest in May 1992 (Figure 17.3). Most remarkable at this station was the presence of insects in the sediments during the entire period, most probably the larvae of *Chironimus salinarius*. In June 1991 *Nereis diversicolor* was the most abundant species in biomass, accounting for 50% of the total (Figure 17.4). Bivalves accounted for less than 50%, the most important species being *Cerastoderma* sp. (35%) and *T. tenuis* (10%). *Gammarus* spp. were the most abundant, with a density exceeding 2000 animals m^{-2}. In April 1992 Nereidae (25%) and the crustacean *Idothea balthica* (20%) were the most important contributors to biomass, with *T. donacina* and the Gammaridae accounting for another 10% of the total. Larvae of *Chironimus salinarius* were the most abundant, with a density of over 2000 animals m^{-2}. In May 1992 *T. donacina* and the Nereidae were dominant in biomass, with respectively, 30% and 35%. The Nereidae were also numerically dominant, with about 800 animals m^{-2}. In July 1992 the Nereidae contributed to 60% of total biomass and were also the most abundant (350 animals m^{-2}). The crab, *Carcinus mediterraneus*, with a biomass of about 12 g ADW m^{-2} and a density of approximately 15 animals m^{-2} (an average biomass of >1 g ADW per crab), accounted for 34% of total biomass. In September 1992 *C. mediterraneus* was the only species present and had a total biomass of about 12 g ADW m^{-2} and a density of about 7 animals m^{-2} (an average biomass of about 2 g ADW per crab).

Station D

The biomass was highest in April 1992 and lowest in May 1992 (Figure 17.3). In April 1992 almost 90% of the total biomass was represented by two species: *Cerastoderma* sp. (56%) and *U. deltaura* (44%) (Figure 17.4). Gammaridae were the most abundant with more than 200 animals m^{-2}. In May 1992 the biomass of *Cerastoderma sp.* was lower than in April, but its contribution to total biomass had increased to 90%. The crustacean *A. gracilis* was the most abundant with a density close to 50 animals m^{-2}. In July 1992 the biomass of *Cerastoderma* sp. was higher than in May but its contribution to total biomass had decreased to 50%. Gammaridae were once again the most dominant in numbers, with a density of 50 animals m^{-2}. In September 1992 the biomass of *Cerastoderma* sp. was similar to that in July, with a contribution of 40% to total biomass. *C. mediterraneus* and *U. deltaura* contributed 25% and 15% respectively, to total biomass. The most abundant species was *A. gracilis*, with a density of about 400 animals m^{-2}.

Environmental factors

Analysis of the response of macrofauna to eutrophication was restricted to the most common species present, with the criteria set at a minimum 10% contribution to total biomass and a gain, at least once during the sampling period, of 1 g ADW m^{-2} or above (Table 17.3).

Table 17.3 Significance of relationships between the biomass of selected macrofaunal species and several environmental parameters

Macrofauna species/families	Mobile/ resident fauna	NH$_4^+$ mmol.l^{-1}	NO$_3^-$ mmol.l^{-1}	Phyto- plankton mg CHLa.l^{-1}	Microphyto- benthos, mg CHLa.m^{-2}	Macroalgae g wet weight m^{-2}	Oxygen saturation %
Actiniidae	r	ns	ns	ns	ns	ns	ns
Cerithium vulgata	m	ns	ns	ns	ns	ns	ns
Gibbula albida	m	ns	ns	ns	ns	ns	ns
Cerastoderma sp.	r	ns	+***	ns	ns	-*	-**
Loripes lacteus	r	ns	ns	ns	ns	ns	ns
Tapes aureus	r	ns	ns	ns	ns	ns	ns
Tapes decussata	r	ns	ns	ns	ns	ns	ns
Tellinus donacina	r	ns	ns	ns	ns	ns	ns
Nereidae	m	ns	ns	ns	ns	ns	ns
Upogebia deltaura	m	+*	+***	ns	ns	ns	ns
Gammaridae	m	ns	ns	ns	ns	ns	ns
Idothea baltica	m	ns	ns	ns	ns	ns	ns
C. mediterraneus	m	ns	ns	+***	ns	ns	ns

ns = non-significant; *-$p < 0.05$; **-$p < 0.01$;***-$p < 0.001$

Practically no significant relationships were found between biomass of macrofaunal species and the various parameters considered (Table 17.3). Biomass of *Cerastoderma* sp. correlated negatively with oxygen and macroalgal biomass and positively with nitrate concentrations. Biomass of some mobile species correlated positively with ammonium and nitrate concentrations (*U. deltaura*) and phytoplankton biomass (*C. mediterraneus*).

DISCUSSION AND CONCLUSIONS

Meiofauna density in the Venice Lagoon is most probably influenced not only by eutrophication but also by other environmental conditions such as sediment composition (Hicks and Coull, 1983; Wieser, 1959), availability of food (Nilsson *et al.*, 1991) and the presence of bioturbators (Reise, 1985) and predators (McIntyre, 1969). Annelids and some crustaceans are known to be very sensitive to low oxygen values (Elmgren, 1975) and may therefore be expected to be low in abundance at those stations where hypoxia, or even anoxia, occurs. This indeed happened with annelids at stations B and C, and with crustaceans at station B.

The density of crustaceans at station C was relatively high compared to other stations and this could have been the result of the stimulating effect of bioturbation by the polychaete *Nereis diversicolor* (Reise, 1979, 1985) which was also abundant at this station. The low density of foraminiferans at stations A and B is probably the result of dissolution of their calcareous shells at pH lower than 7–8 measured at these stations.

The results of the field study were in agreement with those of a mesocosm experiment carried out at the island of Texel (Philippart and Dankers, 1994). Although the environmental conditions of the Venice Lagoon differed in many ways from the mesocosm situation, densities of meiofauna groups and their behaviour were still more or less comparable. During a pilot study in the mesocosms in 1991, food availability following the spring bloom seemed to favour the increase of meiofauna within a few weeks. Macroalgae may subsequently have reduced the densities of meiofauna through a reduction of microphytobenthos. Subsequent decay of macroalgae probably resulted in low oxygen concentrations and a higher nutrient load, followed by a decrease in the density of annelids and crustaceans due to hypoxia, and a decrease of foraminiferans as a result of a decrease in pH.

The mean macrofaunal biomass of 20 g ADW m^{-2} found in the central Venice Lagoon is high compared to several other temperate estuaries (Knox, 1986). The findings on the effects of eutrophication on the macrofauna of the Venice Lagoon roughly agreed with the hypothesis set out earlier (see above). In general, density and biomass of the bivalves were higher at the relatively undisturbed station A than at other stations. The abundance of macrofauna at sites experiencing yearly anoxia events was highest in spring, which conforms to the observed trend in regions undergoing eutrophication.

Bivalves disappeared in July 1992 at station B and in September 1992 at station C, which was probably the result of the incidence of hypoxia and anoxia at these stations. However, the oxygen saturation values considered here were probably not critical for macrofauna, since lowest values are attained at night, and not during daytime when the measurements were made.

Recolonization of station B by bivalves was very fast, and they became important within 2 months. The decrease in their density and biomass coincided with an increase of Gammaridae, polychaetes and *C. mediterraneus*. The Gammaridae may have been associated with macroalgae. Polychaetes are known to be resistant not only to low oxygen values but also to sulphide (Vismann, 1990), and thus are favoured by eutrophication. Crabs occurring at station C probably migrated from other areas of the Lagoon as soon as environmental conditions were conducive, in order to feed on dead remnants of the resident macrofauna; they were found to double their biomass within 2 months at this station.

Several significant relationships between macrofauna abundance and environmental factors in the Venice Lagoon may not be causal but can be interpreted as indirect effects of eutrophication. In contrast to resident species, mobile species may profit from the decay of macroalgae, when nitrogen concentrations are high resulting in a subsequent bloom of phytoplankton; food for these animals, consisting of plant and animal detritus, would be abundant at this time.

Our findings that the biomass (and density) of *C. edule* related negatively to oxygen levels is supported by the results of the mesocosm experiments (Philippart and Dankers, 1994). In the latter experiments, the condition factor of the resident bivalves – *C. edule* and *Macoma balthica* – was negatively correlated with oxygen and positively correlated with phytoplankton biomass, and macroalgal biomass significantly affected the condition factor of *M. balthica*. These results suggest that eutrophication affects resident species in the following way: biomass and density are affected by hypoxia through increased mortality, while the condition factor is affected by restricted food availability.

Biomass of *Nereis diversicolor* correlated positively with nitrogen concentration and with hypoxia, reflecting its high tolerance of low oxygen and the presence of sulphides. The positive relationship of the biomass of *Hydrobia ulvae* with hypoxia may reflect its feeding habits, since these mud snails mainly consume dead organic material, which is more easily digested than living material.

Species composition and biomass of the benthic fauna seem to give an indication of the degree of eutrophication in the different areas of the Venice Lagoon. However, the four different sites cannot be compared solely on the basis of a nutrient gradient. Apart from the differences in nutrient loading, which would affect community structure, there were also differences in salinity, sediment grain size, water residence time, average water temperature and pollutants (Donazzolo *et al.*, 1984; Calvo *et al.*, 1991). Furthermore, the effects of eutrophication on primary producers, and subsequently on macrofauna, are not only the result of the nutrient loading but also of environmental conditions (De Vries *et al.*, 1996). Although the evidence is by no means unambiguous, the benthic fauna of the Venice Lagoon seem to have responded to eutrophication as we hypothesised. However, due to the complexity of environmental factors which play a role in the regulation of biomass and species composition, it is difficult to distinguish between the effects of eutrophication and other factors.

ACKNOWLEDGEMENTS

We are grateful to René Henkens, Anja van Leeuwen, Erwin Winter and Koos Zegers for their help in sampling and analysis of the benthic fauna. We are also grateful to Bruno Pavoni and Adriano Sfriso for their hospitality in the laboratory of the University of Venice and the use of their data on environmental conditions, and to Birthe Bak for linguistic corrections. This project was partly financed by the European Commission DGXII (STEP-CT-91-0120).

REFERENCES

Beukema, J. J. and Cadée, G. C. (1991). Growth rates of the bivalve *Macoma balthica* in the Wadden Sea during a period of eutrophication: relationships with concentrations of pelagic diatoms and flagellates. *Mar. Ecol. Prog. Ser.*, **68**: 249–56

Calvo, C., Grasso, M. and Gardenhi, R. (1991). Organic carbon and nitrogen in sediments and in resuspended sediments of the Venice Lagoon: relationships with PCB contamination. *Mar. Poll. Bull.*, **22**: 543–7

Carlsson, P. and Granéli, E. (1993). Availability of humic bound nitrogen for coastal phytoplankton. *Est. Coast. Shelf Sci.*, **36**: 433–47

De Vries, I., Philippart, C. J. M., De Groot, E. G. and Van Der Toi, M. W. M. (1996). Coastal eutrophication and marine benthic vegetation: A model analysis. In Schramm, W. and Nienhuis, P. H. (eds) *Marine Benthic Vegetation: Recent Changes and the Effects of Eutrophication*. Springer Verlag, Berlin, Heidelberg, Ecological Studies vol. 123: 79–113

Doering, P. H., Oviatt, C. A., Beatty, L. L., Banzon, V. F., Rice, R., Kelly, S. P., Sullivan, B. K. and Fritshen, J. B. (1989). Structure and function in a model coastal ecosystem: silicon, the benthos and eutrophication. *Mar. Ecol. Prog. Ser.*, **52**: 287–99

Donazzolo, R., Orio, A. A., Pavoni, B. and Perin, G. (1984). Heavy metals in sediments of the Venice Lagoon. *Oceanol. Acta*, **7**: 25–32

Elmgren, R. (1975). Benthic meiofauna as indicator of oxygen conditions in the northern Baltic proper. *Merentutkimuslait Julk*, **239**: 265–71

France, R., McQueen, D., Lynch, A. and Dennison, M. (1992). Statistical comparison of seasonal trends for autocorrelated data: a test of consumer – and resource – mediated trophic interactions. *Oikos*, **65**: 45–51

Hicks, G. F. R. and Coull, B. C. (1983). The ecology of marine meiobenthic harpacticoid copepods. *Oceanogr. Mar Biol. A Rev.*, **21**: 67–175

Hofmann, W. and Höfle, M. G. (1993). Rotifer population dynamics in response to increased bacterial biomass and nutrients: a mesocosm experiment. *Hydrobiologia*, **255/256**: 171–5

Kamermans, P. (1992). Food limitation in cockles (*Cerastoderma edule* (L.)): influences of location on tidal flat and of nearby presence of mussel beds. PhD. Thesis, Rijksuniversiteit Groningen, 43–58

Keller, A. A., Doering, P. H., Kelly, S. P. and Sullivan, B. K. (1990). Growth of juvenile Atlantic menhaden, *Brevoortia tyrannus* (Pisces: Clupeidae) in MERL mesocosms: effects of eutrophication. *Limnol. Oceanogr.*, **35**: 109–22

Knox, G. A. (1986). *Estuarine Ecosystems: A Systems Approach*. CRC Press, Boca Raton

McIntyre, A. D. (1969). Ecology of marine meiobenthos. *Biol. Rev.*, **44**: 245–90

Nilsson, P., Jönsson, B., Lindström Swanberg, I. and Sundbäck, K. (1991). Response of a marine shallow-water sediment system to an increased load of inorganic nutrients. *Mar. Ecol. Prog. Ser.*, **71**: 275–90

Oviatt, C. A., Keller, A. A., Sampou, P. A. and Beatty, L. L. (1986). Patterns of productivity during eutrophication: a mesocosm experiment. *Mar. Ecol. Prog. Ser.*, **28**: 69–80

Philippart, C. J. M. and Dankers, N. (1994). Meiofauna and macrofauna dynamics. *Basic Report Benthic Eutrophication Studies*, STEP-CT-91-0120

Reise, K. (1979). Moderate predation on meiofauna by the macrobenthos of the Wadden Sea. *Helgol. Wiss. Meeresunters*, **32**: 453–65

Reise, K. (1985). *Tidal Flat Ecology*. Ecol. Studies 54, Springer Verlag, Berlin

Sampou, P., and Oviatt, C. A. (1991). Seasonal patterns of sedimentary carbon and anaerobic respiration along a simulated eutrophication gradient. *Mar. Ecol. Prog. Ser.*, **72**: 271–82

Sfriso, A., Pavoni, B. and Orio, A. A. (1994). Flora and macroalgal biomass production in different nutrient-enriched areas of the Venice Lagoon. *Basic Report Benthic Eutrophication Studies*, STEP-CT-91-0120

Sfriso, A., Pavoni, B. and Marcomini, A. (1989). Macroalgae and phytoplankton standing crops in the central Venice Lagoon: primary production and nutrient balance. *Sci. Total. Environ.*, **80**: 139–59

Sullivan, B. K. and Banzon, P. V. (1990). Food limitation and benthic regulation of populations of the copepod *Acartia hudsonica* Pinhey in nutrient-limited and nutrient-enriched systems. *Limnol. Oceanogr.*, **35**: 1618–31

Vismann, B. (1990). Sulphide detoxification and tolerance in *Nereis* (*Hediste*) *diversicolor* and *Nereis* (*Neanthes*) *virens* (Annelida: Polychaeta). *Mar. Ecol. Prog. Ser.*, **59**: 229–38

Widbom, B. and Elmgren, R. (1988). Response of benthic meiofauna to nutrient enrichment of experimental marine ecosystems. *Mar. Ecol. Prog. Ser.*, **42**: 257–68

Wieser, W. (1959). The effect of grain size on the distribution of small invertebrates inhabiting the beaches of Puget Sound. *Limnol. Oceanogr.*, **4**: 181–94

CHAPTER 18

FLORA AND MACROALGAL BIOMASS PRODUCTION IN DIFFERENT NUTRIENT-ENRICHED AREAS OF THE VENICE LAGOON

A. Sfriso, B. Pavoni and A.A. Orio

SUMMARY

The composition and production of macroalgae and phytoplankton blooms were studied in three areas (Alberoni, Sacca Sessola and San Giuliano) of the central Lagoon of Venice from 1989–91. Changes of nutrient concentrations in the water column and interstitial waters and their relationships with variations of micro- and macroalgal biomass were also investigated. The studies in the following year (1992) were designed to identify the factors that favour the dominance of macroalgae and phytoplankton at Sacca Sessola and Fusina respectively. This was achieved by a comparison of the growth and decline of *Ulva* under natural conditions and in net cages at different depths in the water column.

INTRODUCTION

The central Lagoon of Venice (Figure 18.1) has undergone major environmental changes during the twentieth century (Pavoni *et al.*, 1992; Sfriso *et al.*, 1992a; Marcomini *et al.*, 1995). These were caused by the lack of a suitable treatment plant for sewage generated by Venice, enormous anthropogenic inputs from Mestre, surrounding villages and the Porto Marghera industrial area and intensive agricultural practices that use large amounts of fertilisers and pesticides. The central basin, with a water surface of ~132 km² and a mean depth of ~1 m, is separated from the southern and northern lagoons by the Malamocco–Marghera canal and the Burano–Torcello islands, respectively. In the past, the central Lagoon suffered marked hydrodynamic changes because of digging of wide (80–150 m) and deep (12–20 m) commercial channels (Vittorio Emanuele, 1919–30 and Malamocco–Marghera, 1961–69). This resulted in rapid tide expansion directly into the innermost parts of the basin and exclusion of vast areas near the watersheds from efficient water renewal. Superimposed on these changes are the effects of high inputs of eutrophic substances which markedly increased the concentrations of nitrogen and phosphorus in water (ammonium from 10–50 to 2000–2500 μmol N l^{-1} and phosphate from <1 to 10–25 μmol P l^{-1}) and in surface sediments (total N and P concentrations greater by 2.4 and 30 times, respectively).

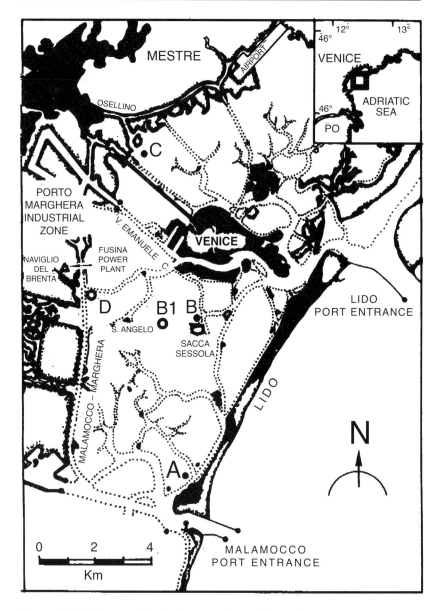

Figure 18.1   The central Venice Lagoon and sampling stations

The flora and phytobenthic associations of the central Lagoon were, as a result, severely affected from the 1970s (Sfriso, 1987) and monospecific populations of *Ulva rigida* C.Ag. progressively replaced the autochthonal rhizophyte and macroalgal populations and outcompeted the phytoplankton bloom. *Ulva* became the dominant species in a few years, attaining biomasses exceeding 20 kg wet weight m^{-2} over vast areas, mostly where water renewal and turbulence were negligible. The uncontrollable spread and growth of *Ulva*, which filled the whole water column up to the water surface for tens of square kilometers (Sfriso *et al.*, 1989), caused frequent and sudden biomass decays, leading to anoxia and the death of benthic populations (Sfriso *et al.*, 1987; 1988; Jak *et al.*, 1994) as well as a marked proliferation of mosquitoes of the genus *Chironomus* which, taking advantage of their ability to resist anaerobic conditions and benefiting from the death of their natural predators, reached densities of over 25,000 larvae per m^{-2} (Ceretti *et al.*, 1985).

Eutrophication and contamination of different areas of the central Venice Lagoon and their effects on benthic populations were studied from 1989. A preliminary study, at Alberoni (station A), Sacca Sessola (station B) and San Giuliano (station C), was carried out from March 1989 to April 1990, to identify the main phytobenthic associations and estimate macroalgal primary production. This study was extended to April 1990–June 1991 and included investigations on the relationship of macroalgal growth with annual trends of the main physicochemical parameters, nutrients (Sfriso *et al.*, 1992a, 1993; Sfriso and Marcomini, 1993) as well as some toxicants (PCBs and heavy metals) in water, sediments and sedimented particulate matter (SPM) (Pavoni *et al.*, 1990; Maroli *et al.*, 1993; Marcomini *et al.*, 1993; Sfriso *et al.*, 1995). Between April and December 1992, the study was extended to two other areas of the lagoon: Fusina (D) and an additional station at Sacca Sessola (B1), to differentiate between the presence and absence of macroalgae and/or phytoplankton and to obtain better estimates of the production and decomposition of algal biomass (Sfriso and Pavoni, 1994).

Over the total period of three years, 84 field studies were done at stations A, B and C, and 34 at stations D and B1, at intervals of 5 to 7 days in spring–summer and 15 to 19 days in winter. About 20 experiments on production and decomposition of *Ulva* at different growth periods were also conducted at stations D and B1 by placing algal fronds of known biomass in net cages at different depths (Sfriso and Pavoni, 1994).

MATERIAL AND METHODS

Physicochemical parameters

Air and water temperatures were measured with a thermocouple electrode (precision 0.1°C) connected to a Delta OHM HD8602 model portable pH meter. Redox potential and pH in water and the surface sediment (top 5 cm layer at stations A, B and C and top 2 cm layer at stations D and B$_1$) were measured with the same instrument, after calibration at pH 7 and pH 9. The photosynthetically active radiation (PAR, 400–700 nm, total incident and reflected irradiance) incident on the surface and the downwelling irradiance at different depths and at the bottom were determined with a portable LI-COR quantum-radiometer-photometer equipped with an air plate quantum sensor LI-190SA and an underwater spherical quantum sensor LI-193SA. Dissolved oxygen concentrations were measured by Winkler titration and chlorinity by Mohr–Knudsen titration.

Nutrients and phytoplankton

Replicate water samples were collected in a homemade Plexiglas bottle (150 cm height, 4 cm ID) and mixed together. Aliquots (100 to 1000 ml) were immediately filtered through Whatman GF/F glass fibre filters in a Millipore swinnex apparatus. The filtrate was analysed in triplicate for nitrate, nitrite, ammonium, phosphate and silicate following standard procedures (Strickland and Parsons, 1992). The particulate matter retained on the filters was analysed for chlorophyll *a* and phaeopigments. Six replicate sediment cores were collected with a 10 cm ID Plexiglas corer and the top 2 or 5-cm layers separated. The replicate cores at any depth were mixed together and compressed in a homemade stainless steel compressor to recover pore water in which ammonium and phosphate concentrations were measured.

Macroalgae

Standing crop

Macroalgal biomass was estimated according to the procedure reported by Sfriso *et al.*, (1991) for shallow water environments. In selected (approximately 20×20 m) areas, ten 1 m^2 subsamples in a predetermined grid were obtained using a $1 \times 1 \times 0.7$ m aluminium box placed on the bottom. Macroalgae fronds trapped inside were cut around the box walls, extracted with a landing net, drained on a sieve, weighed and the average weight calculated. At high biomass (> 5 kg wet weight m^{-2}) levels this procedure had a precision of 2–3% and at lower levels, 5–7%.

In situ macroalgal production and decomposition

The natural rate of net macroalgal increase/decrease was calculated by the method of 'biomass variations' between successive samplings, scaled to obtain mean daily changes in biomass and tested for significance ($p < 0.01$) by ANOVA.

The mean percentage daily biomass variation (DBV%) with reference to the initial biomass was calculated using the formula:

$$\text{DBV}\% = [(B_t - B_0) / B_0] / t \times 100$$

where B_0 is the initial biomass, and
B_t the biomass at day t.

The mean percentage progressive daily biomass variation (PDBV%) was obtained using the formula (Lignell *et al.*, 1987):

$$PDBV\% = [(B_t/B_0)^{1/t} - 1] \times 100$$

By using an average dry/wet weight ratio and the carbon content of young and old fronds in different seasons, the rates of biomass variations can also be expressed in terms of carbon.

Cage studies of macroalgal production, decomposition and grazing

Fluctuations in macroalgal biomass as a function of depths and areas, such as the Fusina station where macroalgae are usually scarce or absent, were measured in seminatural conditions by placing the fronds in 25 cm cubic net cages. At stations B and D, 12 cages, 6 on the surface and 6 at the bottom, were suspended. Each contained 150 g of *Ulva* in wet weight, corresponding to a biomass of about 10 kg m^{-2}, i.e. the biomass found at Sacca Sessola at the beginning of the observations. Variations in biomass were measured in five consecutive surveys during four different seasons: April, May–June, August–September and November–December. The data were processed using the formulae given above and compared with rates of *in situ* biomass variation.

Grazing pressure on macroalgae was evaluated by exposing *Ulva* fronds in net cages of 1 mm and 1 cm mesh size at different depths. The grazing rate was calculated as the difference between the daily biomass variations in the 1 mm and 1 cm net cages.

RESULTS AND DISCUSSION

Physicochemical parameters

Temperature

Surface temperatures at stations A, B and C in 1990–91 ranged between 2 and 28°C with annual averages of 18.3–18.6°C (Table 18.1, Figure 18.2). Average temperatures at stations B1 and D during April–December 1992 were slightly higher than at the other 3 stations, obviously because the lowest temperatures normally occur in January–February. The range and average of the temperature measured at Fusina (Table 18.1, Figure 18.3) were higher than at other stations because of the influence of warm waters from the cooling system of a nearby electric power plant. Temperatures near the mouth of the Naviglio del Brenta canal, at about 300 m from station D, were up to 10°C higher than those measured in other areas during the cold season. In contrast, the shallow Sacca Sessola area, located far from the sea and from freshwater inputs, showed the lowest temperatures.

Table 18.1 Physicochemical parameters and nutrients in the sampling stations from 16 April 1990 to 25 June 1991. Pore water nutrients were sampled at 5 cm in 1990–91 and at 2 cm in 1992

STATION		BIOMASS (wet wt) g/m²y	Temp. (water) °C	PHYSICOCHEMICAL PROPERTIES						PIGMENTS		WATER NUTRIENTS (µmol/l)						
				pH water	sed.	Eh water mV	sed.	Chlor g/l	Oxygen sat. %	Chl a	Phaeo a mg/m³	SURFACE RP	Am	Nit	Nitrate	TIN	PORE RP 5 cm	Am
1990–91																		
Alberoni (A)	Mean	580	18.6	8.42	7.78	379	55	17.2	150	1.3	0.9	0.59	6.66	0.58	6.80	14.04	4.0	79
	max	2855	28.2	9.26	8.20	445	228	18.9	281	11.4	9.9	1.50	42.81	2.47	48.51	92.25	9.5	261
	min	17	3.6	7.98	7.43	312	−114	15.8	96	0.1	0.01	0.10	0.78	0.07	0.12	1.88	1.0	18
Sacca Sessola (B)	Mean	1452	18.3	8.39	7.71	360	−2	16.7	139	1.7	1.4	0.84	9.46	1.06	8.27	18.79	4.7	169
	max	10652	27.0	9.56	8.15	432	142	18.6	254	7.4	5.6	3.00	31.67	3.38	44.74	61.35	17.7	618
	min	0	2.0	7.60	7.23	250	−203	13.0	60	0.04	0.07	0.05	2.94	0.26	0.69	5.09	1.4	34
San Giuliano (C)	Mean	1589	18.5	8.36	7.70	360	−85	14.7	130	8.0	9.4	2.72	11.93	1.76	11.40	25.09	11.3	208
	max	8228	27.6	9.35	8.21	419	62	18.3	243	86.0	109.8	16.25	44.73	5.34	47.58	94.20	31.8	581
	min	0	4.0	7.28	7.05	−110	−229	9.1	0	0.19	0.28	0.28	1.70	0.21	0.37	5.10	2.3	16
1992																	2 cm	
Sacca Sessola (B1)	Mean	4207	20.6	8.34	7.44	268	−132	17.0	117	3.2	0.7	0.86	20.83	1.29	9.31	31.43	53.5	1114
	max	19785	28.6	9.42	7.90	400	49	18.0	237	12.1	15.8	6.25	107.75	3.62	48.53	143.91	209	2508
	min	0	9.5	7.42	7.04	−18	−243	14.7	0	0.28	0.12	0.05	3.20	0.10	0.13	4.42	2.5	222
Fusina (D)	Mean	0	22.4	8.28	7.60	293	12	16.0	139	9.1	16.5	1.61	20.48	2.19	25.89	48.56	21.4	552
	max	0	30.2	8.95	8.12	397	84	17.8	274	57.8	47.5	7.05	98.80	4.93	61.15	144.81	153.5	1633
	min	0	9.8	7.48	7.06	31	−119	14.5	86	0.43	2.00	0.10	1.89	0.31	0.65	3.39	7.1	73

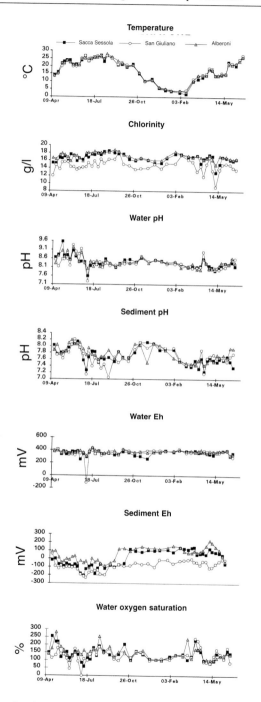

Figure 18.2 Physicochemical parameters at stations A (Alberoni), B (Sacca Sessola) and C (San Giuliano) from April 1990 to July 1991

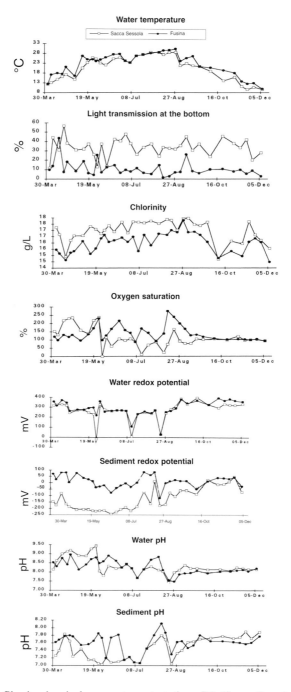

Figure 18.3 Physicochemical parameters at stations B1 (Sacca Sessola) and D (Fusina) in 1992

Chlorinity

Chlorinity ranged between 9.1 mg g^{-1} at San Giuliano near the mouth of the Osellino river to 18.9 mg g^{-1} at Alberoni near the Malamocco channel entrance. The highest values were found in summer.

Dissolved oxygen

Percentage saturation of dissolved oxygen and redox potential (E_h) and pH of water and top sediment layers were strictly related to fluctuations in micro- and macroalgae and zoobenthos. During fluctuations of algae, oxygen satura- tion varied from a peak 281% during the highest algal growth to 0% during their decomposition. Frequent hypoxic and anoxic conditions were observed at Sacca Sessola (B and B$_1$) and San Giuliano (C) stations between late April and July (Figures 18.2 and 18.3). In contrast, oxygen levels remained at ~100% saturation from September to December. Similarly, E_h fluctuated bet- ween 200 and 300 mV in March–August and between 300 and 400 mV during other periods (Figures 18.2 and 18.3). Negative peaks down to 0 mV were detected during anoxia. No significant differences in E_h between different areas were found but marked E_h variations were detected on the surface sedi- ment. When high biomass was present, as at Sacca Sessola and San Giuliano, the redox potential measured at the sediment top layer (2 or 5cm) fluctuated between −150 and −250 mV. When macroalgae completely disappeared, E_h values progressively increased and ranged between 0 and 150 mV in October–December. E_h at station A was rather higher than at other stations and ranged between −50 and 150 mV (av. 55 mV, Table 18.1).

Irradiance

The surface total and incident PAR and its downwelling percentages were measured only during 1992 at stations D and B$_1$ (Table 18.2, Figure 18.3). PAR varied depending on the sampling hour and cloudiness. The maximum photon flux (incident + reflected irradiance) measured was ~3000 μE m^{-2} s^{-1}, corresponding to ~150 Klux. The irradiance measured at a depth of 5 cm was similar at the two stations and was ~60% of that measured in air. About 33% and 11% of incident radiation reached the bottom at the Sacca Sessola and Fusina stations (Table 18.2, Figure 18.3). The low light intensity (av. ~82 μE m^{-2} s^{-1}, i.e. ~4.2 Klux) measured at the bottom at station D was, however, sufficient for algal growth. The compensation irradiance, at which respiration rate equals photosynthetic rate, for *Ulva* is 9.4 μE m^{-2} s^{-1} and the saturation irradiance, at which the growth is maximal, is ~412 μE m^{-2} s^{-1} (Keith and Murray, 1980). Hanisak (1987) measured a saturation irradiance of ~50 μE m^{-2} s^{-1} for *Gracilaria*.

Table 18.2 Photosynthetically active radiation (PAR) in air and in the water column

Station		*PAR($\mu E/m^2/s$)*				*Transmission(%)*	
		Incident	*Total*			*Total*	
		air	*air*	*water*		*water/air*	
				–5cm	*bottom*	*–5cm*	*bottom*
Sacca	mean	1203	1981	1362	673	67	33
Sessola	max	1840	2900	1980	1170	97	57
	min	93	398	95	45	24	11
	mean	1173	1924	1202	205	62	11
Fusina	max	1900	2830	2050	850	105	44
	min	30	52	30	2	36	1

Nutrients and phytoplankton

Nutrient concentrations in the water column

Annual cycles of nitrate, nitrite, ammonium and their sum (TIN) are shown in Figure 18.4. The highest concentrations (60–144 μmol N l^{-1}) were found in winter or during stormy weather and/or after anoxic periods. Marked differences in N concentrations were found between 1990–91 and 1992 at all stations. The annual average TIN concentrations at station A in 1990–91 (~14 μmol N l^{-1}) were 25% and 45% lower than those found at stations B (~19 μmol N l^{-1}) and C (~25 μmol N l^{-1}) (Table 18.1). The average concentrations measured in 1992 in the areas selected for the highest macroalgae and phytoplankton abundance (stations B1 and D) were 31–49 μmol N l^{-1}. Nitrate was the first nitrogen species which decreased during algal production; its concentrations, hence, were high only in autumn–winter when the macroalgae were absent. At stations A and D, characterized by low abundance or total absence of macroalgae and influenced by sea- and freshwaters of the Naviglio del Brenta, respectively, nitrate was dominant over ammonium. On the other hand, at Sacca Sessola station which sustained marked macroalgal and phytoplankton blooms and their decomposition, ammonium concentrations were equal to, or even twice as high, as nitrate (Table 18.1, Figure 18.4). Nitrite concentrations were always about an order of magnitude lower than those of ammonium and nitrate.

The annual average phosphate concentration (Figure 18.5) increased from station A (~0.6 μmol P l^{-1}) toward the other areas (up to ~2.7 μmol P l^{-1}). It was only slightly higher at station B (~0.8 μmol P l^{-1}) than at station A, whereas at stations D and C, it was 2.5 and 4.5 times higher. Phosphate changes, in general, showed a trend opposite to those of nitrogen compounds. The highest concentrations were found in spring–summer during macroalgal decomposition (up to ~16 μmol P l^{-1}) and during stormy weather (~3-7 μmol P l^{-1})

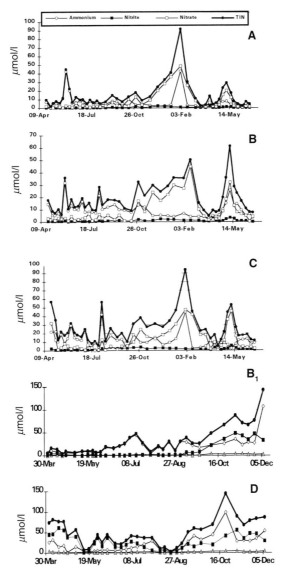

Figure 18.4 Changes in total inorganic nitrogen (TIN) species at stations A, B and C sampled in 1990–1991 and at station B$_1$ and D sampled in 1992

Silicate was only measured in 1992. Its concentrations were not significantly different between the two stations (Figure 18.5). As expected, silicate was not related to macroalgal growth but only to phytoplankton blooms, being a major diatom component, and the lowest concentrations, hence, were found after decomposition of macroalgae (between July and September – Figure 18.6) when phytoplankton became dominant.

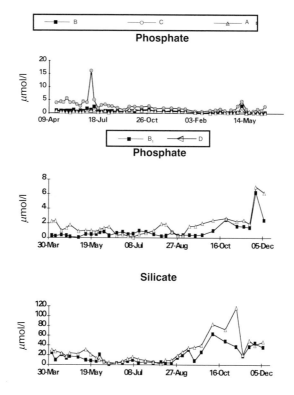

Figure 18.5 Changes in phosphate and silicate concentrations at stations sampled in 1990–1991 (top) and 1992 (middle and bottom)

Nutrient concentrations in pore water

Pore water nutrient concentrations at stations A, B and C (Table 18.1) were markedly lower than those found at stations D and B_1 because of a sampling difference: interstitial waters from station A, B and C were extracted from the top 5 cm sediment layer and not from the top 2 cm layer, as at stations D and B1. Among the first set of stations, the phosphate concentration measured at station A (~4.0 μmol P l^{-1}) was only slightly lower than that at station B (~4.7 μmol P l^{-1}) and about a third of that at station C (~11.3 μmol P l^{-1}). The phosphate concentration in the pore water of the first 2 cm sediment layer at Sacca Sessola (1992) was nearly twelve times higher than that found in the 5 cm layer (1989–91) showing the marked influence of decomposing biomass at the sediment–water interface. In this case, the concentration of phosphate measured at Sacca Sessola was ~2.5 times higher than that at Fusina. A similar trend was also found for ammonium. Its concentrations at stations C and B were about 2.7 and 2.1 times higher than at station A. Similarly, the ratio of its concentrations between the 2 and 5 cm layer pore waters at Sacca Sessola was

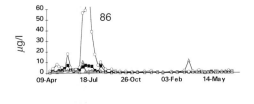

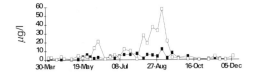

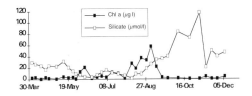

Figure 18.6 Trends in chl *a* concentrations in 1990–1991 (top) and 1992 (middle) and their relationship with silicate at Fusina (bottom). Symbols for stations as in Figure 18.5

~14, and the average concentration here was twice as high as at Fusina. Further, during macroalgal decomposition both ammonium and phosphate concentrations in the Sacca Sessola pore waters were up to ten times higher than at Fusina (maximum concentrations: 209 μmol P l^{-1} and 2508 μmol N l^{-1}). These data agree with the measurements of organic content (Jak *et al.*, 1994) in the surface sediments; it increased from station A (2–5%) to station C (6–9.5%).

Phytoplankton concentrations

As expected, phytoplankton producers were of secondary importance in autotrophic biomass and production. They were more important than macroalgae only at certain areas, such as Fusina, but elsewhere they were important producers only in summer (Figure 18.6), after decline of the macroalgal bloom. Average chl *a* at stations A and B in 1990–91 was, respectively, ~1.3 and ~1.7 mg m^{-3} (Table 18.1). Average chl *a* at San Giuliano was ~8 mg m^{-3}; weak phytoplankton blooms (Chl *a*: ~3-8 mg m^{-3}) occurred in this area even in the presence of high macroalgal densities (~3–8 kg wet weight m^{-3}) because of the influence of the Osellino freshwater rich in nutrients. The highest chl *a* measured at this station was ~86 mg m^{-3}, after total decomposition of macro-algae in July. Average chl a values at stations B1 and D in 1992 were only slightly higher than the 1990–91 averages (Figure 18.6). No phytoplankton blooms were observed at Sacca Sessola in either year and the maximum chl a measured at this station was only ~12 mg m^{-3}. Fusina, from where macroalgae were absent, showed the highest annual average chl a concentration (~9.1 mg m^{-3}) with the largest blooms in August and September (Chl a up to ~60 mg m^{-3}). Still larger blooms (up to ~150 mg m^{-3}) were observed outside the sampling areas, i.e. between Sacca Sessola and Fusina, near the island of S. Angelo della Polvere at the edges of areas covered by decomposing macro-algae. Here, the influence of tidal flow coming from the Malamocco–Marghera canal was negligible.

Phaeopigment levels at stations A and B were, on average, lower than chl a concentrations, but exceeded chl a at stations C and D, probably because of enhanced mortality of phytoplankton due to mixing of waters of different densities.

Macroalgae

Macroalgal typology and standing crop

One of the main objectives of this study was to describe composition, standing crop and production of macroalgae and their responses to the degree of eutrophication. With this purpose, the dominant macroalgal associations were studied at stations A, B and C over three successive years (1989–91). In the subsequent year, their growth and competition with phytoplankton were inves-tigated at stations B1 and D.

Between 1989 and 1991, two major macroalgal associations, one charac-terized by the dominance of *Ulva rigida* and the other by the growth of a few species at different times of the year with a nearly equivalent share in annual production, were identified in the central Lagoon of Venice. San Giuliano and the two Sacca Sessola stations were dominated by the first association and Alberoni, by the second one. At the latter station, 5 taxa – *Enteromorpha* (~30%), *Ulva rigida* C.Ag. (~23%), *Gracilaria verrucosa* (Huds.) Papenf. (~17%), *Dictyota dichotoma* (Huds.) Lamour. (~8%) and *Porphyra leucosticta* Thur. (~8%) – each contributed > 5% of the total annual biomass, while two others – *Ceramium rubrum* (Huds.) C.Ag. and *Polysiphonia sanguinea* (C. Ag.) Zanard. + *P. breviarticulata* (C. Ag.) Zanard. – accounted respectively for ~3.4 and ~1.6% (Figure 18.7). However, with regression of *Ulva* patches and a general increase in the species considered to be more sensitive to eutrophication, the percentage composition changed in the two subsequent years. In the infralittoral zone (i.e. 0.8 m below the mean tide level) of this area, many specimens of a new Rhodophycea, viz. *Grateloupia doryphora* (Mont.) Howe, and a marked repopulation by *Zostera marina* L., which had disappeared in the 1970s, were seen in 1990–1991.

In the same period both the Sacca Sessola stations had almost monospecific *Ulva rigida* populations accounting for ~95% of the biomass and ~99% of the total macroalgal production. Until 1990, this nitrophilic species covered ~66, ~46 and ~26 km² of the central Lagoon water surface with a mean wet biomass of 3.0, 7.5 and 12.5 kg m⁻², respectively. Later the areal coverage and biomass of *Ulva* began to decrease. In fact, within the 3 years (1989–91), the highest biomass measured decreased by half at Sacca Sessola (from ~20 to ~10 kg m⁻²) and at San Giuliano (from ~8 to ~4 kg m⁻²) stations (Table 18.3, Figure 18.7) and the growth period, from ~4 (1989) to ~2 (1990, 1991) months at Sacca Sessola and from ~5 (1989) to ~3.5 (1990) and ~1 (1991) month at San Giuliano (Sfriso *et al.*, 1993). Because of this, a new station (B1) at Sacca Sessola, where macroalgae were abundant, was selected in 1992. At this station, macroalgal biomass was > 10 kg wet weight m⁻² in April and increased to ~20 kg wet weight m⁻² at the end of May (Figure 18.7). Subsequently, the biomass decreased by half within 4 days, and was down to a few g m⁻² in July. After feeble regrowth, up to ~0.7 kg wet weight m⁻² in early August, the macroalgae disappeared completely at the beginning of September. In 1993 the macroalgae were virtually absent from this new station.

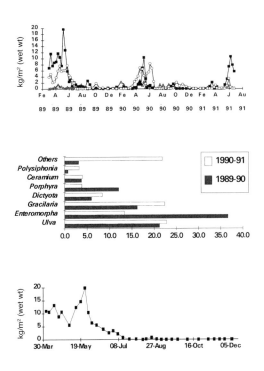

Figure 18.7 Macroalgal biomass trends at stations A, B and C (top) and B1 (bottom) and percentage biomass composition of different species at Alberoni (middle). Symbols as in Figure 18.5

In situ macroalgal production and decomposition

Biomass variations in terms of percent changes in weight were studied both under field and experimental conditions (Sfriso and Pavoni, 1994). The results obtained by two methods were very different. The true biomass variations (net production and decomposition) could be estimated only from *in situ* studies (Sfriso *et al.*, 1991; 1993) since growth in the cages was markedly different from that observed in the field (Figure 18.8).

The rate of net macroalgal production and/or decomposition was estimated by summing the significant increases and losses of biomass during an entire vegetative year, i.e. between the highest standing crops of consecutive years (Sfriso *et al.*, 1993). Production differed between stations, seasons and years (Table 18.3). During 1989–90, biomass production at stations A, B and C were ~9.0, ~16.1 and ~16.5 kg wet weight m^{-2} y^{-1} but decreased to ~6.1, ~15.5 and ~6.8 kg m^{-2} y^{-1} in the next year. Production at Sacca Sessola in 1992 was ~20 kg m^{-2} y^{-1}, equivalent to ~132–435 g C m^{-2} y^{-1}. The peak production

Table 18.3 Macroalgal standing crop (wet weight, g/m^2) and production (g/m^2y) at the sampling stations

		A	B	B1	C	D
Biomass	n	84	84	–	84	–
1989–91	range	0–2855	0–19638	–	0–8228	–
	$\bar{x}$	359	1945	–	1428	–
1989–90	n	36	33	–	35	–
	range	0–2205	0–12850	–	0–8228	–
	$\bar{x}$	385	1465	–	1452	–
Production						
1990–91	biomass	9047	16089	–	16455	–
	carbon	196	350	–	358	–
ABP/HMB		4.10	1.55	–	2.00	–
Biomass						
1990–91	n	35	38	–	33	–
	range	17–1436	0–10652	–	0–8228	–
	$\bar{x}$	269	736	–	589	–
Production						
1992	biomass	6074	15507	–	6833	–
	carbon	132	337	–	149	–
ABP/HMB		4.89	1.46	–	2.08	–
Biomass						
1992	n	–	–	34	–	34
	range	–	–	0–19785	–	0
	$\bar{x}$	–	–	4207	–	0

ABP: Annual biomass production; HMB: Highest measures biomass

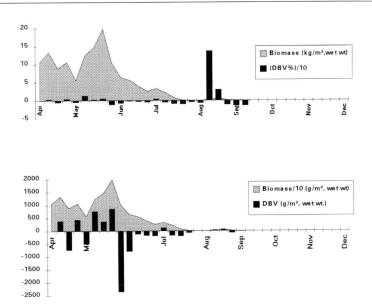

Figure 18.8 Relationship between macroalgal biomass trends at Sacca Sessola and growth rates as percentage DBV (top) and wet weight DBV (bottom)

(~1500 g wet weight m^{-2} d^{-1}, ~33 g C m^{-2} d^{-1}) period lasted for about 10 days in May–June and the highest decomposition rates (2363 g wet weight m^{-2} d^{-1}, ~51.3 g C m^{-2} d^{-1}) were usually observed very close to the period of highest production. From the data at different stations, the ratios of annual biomass production to the highest measured biomass (ABP/HMB, Thorne-Miller and Harlin, 1984) at different biomass levels were calculated. Despite the marked decrease in biomass, these ratios remained quite similar between 1989–90 and 1990–91, and varied from 1.51 (1.55 in 1989–90 and 1.46 in 1990–91) at Sacca Sessola (maximum biomass ~20 kg wet weight m^{-2}) to 2.0 (2 in 1989–90 and 2.08 in 1990–91) at San Giuliano (maximum biomass ~8 kg wet weight m^{-2}) and to 4.5 (4.1 in 1989–90 and 4.9 in 1990–91) at Alberoni (biomass less than ~3 kg wet weight m^{-2}) (Sfriso *et al.*, 1993). The ratios calculated at Sacca Sessola were similar to those (~1.59) found at the Lido station (biomass ~10–15 kg wet weight m^{-2}) in 1985–86 (Sfriso *et al.*, 1988). Thus, these ratios can be applied for a quick estimate of net production with only a need to measure the highest standing crop which, for a given species, occurs at about the same period every year. For example, the highest standing crop of *Ulva* usually occurs between May and June. By applying these ABP/HMB ratios to the highest standing crop found among 178 samples in 1987 in the whole central Lagoon, a net production of ~1,541,600 tonnes wet weight, equivalent to ~34,320, ~3581 and ~315 tonnes of C, N, and P, respectively, was calculated (Sfriso *et al.*, 1993).

The biomass production and/or degradation can also be expressed as DBV% (Figures 18.8 and 18.9) or as PDBV% (see above). The highest DBV% (up to ~200%) were measured at station A throughout the year and at other stations when biomass was < 100 g m⁻², because of the sensitivity of the measurement (5–10 g m⁻²) and the biomass contribution of new thalli from spores or gametes which could be added to the previous ones. On the other hand, higher biomass levels (5–15 kg m⁻²), even under optimal growth conditions, gave very low or negative DBV% (Sfriso *et al.*, 1993). However, as expected, the highest biomass yields, i.e. production in terms of wet weight, were obtained at biomass levels of ~5–7 kg m⁻².

Cage studies of macroalgal production, decomposition and grazing

The maximum DBV% measured for *Ulva* fronds in net cages at the two stations was only ~23% (Figure 18.10). This value was obtained with ~150 g of *Ulva*, corresponding to ~10 kg wet weight m⁻², and is about the highest biomass increase rate expected from the vegetative growth of macroalgal fronds when exposed to optimal growth conditions, even starting from lower biomass (Sfriso and Pavoni, 1994).

In contrast with the data from Sacca Sessola, DBVs in cages were highly positive even during periods of anoxia (Figure 18.8) when macroalgae

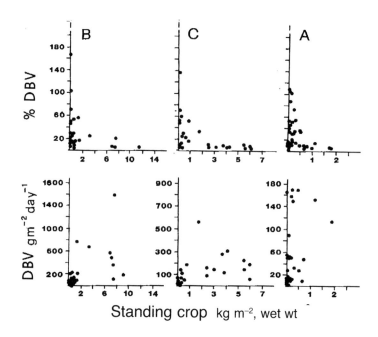

Figure 18.9 DBV% and wet weight DBV at stations A, B and C

decayed rapidly *in situ*. This is probably because algal fronds in the cages were spatially separated both from the markedly reduced sediment surface (Figures 18.2 and 18.3) and the decay process at the sediment–water interface.

On average, macroalgal growth was higher at Fusina than at Sacca Sessola (Figure 18.10) and, obviously, higher in the surface than in the bottom cages. At Fusina the DBV% in the surface cages during the first two surveys were up to three times higher than at Sacca Sessola. On the other hand, in August–September and November–December the growth rate at Sacca Sessola was higher than at Fusina, where macroalgae decreased both in surface and bottom cages. In these surveys many benthic crustaceans such as the shrimps *Palaemon serratus* Penn. and *Palaemon elegans* Rathke, the isopods *Idotea baltica* Pall. and *Sphaeroma serratum* Fabr., and the amphipod *Gammarus aequicauda* Martinov were found to feed on exposed macroalgae. In particular, *Gammarus* densities were from 2000 to 3000 individuals at any one cage over an exposed biomass of only 100–300 g wet weight. The hypo-anoxic conditions occurring in a large part of the central Lagoon between May and August caused the death of sessile and non-migratory benthic species and a migration of fish and crustaceans toward more oxygenated areas, such as Fusina. The effect of grazing pressure, calculated from the differences in DBV% in cages with 1 mm and 1 cm mesh, was ~4–15% of the initial biomass day^{-1}.

During the whole study period, growth in the bottom cages was always higher at Sacca Sessola than at Fusina (Figure 18.10). As light reaching the bottom was sufficient for algal growth, this difference was due to turbulence caused by tidal currents in the Malamocco–Marghera canal, and wave motion produced by commercial ships in transit. Water turbulence resuspended surface sediments and caused a reduction of transparency as well as deposition of a thin microlayer of particulate matter over the substrate. Under these conditions macroalgal fronds were completely shadowed and algal spores and gametes would presumably not be able to stick to any substrate. During 1992, the amount of settled particulate matter (SPM) collected by $20 \times 20 \times 10$ cm traps, especially designed for this shallow lagoon environment (Sfriso *et al.*, 1992b), was ~850 kg m^{-2} y^{-1}, about 3.1 times the SPM collected in the same period at Sacca Sessola (B1).

CONCLUSIONS

The data from three years of samplings in the Venice lagoon between 1989 and 1991, and then in 1992, led to the following conclusions:

(1) The central Lagoon is populated by two main phytoassociations, one characterized by the dominance of *Ulva* and the other by seasonal dominance of *Enteromorpha*, *Ulva*, *Gracilaria*, *Dictyota* and *Porphyra*. Contribution from each of the the latter five genera to the annual average biomass ranged from 8 to 32%.

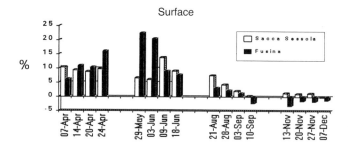

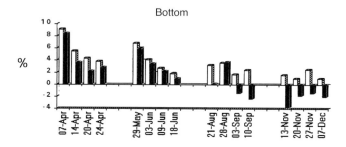

Figure 18.10   DBV% in cages at stations D and B1

(2) At the beginning of the sampling programme (1989), the *Ulva* association extended over about half (66 km²) of the central Lagoon, with a net production estimated at above 1.5 million tonnes wet weight. Later, this association began to decrease, halving its expanse and production in favour of the multigenera association that included new species and a marked rhizophyte regrowth.

(3) *In situ* production was strictly related to biomass levels. Under favourable conditions and with a biomass below 100 g wet weight m⁻², the daily production could be above 100% per day. On the contrary, at biomass levels higher than 3 kg wet weight m⁻², on average the production rate was below 5%. However, the highest biomass yield, in terms of wet weight, occurred at biomass levels ranging from 5 to 7 kg wet weight m⁻². Production in net cages never exceeded 23% per day.

(4) Phytoplankton played an important role only after macroalgal decomposition, during anoxic events, and in limited areas such as Fusina where water turbulence, sediment resuspension and grazing by crustaceans were high, hampering macroalgal growth.

(5) Differences in nutrient concentrations in water and pore water as well as the physicochemical parameters, such as underwater light transmission and temperature, played a minor role in the distribution of algae and their productivity.

ACKNOWLEDGEMENTS

The BEST project was partly funded by the European Commission (DG XII) within the framework of the STEP environmental research programme (contract STEP-CT-91-0120).

REFERENCES

Ceretti, G., Ferrarese, U. and Scattolin, M. (1985). I Chironomidi nella laguna di Venezia. *Arsenale, Venezia*, pp. 59

Hanisak, M. D. (1987). Cultivation of *Gracilaria* and other macroalgae in Florida for energy production. In Bird, K. T. and Benson, P. H. (eds), *Seaweed Cultivation for Renewable Resources*. Elsevier, Amsterdam, 191–218

Jak, R. G., Scholten, M. C. Th., van het Groenewoud, H. (1994). Nutrient, algal production and zooplankton control. *TNO report*, IMW-R94/049 A

Keith, E. A. and Murray, N. M. (1980). Relationship between irradiance and photosynthesis for marine benthic green algae (*Chlorophyta*) of differing morphologies. *J. Esp. Mar. Biol. Ecol.*, **43**: 183–92

Lignell, A., Ekmann, P. and Pedersén, M. (1987). Cultivation technique for marine seaweeds allowing controlled and optimized conditions in the laboratory and on a pilot scale. *Botanica Marina*, **30**: 417–24

Marcomini, A., Sfriso, A., Pavoni, B. and Orio, A. A. (1995). Eutrophication of the lagoon of Venice: nutrient loads and exchanges. In McComb, A. J. (ed.) *Eutrophic Shallow Estuaries and Lagoons*. CRC Press, Boca Raton, Fl., USA, 59–80

Marcomini, A., Sfriso, A. and Zanette, M. (1993). Macroalgal blooms, nutrient and trace metal cycles in a coastal lagoon. In Rijstenbil, J. W. and Haritonidis, S. (eds) *Macroalgae, Eutrophication and Trace Metal Cycling in Estuaries and Lagoons*. EC Report BRIDGE – DG XII, 66–90

Maroli, L., Pavoni, B., Sfriso, A. and Raccanelli, S. (1993). Concentrations of polychlorinated biphenyls and pesticides in different species of macroalgae from the lagoon of Venice. *Marine Pollution Bulletin*, **26**: 553–8

Pavoni, B., Calvo, C., Sfriso, A. and Orio, A. A. (1990). Time trend of PCB concentrations in surface sediment from a hypertrophic, macroalgae populated area of the lagoon of Venice. *Science of the Total Environment*, **91**: 13–21

Pavoni, B., Marcomini, A., Sfriso, A., Donazzolo, R. and Orio, A. A. (1992). Changes in an estuarine ecosystem. The lagoon of Venice as a case study. In Dunnett, D. A. and O'Brien, R. J. (eds), *The Science of Global Change*, American Chemical Society, Washington, DC, USA, 287–305

Sfriso, A. (1987). Flora and vertical distribution of macroalgae in the lagoon of Venice: a comparison with previous studies. *Giornale Botanico Italiano*, **121**: 69–85

Sfriso, A., Pavoni, B., Marcomini, A. and Orio, A. A. (1988). Annual variation of nutrients in the lagoon of Venice. *Marine Pollution Bulletin*, **19**: 54–60

Sfriso, A., Pavoni, B. and Marcomini, A. (1989). Macroalgae and phytoplankton standing crops in the central Venice lagoon: primary production and nutrient balance. *Science of the Total Environment*, **80**: 139–59

Sfriso, A., Raccanelli, S., Pavoni, B. and Marcomini, A. (1991). Sampling strategies for measuring macroalgal biomass in the shallow waters of the Venice lagoon. *Environment Technology*, **12**: 263–9

Sfriso, A., Pavoni, B., Marcomini, A. and Orio, A. A. (1992a). Macroalgae, nutrient cycles and pollutants in the lagoon of Venice. *Estuaries*, **15**: 517–28

Sfriso, A., Pavoni, B., Marcomini, A., Raccanelli, S. and Orio, A. A. (1992b). Particulate matter deposition and nutrient fluxes onto the sediments of the Venice lagoon. *Environment Technology*, **13**: 473–83

Sfriso, A., Marcomini, A., Pavoni, B. and Orio, A. A. (1993). Species composition, biomass, and net primary production in shallow coastal waters: the Venice lagoon. *Bioresources Technology*, **44**: 235–50

Sfriso, A., Marcomini, A. and Zanette, M. (1995). Heavy metals in sediments, settled particulate matter and phytozoobenthos of the Venice lagoon. *Marine Pollution Bulletin*, **30**: 116–24

Sfriso, A. and Marcomini, A. (1994). Gross primary production and nutrient behaviours in shallow lagoon waters. *Bioresources Technology*, **47**: 59–66

Sfriso, A. and Pavoni, B. (1994). Macroalgae and phytoplankton competition in the central Venice lagoon. *Environment Technology*, **15**: 1–14

Strickland, J. D. H. and Parsons, T. R. (1992). A Practical Handbook of Seawater Analysis, *Fish. Res. Board of Canada, Ottawa,* pp. 310

Thorne-Miller, B. and Harlin, M. M. (1984). The production of *Zostera marina* L. and other submerged macrophytes in a coastal lagoon in Rhode Island, USA. *Botanica Marina*, **27**: 539–46

Section 4

Biological Effects of
Environmental Pollution

CHAPTER 19

DETERMINATION OF WATER QUALITY IN THE VENICE LAGOON UTILIZING THE EARLY LIFE STAGES OF A FISH (SPARUS AURATA) AND AN ECHINODERM (PARACENTROTUS LIVIDUS)

I.R.B. McFadzen

SUMMARY

Industrial and agricultural activities around the Venice Lagoon contribute contaminants and waste products to the basin via effluent discharge and runoff. In addition to this, urban sewage with its associated toxic components (e.g. faecal sterols) is discharged without treatment directly into the canals of Venice and coastal waters. Sediment elutriates from the inner city canals and surface microlayer samples from the Lagoon were highly toxic and produced deleterious biological effects when tested with marine organisms. Early life stages of the fish (*Sparus aurata*) and the sea urchin (*Paracentrotus lividus*) species utilised in this study show potential as test organisms for biological effects monitoring in the Venice Lagoon.

INTRODUCTION

Rationale for the application of bioassays to the Venice Lagoon

Water quality is often assessed chemically in terms of concentrations of known (or detectable) toxic contaminants. This method may be satisfactory when there are a limited number of contaminants whose biological effects are known and predictable but effluents are often extremely complex and may contain numerous synthetic organic compounds. Effluents from industry entering a harbour or an estuary may include thousands of individual elements and compounds, making it impractical to define, and costly to monitor, the water quality by chemical analysis alone. Many factors influence the toxicity of contaminants on their entry into seawater. Some contaminants may degrade quickly into harmless products, some may become bound by organic material, with attendant changes in their biological availability and some others may interact chemically to become more, or less, toxic in combination than separately.

Since the ultimate concern is the capacity of the sea to support life, it is desirable to measure water quality in terms of biological response. The response of suitable test organisms in the laboratory to water samples taken from polluted areas can be used as an index of water quality. In this way all of the biologically relevant variables that affect water quality are integrated in terms of the responses of individual organisms. Among the advantages of this approach is the integration it provides, not only of the effects of toxicants and modifying variables that are known and understood, but also those that are not yet known (Stebbing, 1985).

The early life stages of marine organisms, especially pelagic embryos and larvae, represent the most sensitive stages in their life cycle. Exposure to anthropogenic contaminants during these extremely sensitive developmental stages can have deleterious results on the population of the respective species. It is the sensitivity to variations in ambient water quality that makes the early life stages suitable for short-term exposure experiments. Many embryonic/larval stages have the advantage that they are lecithotrophic (nourishment is gained from endogenous feeding on yolk) for up to several days, thus avoiding the problems and inconvenience of feeding them for the duration of the test. Fish and echinoderm embryos and larvae have previously been used in water quality assessment, due to their ease of culture, distinct developmental stages and relative sensitivity to environmental variables. Exposure to environmental contaminants can result in an impact on the early life stages which is detectable using simple bioassay techniques. Vertebrate and invertebrate species were identified as potential test organisms, and were used to bioassay the waters of the Venice Lagoon. Gilthead sea bream (*Sparus aurata*) embryos were obtained from a commercial hatchery and native sea urchins (*Paracentrotus lividus*) were collected from the Adriatic coast of the Lagoon (Lido). The aims of the bioassay component of the programme were to:

(1) Identify an invertebrate species in a gravid condition for subsequent application to a larval bioassay;
(2) Assess the application of hatchery-reared finfish for bioassays;
(3) Conduct short-term static toxicity tests on sediment elutriates from S. Elena and S. Margherita in the City of Venice; and
(4) Conduct a preliminary toxicity assessment of the sea-surface (micro-layer) waters of the Venice Lagoon.

MATERIALS AND METHODS

Sediment elutriates

Surface sediments were taken by a Van Veen grab from two stations (Figure 19.1). The first was located at S. Elena and the second, at Carmini in the rio of S. Margherita (stations 12 and 22, respectively, according to Van Vleet *et al.*, 1987). The sediments were carefully drained and preserved air-free in nylon bags. Three to four hours after sampling, 1 kg of sediment was transferred to a 10 l pyrex bottle containing 8 l of seawater taken from the Gulf of Venice that was filtered and eluted through two columns packed with amberlite XAD-2 and 8-hydroxyquinoline, to retain organic pollutants and trace metals, respectively.

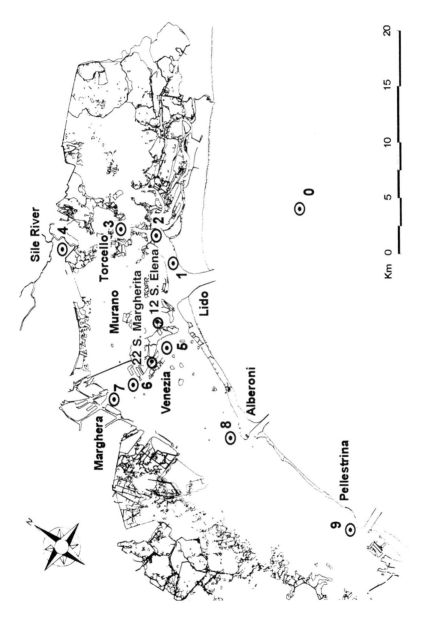

Figure 19.1   Sampling stations in the Venice Lagoon

For oxic elutriate tests, sediment–water mixtures were mixed by vigorously bubbling compressed air through a gas diffusion tube for 30 min. This was then left to settle for 30 min, mixed again for 30 min with air and finally allowed to settle overnight. Six litres of supernatant were filtered through a glass fibre filter (10 μm pore size, 142 mm dia) and used for chemical analyses and bioassays.

Sea-surface microlayer and subsurface water samples

Microlayer samples (2 l) were collected by the Garrett screen method (Garrett, 1965), using a stainless steel screen for organotins and a nylon screen for metals. Further details of sampling protocols and tidal conditions are given in Cleary *et al.*, Chapter 8. A summary of sampling times and tidal conditions is given in Table 19.1. Each station was sampled and assayed twice, once during the ebb tide and once again during the flood tide (Figure 19.1).

Unfiltered water samples were tested using pelagic embryos of *Sparus aurata*. The degree of particulate contamination among samples was highly variable, but due to the static tests and the period of temperature acclimation prior to exposure, all particulates settled to the bottom of the exposure containers. Surface microlayer samples, once sampled and dispensed into test receptacles, do not repartition into another microlayer and subsurface waters (see Cleary *et al.*, Chapter 8).

Table 19.1   Sample collection and tidal conditions. Ebb tide (E), Flood tide (F)

Station	Date	Time	Sample	Tide	Salinity (PSU)
Lio Grande	10/5/93	1600	microlayer	F	35.2
	"	1640	subsurface	F	34.8
	11/5/93	1030	microlayer	E	33.5
	"	1100	subsurface	E	33.5
Punta Salute	12/5/93	1000	microlayer	E	31.6
	"	1030	subsurface	E	31.5
	"	1715	microlayer	F	33.5
	"	1745	subsurface	F	33.3
CVE	12/5/93	0900	microlayer	E	27.4
	"	0930	subsurface	E	27.6
	"	1600	microlayer	F	30.3
	"	1630	subsurface	F	30.6
Platform	11/5/93	1430	microlayer	*	36.3
	"	1500	subsurface	*	36.3

*Offshore station with minimal tidal influence

BIOASSAY

Sea urchin (*Paracentrotus lividus*)

Adult sea urchins were collected by divers from the port entrance of the Lido (Figure 19.1), on the Adriatic coast of the Venice Lagoon, one week prior to the commencement of the experiment. Adults were acclimated in the laboratory in 150 l glass aquaria at 18 °C in clean aerated seawater and maintained on macroalgae (*Ulva rigida*).

Individual adults were induced to spawn, as and when required, by injecting 1 ml of 0.75 M KCl solution into the coelum, as described by Pagano *et al.* (1986). Five minutes after injection, the quality of released gametes was assessed by microscopic examination. Adults failing to release gametes after 5 min were discarded. Sperm quality was assessed from the relative motility of released gametes. Males with poor quality sperm (low motility) were discarded while selected males were retained. Egg quality was dependant on regularity of shape (spherical) and uniformity of size. Only females releasing large regular oocytes were selected for subsequent matings. Egg size (maximum diameter) was measured using a Wild M5 stereomicroscope fitted with an eye-piece graticule.

Fertilization

Egg density was estimated by eye and then each suspension was adjusted to give approximately equal densities. Single male/female crosses were achieved by adding 1 ml of motile sperm to 100 ml of oocyte suspension. This was repeated with separate individuals to yield three distinct parental crosses (crosses A, B and C). All fertilisations were conducted at 18 °C ± 1 °C under static conditions, using seawater collected from a reference station 5 miles offshore, filtered through 0.45 μm and kept aerated. Fertilisations were done within 20 min of the first signs of gamete release. Fertilised eggs from each cross were held on a 53 μm nytex mesh to enable subsequent rinsing of the embryos.

Toxicity test

Twenty minutes after the addition of sperm to the eggs, the sample was carefully rinsed thrice in clean reference seawater to remove excess sperm and reduce the risk of polyspermy. Gentle rinsing of the eggs removes redundant sperm and any excess coelomic fluids liberated during gamete release, thus reducing the chances of bacterial infection. Thirty minutes after the addition of sperm, the fertilization success was determined as the percentage of eggs with a visible fertilisation membrane. Simultaneously, the total numbers of embryos were measured for each parental cross using a Sedgewick Rafter Cell.

One hour prior to inoculation, all water samples were vigorously aerated for 15 min to obtain maximum oxygen saturation. Six 100 ml aliquots were then withdrawn and dispensed into clean 250 ml glass beakers. All the beakers were arranged at random in a thermostatically controlled room and allowed to equilibrate to the incubation temperature (18°C ± 1°C) prior to the addition of embryos. The embryo density was adjusted to a final density of 50 embryos ml^{-1} and then carefully transferred to the test water via Oxford pipette.

The three separate parental crosses were each tested in triplicate for all samples. One sample was used to assess the development stage after 2 and 3 h, the remaining two samples were then used to assess survival, pluteus growth and developmental stages reached after 48 h. Vials were maintained under static conditions with a loose-fitting top for 48 h at 18°C ± 1°C. Each vial was then added, with 4% buffered formalin to preserve the larvae for subsequent analysis. Developmental stages reached after the 48-h period were recorded, based on observations of the first 200 larvae encountered from a random subsample. Growth of the first 20 echinopluteus larvae encountered were taken by measuring the length of the post-oral arm.

Gilthead seabream (*Sparus aurata*)

One-day-old embryos were obtained from a single male/female parental cross, 18–20 h after fertilization. Five thousand embryos were transported from a commercial hatchery (Maricoltura, Italia) to the laboratory in aerated 10 l temperature-controlled (17–18°C) vessels.

Toxicity test

Preparation of water samples was done in the same way as in the previous case, with each water sample tested in triplicate. Each embryo was then checked visually using a stereodissecting microscope and abnormal embryos were discarded. Fifty morphologically normal embryos were then carefully transferred to the test water via wide-bore pipette (effective density of 500 embryos 1^{-1}). Samples were maintained under static conditions with a loose-fitting top at 18°C ± 1°C, for 48 h. One ml of 4% buffered formalin was then added to each beaker to arrest development. Specimens were then transferred into Baker's formol calcium for subsequent analysis and long-term preservation.

Hatching success was recorded and subsequent larval survival determined. Embryos remaining in the egg capsule at the termination of the exposure period appeared opaque (abnormal) and were classified as dead. Morphological appearance in fixed specimens was used to assess post-hatch survival. Those specimens maintaining a regular straight notochord, an intact primordial marginal fin and no obvious tissue deterioration were deemed to be normal, surviving, larvae.

This procedure was repeated one week later, with all elutriate samples being housed at ambient temperature prior to the onset of the second experiment.

RESULTS

Sediment elutriate toxicity

Paracentrotus lividus

Artificially induced spawning of *P. lividus* collected from the field was successfully achieved in the laboratory. Fertilization success was > 98% for all three parental crosses with rapid progression through all developmental stages. During early embryogenesis of *P. lividus*, the symmetrical cleavages, large blastomere size and distinct morphological stages of development, provided unequivocal sublethal end-points. Egg size was similar to those reported in the literature (82.8 μm ± 5.7 μm; Cellario and George, 1990), with the smaller eggs found in female C (Figure 19.2).

This slight reduction in egg size becomes significant after the growth phase (maternal, rather than paternal, influences being the size determinant for larvae) and becomes emphasised in the significantly smaller echinopluteus larvae observed in the purified water samples (where maximum growth would be expected) for cross C, compared to the other two matings (Figure 19.3).

Embryonic deformities　Developmental stages reached following 2- and 3-h exposure to sediment elutriates were assessed for a single parental cross (B)

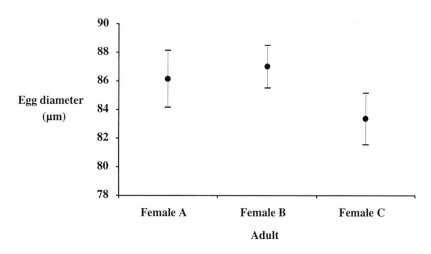

Figure 19.2　Egg size from induced spawnings of each female *P. lividus*. Mean values ± 95% CI limits for each female (A,B,C) are based on 50 eggs

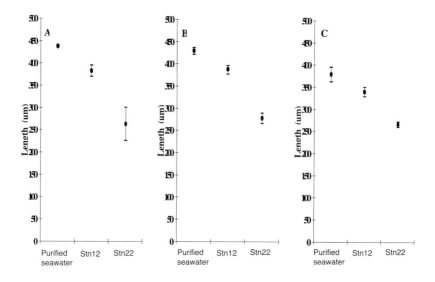

Figure 19.3 Size of *P. lividus* echinopluteus larvae following 48 h exposure to sediment elutriates taken from the canals of Venice. Each value is based on the first 20 embryos encountered in a live subsample. Three parental crosses (A,B,C) are represented. Station 12 – S. Elena, station 22 – S. Margherita

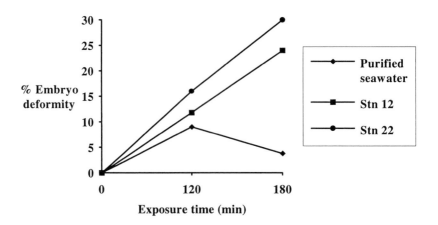

Figure 19.4 Percentage deformity of *P. lividus* embryos following exposure to sediment elutriates taken from the canals of Venice. Parental cross B. Each value is based on the first 200 embryos encountered in a live subsample. Station 12 – S. Elena, station 22 – S. Margherita

and are presented in Figure 19.4. Abnormal embryonic development, consisting predominantly of asymmetrical blastomere formation, gut evagination, and, in older larvae, skeletal deformities, were commonly observed in all water samples.

Developmental rate was assessed for a single cross (B), with each test water being sampled at approximately 2 and 3 h after fertilisation. Cell stages reached were assessed in live samples. Each sample required approximately 4 min to process, hence the time intervals were not precisely 2 and 3 h. However, the general trend in the increase in cell number with time can be followed (Figure 19.5). After 2 h development, 2- and 4-cell stages were dominant in each water sample. A further 1 h exposure led to an increase in cell number, peaking at 8 cell and entering a further division to 16 cell. Only the purified water samples showed a lower deformity rate. Differences in embryo development rates were observed between sites, with slower rates at the inner-city site, station 22. Developmental success was lower for embryos reared in sediment elutriates compared with those reared in purified water.

Figure 19.6 shows the developmental stages reached following 48 h exposure. Significant differences in the number of echinopluteus larvae that reached different developmental stages were found among all three water samples. The highest developmental success was observed in the purified water and the lowest (< 50%), in the elutriates from station 22. The embryos/larvae that did not develop to echinopluteus larvae were predominantly in the late blastula and early gastrula stages of development (similar results are reported by Bressan *et al.*, 1991).

Sparus aurata

Results of the hatching success of the fish embryos showed a trend similar to that for sea urchins (Figure 19.7). In spite of transportation, the embryos received in the laboratory were in excellent condition. The cohorts were homogeneous both in size and developmental stage. Prior to the exposure period, all embryos were individually checked for abnormalities and poor quality material was discarded. The high quality of the spawn allowed this stage of the experiment to pass quickly and careful handling procedures ensured high survival. It is worth noting that rough or excessive handling of the embryos at this stage will introduce stress factors which may bias the results or cause artificially high levels of mortality.

Two separate cohorts of *S. aurata* embryos were exposed to all three elutriate samples. Despite storing the elutriates in the refrigerator for one week (4°C) between consecutive tests, the results from the bioassays were similar. It is likely that some of the more volatile contaminants were lost in the second test (especially after a second period of aeration) but no differences between tests were evident.

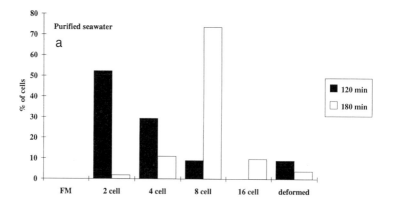

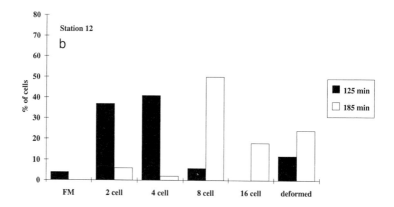

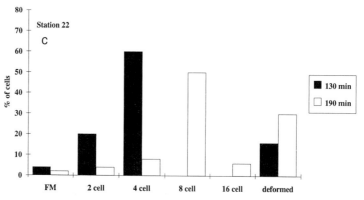

Figure 19.5 Developmental stages of *P. lividus* embryos following exposure to sediment elutriates taken from the canals of Venice. Parental cross B. Each value is based on the first 200 embryos encountered in a live subsample. Station 12 – S. Elena, station 22 – S. Margherita. Note the increase in cell number with time

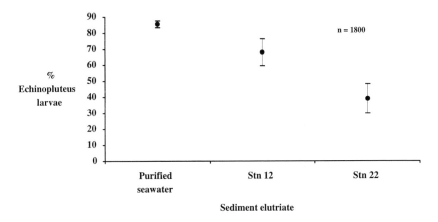

Figure 19.6 Percentage of all echinopluteus larvae of *P. lividus* reached following 48 h exposure to sediment elutriates taken from the canals of Venice. Mean values ± 95% CI limits are based on the 3 parental crosses. 200 random larvae were counted at random for each cross. Station 12 – S. Elena, station 22 – S. Margherita

Sea-surface microlayer

Sparus aurata

Assessment of microlayer toxicity showed a general increase in toxic effects attributable to the microlayer compared to subsurface waters and also increased toxicity during ebb tides (Figure 19.8). Microlayer and subsurface waters were sampled for the selected stations as given in Table 19.1. All samples were tested twice in duplicate, one week apart, using two separate cohorts of larvae.

Salinity was highly variable between the stations sampled, ranging from 27.4 PSU at CVE (upper Lagoon) to 36.3 PSU at the offshore platform station located in the northern Adriatic (see Table 19.1 for further details on salinity).

Hatching success and subsequent survival of larvae were assessed after a 48 h static exposure and the results are presented in Figure 19.8. Hatching success and survival were highest for the offshore platform (Figure 19.8a,b) and lowest for the upper Lagoon station CVE (Canale Vittorio Emanuele) (Figure 19.8g,h). When hatching success and survival were extremely high or low, the variability was low (Figure 19.8a,b,g,h) but when survival was moderate, there was a greater variability, as with the stations Punta Salute and Lio Grande (Figure 19.8c–f).

In general, hatching and survival were better during flood tide conditions than in ebb tides (Figure 19.8). Salinity was lower during ebb than during flood (Table 19.1). Salinity stress may be one factor contributing to toxicity at the stations, perhaps having a synergistic effect with the contaminants found, in spite of *Sparus aurata* being a euryhaline species.

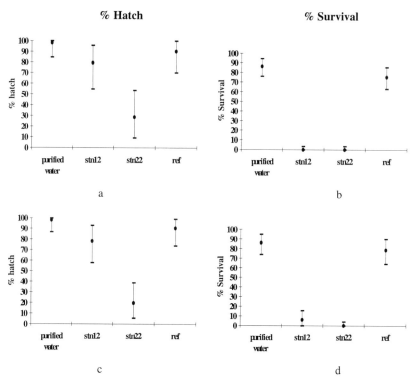

Figure 19.7 Percentage of embryos of *S. aurata* that hatched (a,c) and percentage survival of those larvae that hatched (b,d) following 48 h exposure to sediment elutriates taken from the canals of Venice. Means ± 95% CI limits were derived after arc-sin transformation followed by a back transformation. (Note: A pooled variance was calculated on the transformed scale, df = 8). Charts a and c were experiment 1, and charts b and d were from experiment 2, one week later on the same stock water samples

DISCUSSION

Toxicity of inner canal sediment elutriates

Industrial and agricultural activities around the Venice Lagoon contribute contaminants and waste products (e.g. hydrocarbons, pesticides, fertilisers) to the basin via effluent discharge and runoff. In addition to these, urban sewage with its associated pathogenic and toxic components (e.g. faecal sterols) is discharged without treatment directly into Venice's canals and coastal waters (Fossato *et al.*, 1988). Although the main canals experience partial tidal flushing (exchange with northern Adriatic water), most inner-city canals do not receive significant water exchange. This inadequate flushing extends to the Lagoon as a whole, with contaminated Lagoon waters being only partially replaced by the relatively cleaner off-shore waters, resulting in the accumulation of chemical pollutants in the water column, sediments and marine organisms (Van Vleet *et al.*, 1987).

Several studies have investigated the distribution of contaminants in sediments and organisms (Rodgers *et al.*, 1987, Cofino *et al.*, 1992) and in particular from the Venice Lagoon (petroleum hydrocarbons – Fossato and Siviero, 1974; chlorinated hydrocarbons – Nasci and Fossato, 1982; faecal sterols – Fossato *et al.*, 1988, 1989; metals). Few studies, however, have investigated the biological effects of the sediments taken from the inner-city canals.

The sediment elutriate stations (Figure 19.1) were selected as typical canal sites from the city of Venice (Van Vleet *et al.*, 1987). Station 12, on the relatively clean seaward side of the city, was taken as representative of the deeper (2–4 m) canals around the city. Outer city canals receive a favourable water exchange, due to tidal flushing and strong currents. Station 22 was considered representative of the shallower (1 m) canals of the city interior, having limited water circulation and receiving untreated urban sewage. Seawater taken from the control station at 15.6 km offshore in the Adriatic Sea (Van Vleet *et al.*, 1987) was purified and used for eluting the sediments.

In order to ascertain the toxicity of sediment elutriates, oxic testing is necessary, as oxygen is a requirement of marine organisms used in ecotoxicology testing. This also provides a more realistic assessment of potential toxicity to the test organism as oxic conditions prevail in most sediment dumping sites (Lee and Jones, 1987).

The inner city sites were highly toxic, resulting in reduced hatching success and very poor survival in fish (Figure 19.7). Similar results were obtained with the echinoderm embryos. Deformities in the early developmental stages were highest at the most inner site (station 22) and relatively low for the reference water (Figure 19.4). The ecological consequences of the elutriate toxicity cannot be ascertained from these preliminary results. However, the severe effects on development, hatching and growth are serious enough to warrant further study. The biological consequences of the resuspension of sediments within the inner-city canals resulting from tidal mixing, storm events and dredging activity should be a priority in future studies.

The sea-surface microlayer

Air–sea transfer of contaminants leads to inputs and outputs of contaminants through the surface of sea- or freshwaters. Dehairs *et al.* (1982) calculated that, on the scale of the North Sea, atmospheric fallout of metals such as copper, zinc, lead and cadmium exceeds by one order of magnitude the combined input of the rivers Rhine, Scheldt and Meuse. Similarly, after deposition in sediments, chlorinated hydrocarbons are then released into the water column, then through the surface into the atmosphere and are then transported over wide areas through marine aerosols (Franzen, 1990). The sea-surface microlayer is approximately 50 μm thick and serves as a concentration site for anthropogenic contaminants such as heavy metals, organochlorines, and polyaromatic hydrocarbons. The sea-surface microlayer represents a unique physical and chemical environment, with enriched levels of contaminants compared to the subsurface water mass. Embryonic and larval stages of marine organisms can reside or visit the microlayer and are, therefore, vulnerable to accumulated contaminants. Considering the importance of surface waters for the accumulation and transfer of pollutants, water quality bioassays were conducted to ascertain the biological effects of microlayer samples from several sites (Figure 19.1) in the Venice Lagoon.

The literature indicates that contaminants are concentrated in the microlayer compared with subsurface waters, demonstrating enhanced toxicity, but the evidence for this in the present study was variable (Figure 19.8). Greater differences in toxicity were demonstrated between tidal conditions than between surface and subsurface samples (Figure 19.8). High boating activity and strong seasonal winds causing surface mixing (Langmuir circulation) can disrupt the microlayer and this could be a possible explanation for the lack of enhanced toxicity in the microlayer samples in this study.

Results from the exposure of fish embryos (Figure 19.8) to water samples from around the Lagoon showed distinct differences between sites. Significant differences were observed between stations but toxicity within a station was dependent on the tidal condition. Such dramatic discrepancies in toxicity within a site highlights the need for caution when interpreting data from regional surveys.

High microlayer concentrations of metals, organotin and TBT in the sea-surface microlayer (see Cleary *et al.*, Chapter 8) have been reported previously in nearshore (Hall *et al.*, 1986; Cleary, 1991) and offshore (Hardy and Cleary, 1992) waters. Contaminant enrichment occurred in this study and microlayer concentrations of TBT and copper at ebb tide were higher than water quality criteria established for marine life protection, i.e. the UK Environmental Quality Standard value of 2 ng TBT l^{-1} and the US Environmental Protection Agency chronic water quality limit of 2.9 μg Cu $l^{-1.}$

In order to ascertain the significance of both sediment and microlayer toxicity on the early life stages of marine organisms, future studies should consider the possible pathways of contaminant uptake. The exposure of early life stages to contaminants could be by direct uptake from the water column, sediments and microlayer, or possibly indirectly through the accumulation of contaminants in maternal gonads during oogenesis. Additional information on the occurrence and abundance of embryos and larvae in the water column, and the frequency of contact with the microlayer would aid extrapolation of results from laboratory studies to the field.

CONCLUSIONS

Results obtained for the sediment elutriates taken from the inner-city canals, using early life stages of both sea urchins and larval fish, were similar (Figures

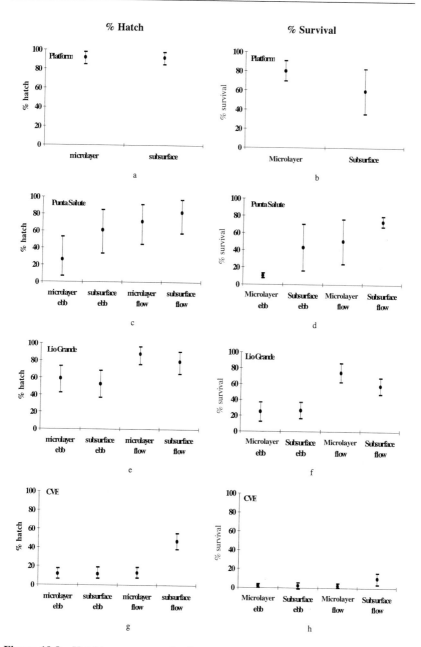

Figure 19.8 Hatching success and subsequent survival of *S. aurata* following 48 h exposure to surface microlayer and subsurface waters, taken on ebb and flood tides from the Venice Lagoon. Platform reference site was at 15 km offshore with no tidal variation. Means ± 95% CI limits were derived after arc-sin transformation followed by a back transformation. (Note: A pooled variance was calculated on the transformed scale, df = 17)

19.6 and 19.7). Elutriates are highly toxic and their biological effect can be easily detected using early life stages of marine invertebrates and vertebrates. Both species utilised in this study show potential as test species for bioassay application and biological monitoring of the Venice Lagoon.

The sensitivity and reproducibility of the test systems (*S. aurata* and *P. lividus*) are sufficiently high to detect differences between stations and tidal conditions. Further studies are warranted to assess the sensitivity of larvae to hydrobiological parameters in conjunction with environmental contaminants. Comprehensive seasonal studies should be carried out in the Lagoon to assess the formation, stability, duration and contamination of the sea-surface microlayer as a reservoir and vehicle for contaminant exposure.

Many biological effects respond at various levels of organisation or complexity to changes in environmental contamination. The choice of biological effect monitoring methods depends on the objectives of the monitoring programme. In many organisms cellular responses to environmental stress may be detected before stress becomes apparent in the physiology of the whole organism. Biochemical/cellular stress responses may provide a rapid and sensitive indicator of environmental alterations and can potentially provide direct evidence of biological deterioration and the probable aetiology of the biological damage. At the other end of the scale of biological complexity, analysis of community structure offers a retrospective picture of responses to general environmental changes. The results from the present programme are encouraging, but they demonstrate the need for a greater level of integration between disciplines to interpret fully the results of environmental toxicology. The use of sublethal biomarkers in conjunction with a range of conventional bioassays, such a those proposed here, can be a powerful tool for biological effect monitoring in the future.

REFERENCES

Bressan, M., Marin, M. G. and Brunetti, R. (1991). Effects of linear alkylbenzene sulphonate (LAS) on skeletal development of sea urchin embryos (*Paracentrotus lividus*). *Water Research.*, **25**: 613–6

Cellario, C. and George, S. (1990). Second generation of *Paracentrotus lividus* reared in the laboratory: Egg quality tested. In De Ridder, Dubois, Lahaye Jangoux (eds). *Echinoderm Research,* Balkema Rotterdam

Cleary, J. J. (1991). Organotin in the marine surface microlayer and subsurface waters of southwest England: relation to toxicity thresholds and the UK environmental quality standard. *Mar. Env. Res.*, **32**: 213–22

Cofino, W., Smedes, F., De Jong, A. S., Abarnou, A., Boon, J. P., Oostingh, I., Davies, I., Klungsoyr, J., Wilhelmsen, S., Law, R., Whinnett, J. A., Schmidt, D. and Wilson, S. (1992). The chemistry programme. *Mar. Ecol. Prog. Ser.*, **91**: 47–56

Dehairs, F., Dedeurwaerder, H., Dejongie, M., Decadt, G., Gillain, G., Baeyens, W. and Elskens, I. (1982). Boundary conditions for metals in the air sea interface. *ICES.*, C.M.E 33

Fossato, V. U. and Siviero, E. (1974). Oil pollution monitoring in the Lagoon of Venice using the mussel *Mytilus galloprovincialis*. *Mar. Biol.*, **25**: 1–6

Fossato, V. U., Van Vleet, E. S. and Dolci, F. (1988). Chlorinated hydrocarbons in sediments of Venetian canals. *Archo. Oceanogr. Limnol.*, **21**: 151–61

Fossato, V. U., Campesan, G., Craboledda, L. and Stocco, G. (1989). Trends in chlorinated hydrocarbons and heavy metals in organisms from the Gulf of Venice. *Archo. Oceanogr. Limnol.*, **21**: 179–90

Franzen, L. G. (1990). Transport, deposition and distribution of marine aerosols over southern Sweden during dry westerly storms. *Ambio*, **19**: 180–8

Garrett, W. D. (1965). Collection of slick forming metals from the sea surface. *Limnol. Oceanogr.*, **10**: 602–5

Hall, L. W., Lenkevich, M. J., Hall, W. S., Pinkney, A. F. and Bushong, S. J. (1986). Monitoring organotin concentrations in Maryland waters of Chesapeake Bay. In *Organotin Symposium Proceedings, Oceans 86*. Marine Technology Society. Washington, Vol. 4., D.C., 1275–9

Hardy, J. T. and Cleary, J. J. (1992). Surface microlayer contamination and toxicity in the German Bight. *Mar. Ecol. Prog. Ser.*, **91**: 203–10

Lee, G. F. and Jones, R. A. (1987). Water quality significance of contaminants associated with sediments: An overview. In Dickson, K. L., Maki, A.W. and Brungs, W.A. (eds). *Fate and effects of sediment-bound chemicals in aquatic systems*, Pergamon, New York, 1–34

Nasci, C. and Fossato, V. U. (1982). Studies on physiology of mussels and their ability in accumulating hydrocarbons and chlorinated hydrocarbons. *Environ. Technol. Lett.*, **3**: 273–80

Pagano, G., Cipollaro, M., Corsale, G., Esposito, A., Ragucci, E., Giordano, G. and Trieff, N. (1986). The sea urchin: Bioassay for the assessment of damage from environmental contaminants. In Cairns, J. (ed.). *Community Toxicity Testing*. Association for Standard Testing and Materials, Philadelphia, 67–92

Rodgers, J. H., Dickson, K. L., Saleh, F. H. and Staples, C. A. (1987). Bioavailability of sediment-bound chemicals to aquatic organisms – some theory, evidence and research needs. In Dickson, K.L., Maki, A.W. and Brungs, W.A. (eds). *Fate and effects of sediment-bound chemicals in aquatic systems*. Pergamon, New York, 245–66

Stebbing, A. R. D. (1985). Bioassay. In Bayne, B. L., Brown, D. A., Burns, K., Dixon, D. R., Ivanovici, A., Livingstone, D. R., Lowe, D. M., Moore, M. N., Stebbing, A. R. D. and Widdow, J. (eds). *The effects of stress and pollution on marine animals*. Praeger, New York, 133–7

Van Vleet, E. S., Fossato, V. U., Sherwin, M. R., Lovett, H. B. and Dolci, F. (1987). Distribution of coprostanol, petroleum hydrocarbons and chlorinated hydrocarbons in sediments from the canals and coastal waters of Venice, Italy. *Org. Geochem.*, **13**: 757–63

CHAPTER 20

BIOTRANSFORMATION AND ANTIOXIDANT ENZYMES AS POTENTIAL BIOMARKERS OF CONTAMINANT EXPOSURE IN GOBY (ZOSTERISSOR OPHIOCEPHALUS) AND MUSSEL (MYTILUS GALLOPROVINCIALIS) FROM THE VENICE LAGOON

D.R. Livingstone and C. Nasci

SUMMARY

Induction of hepatic cytochrome P4501A (CYP1A) was examined in goby (*Zosterissor ophiocephalus*) as a biomarker for exposure to polynuclear aromatic hydrocarbons (PAHs), PCBs and related compounds. Marked induction was found at Porto Marghera (station 7) compared to other sites (stations 1–3). This is consistent with higher levels of tissue contaminants and indicates a potential for pollutant-mediated biological damage. Hepatic CYP1A is recommended for use in goby as a specific biomarker of organic pollution.

The responses of a CYP1A-like enzyme were also examined in digestive gland of mussel (*Mytilus galloprovincialis*). The results indicate a potential for its use as a biomarker of exposure to PAHs and PCBs. Mussels in the Venice Lagoon (stations 1–3, 5, 6) were more impacted than those in the Adriatic Sea (station 0), and those at CVE (station 6) were possibly more impacted than those at Crevan (station 2).

Few or no differences were seen in goby and mussel antioxidant enzyme (catalase, superoxide dismutase, DT-diaphorase) and mussel glutathione S-transferase activities at different stations, or in cytochrome P450 in cockles (*Cerastoderma edule*) at Lio Grande (station 1) compared to Canale di Torcello (station 3). However, some oxidative stress in male goby at Porto Marghera station was evident.

INTRODUCTION

Lipophilic organic contaminants, such as polynuclear aromatic hydrocarbons (PAHs), polychlorobiphenyls (PCBs) and dioxins, are readily taken up into the tissues of marine organisms, both from water and sediment interfaces, and through the food chain (Livingstone, 1991a; Walker and Livingstone, 1992). The integrated use of biomarkers and chemical contaminant levels has been advocated as an effective means of monitoring the impact of such pollution (McCarthy and Shugart, 1990; Stegeman *et al.*, 1992, Livingstone, 1993). This approach was applied to a study of organic pollution in the Venice Lagoon as part of the UNESCO-MURST Venice Lagoon Ecosystem Project, using the goby (*Zosterissor ophiocephalus*), mussel (*Mytilus galloprovincialis*) and cockle (*Cerastoderma edule*) as sentinel organisms.

Goby

Cytochrome P4501A1 (CYP1A1) of the liver of vertebrates metabolises and detoxifies organic chemicals such as PAHs, PCBs, dioxins and others. It is specifically induced by these compounds and this induction has been used worldwide as a biomarker for impact by organic pollution in fish, birds and mammals (note that it has been recommended that unless the enzyme has been catagorically identified as CYP1A1, i.e. completely sequenced, the more general CYP1A term should be used – Stegeman, 1992). To date some 20 or more field studies have been reported in the literature where this biomarker has been successfully used on fish (Livingstone, 1993). Its application in biomonitoring can, therefore, be considered routine. CYP1A can be measured in terms of its enzyme activity (7-ethoxyresorufin *O*-deethylase (EROD) or benzo[a]pyrene (BaP) metabolism), protein amount (immunochemical detection by Western blotting or ELISA using antibody which is specific for the enzyme) and the amount of messenger RNA (mRNA) responsible for its synthesis (Northern blotting with detection by specific cDNA probe) (note that in the recommended convention, the enzyme is represented as CYP1A and the gene or mRNA as *CYP1A* – Nebert *et al.*, 1991).

Reactive oxygen species (ROS) are continually produced in biological systems as unwanted by-products of normal oxidative metabolism. They include the superoxide anion radical (O_2^-) and hydrogen peroxide (H_2O_2). Production of ROS can be increased by exposure to foreign compounds (xenobiotics), such as pollutants, via a number of mechanisms, including induction of cytochrome P450s, redox cycling and other free radical interactions of certain metals (Fe, Cu, Va, Co, Ni, Cr) and organic xenobiotics (nitroaromatics, quinones) (Lemaire and Livingstone, 1993). ROS are removed by antioxidant enzymes such as superoxide dismutase (SOD) (EC 1.15.1.1 – converts O_2^- to H_2O_2) and catalase (EC 1.11.1.6 – converts H_2O_2 to water and oxygen) which may be induced by conditions of enhanced ROS generation. Failure to detoxify and remove ROS can result in oxidative damage to key molecules, including lipid and DNA, which in turn has been linked to disease processes, including cancer (Goldstein and Witz, 1990; Trush and Kensler, 1991).

A similar scenario of pollutant-mediated pro-oxidant (ROS generation) and antioxidant (defence) events, as has been described for mammals (see above), has been shown in fish and marine invertebrates (Livingstone *et al.*, 1990; Livingstone, 1991a; Winston and Di Giulio, 1991). It is therefore possible that pollution might lead to the elevation of hepatic SOD and catalase activities and that these enzymes also can be used as biomarkers for such a situation (Stegeman *et al.*, 1992; Livingstone, 1993). In contrast to CYP1A, this, therefore, represents an investigative study rather than a routine application of a well-established biomarker. Another antioxidant enzyme investigated which is thought to offer biomonitoring potential was DT-diaphorase (EC 1.6.99.2 – also called quinone oxidoreductase). This enzyme metabolises quinones to hydroquinones, thus preventing redox cycling of the quinones and O_2^- production (Lind *et al.*, 1982; Smith *et al.*, 1985). In mammals it is part of the same gene battery containing CYP1A1 (so-called [*Ah*]-gene battery) and is induced by the same xenobiotics, including PAHs and PCBs (Nebert *et al.*, 1990).

Mussels and cockles

Use of the CYP1A biomarker in fish is based on (i) a fundamental understanding of the properties and regulation of the enzyme (Stegeman, 1989; Goksøyr and Förlin, 1992) and (ii) numerous successful field studies with a wide variety of species (see above). Much less is known of this enzyme in molluscs, and the success of field results using various general measurements of cytochrome P450, or the mixed-function oxygenase (MFO) system (cytochrome P450 is the terminal catalytic component of the pollutant-metabolizing MFO system), has been variable (Livingstone, 1991a,b). However, recent studies indicate the existence of a CYP1A-like enzyme in *Mytilus* sp. (Porte *et al.*, 1995; Wootton *et al.*, 1995) with possible biomonitoring potential (Narbonne *et al.*, 1991). Measurements of a CYP1A-like enzyme were therefore made at the levels of mRNA, protein and putative catalytic activity (BaP metabolism to free polar metabolites and protein adducts) in digestive gland of mussel. In addition, a number of other MFO system, antioxidant enzyme and biotransformation enzyme variables, which in past studies have been observed to be elevated with exposure to organic contaminants, were also measured in digestive gland of various mussels and cockles. These were total cytochrome P450 and '418-peak' (putative denatured P450) contents and NADPH-cytochrome c (P450) reductase activity (component of MFO system) (Moore *et al.*, 1987; Livingstone, 1988, 1991b; Yawetz *et al.*, 1992), SOD, catalase, DT-diaphorase (Livingstone *et al.*, 1990; Porte *et al.*, 1991) and glutathione S-transferase (GST) (Sheehan *et al.*, 1991) activities.

MATERIALS AND METHODS

Some measurements were carried out collaboratively. The laboratories and personnel involved were: (i) D.R. Livingstone, P. Lemaire, A. Matthews, L. Peters and C. Porte (NERC Plymouth Marine Laboratory, Plymouth, UK); (ii) C. Nasci (Istituto di Biologia del Mare CNR, Venice, Italy); (iii) A.N. Wootton and P. Goldfarb (School of Biochemistry, University of Surrey, UK – *CYP1A* mRNA studies); (iv) P.J. Fitzpatrick (Dept. Biochemistry, University College, Cork, Ireland – GST studies); and (v) L. Förlin (Dept. Zoophysiology, University of Göteborg, Sweden – CYP1A protein studies).

Male and female gobies, mussels and cockles were collected at several times of the year (December 1991, March–April 1992, June 1992, May 1993) and during a Training campaign (material collected in October 1992). The collection stations are shown in Figure 20.1, *viz.* goby: stations 1 (Lio Grande), 2 (Isola di Crevan), 3 (Canale di Torcello) and 7 (Porto Marghera); mussel: stations 0 (CNR platform), 1, 2, 3, 5 (Punta Salute) and 6 (Canale V. Emanuele); cockle: stations 1 and 3. Individual goby livers or pooled mussel or cockle digestive glands were used. The numbers of samples taken are given in the section on results. Tissues were immediately dissected out, frozen in liquid nitrogen, and stored at 70°C. Biochemicals were generally obtained from Sigma Co. Ltd., and general chemicals were of Analar grade (Merck, UK) or equivalent. All procedures for preparation of biochemical samples from the frozen tissues were carried out at 4°C.

Cytosolic (9000 g or 100,000 g) and microsomal (100,000 g resuspended pellet) fractions for enzymatic analysis were prepared by standard procedures in 0.15 M KCl/1 mM EDTA pH 7.5 (goby) or 10 mM Tris-HCl/0.5 M sucrose/ 0.15 M KCl pH 7.6 (mussel) (20% w/v glycerol was included in microsomal buffer) (Livingstone, 1988; Livingstone *et al.*, 1992a). EROD activity was measured fluorometrically (Burke and Mayer, 1974) in hepatic 9000 g or microsomal fractions of goby. Metabolism of BaP to free metabolites (dihydrodiols, phenols, diones) and putative protein adducts was measured radiometrically, in the presence of NADPH using ^{3}H-BaP, in digestive gland microsomes of mussel (Lemaire *et al.*, 1993). Free polar metabolites were resolved by reverse-phase HPLC and quantified by on-line radiometric counting. Putative BaP-protein adducts were determined by protein precipitation, washing, dissolution and scintillation counting of bound radioactivity. Western blotting for CYP1A was carried out on goby and mussel microsomes as described by Towbin *et al.* (1979) using polyclonal anti-CYP1A from perch (*Perca fluviatilis*) (alkaline phosphatase visualization, quantified by image analysis). Total cytochrome P450 and '418-peak' contents and NADPH-cytochrome c reductase activity were determined spectrophotometrically on digestive gland microsomes of mussel and cockle as described in Livingstone (1988). Antioxidant enzymes were measured spectrophotometrically, mainly in the 9000 or 100,000 g cytosolic fractions, as described in Livingstone *et al.* (1992 a, b) (SOD, catalase), Livingstone *et al.* (1990) (putative DT-diaphorase in mussel and cockle, *viz.* dicumarol-inhibitable NAD(P)H-dependent dichlorophenolindophenol (DCPIP) reductase activity) and Nims *et al.* (1984) (putative DT-diaphorase in goby, *viz.* NADH-dependent resorufin reductase activity). GST activity was measured spectrophotometrically (substrate: 1-chloro-2,4-dinitrobenzene) in mussel digestive gland cytosol as described in Fitzpatrick *et al.* (1995). The presence of different isoenzymes of GST in mussel digestive gland was also examined by HPLC and other techniques (see Fitzpatrick *et al.*, 1995).

Analysis of *CYP1A1*-like mRNA in mussel digestive gland was carried out by RNAzol B extraction of RNA and Northern blotting using ^{32}P-radiolabelled

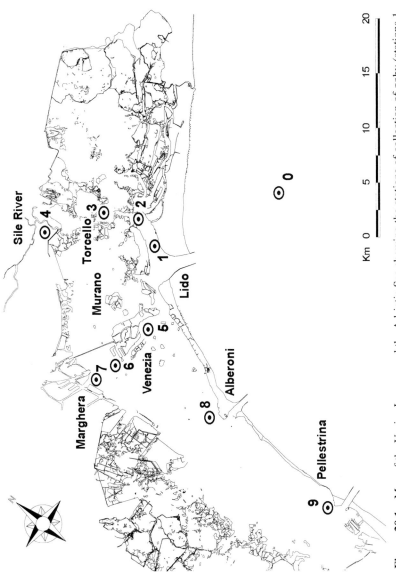

Figure 20.1 Map of the Venice Lagoon and the Adriatic Sea showing the stations of collection of goby (stations 1, 2, 3 and 7), mussel (stations 0, 1, 2, 3, 5 and 6) and cockle (stations 1 and 3) for biochemical biomarker studies

cDNA probe for *CYP1A1* (pfP$_1$450-3' from rainbow trout – *Oncorhynchus mykiss*) as described in Wootton *et al.* (1995).

In addition to the sampling programme, experimental induction of goby hepatic CYP1A with a classical P4501A1-inducer (β-naphthoflavone; single i.p. injection, 20 mg per kg body weight, 30 h) was also carried out to investigate the responsiveness of the enzyme.

Values are expressed as mean ± SE and were compared by one-way analysis of variance ($p < 0.05$).

RESULTS

Detailed results are presented for the major findings only.

Goby

The results for hepatic EROD activity in goby are given in Table 20.1. The activity varied with season and, to a lesser extent, with sex. Despite seasonal variation, the activity in both sexes was consistently higher (3 to 57 times) at Porto Marghera (station 7) than at Isola di Crevan (station 2) and, on most occasions, was also higher (up to 71 times) at Porto Marghera (station 7) than at Lio Grande (station 1). Activity was generally higher (2 to 13 times) at Lio Grande (station 1) than at Isola di Crevan (station 2). Occasionally, hepatic EROD activity in goby was also measured at Canale di Torcello (station 3); the activity measured was lower than at Porto Marghera but higher than at Isola di Crevan (data not shown). A similar pattern of differences was also

Table 20.1 Hepatic 7-ethoxyresorufin *O*-deethylase (EROD) activities (pmol per min per mg protein) in goby (*Z. ophiocephalus*) from Venice Lagoon stations at different times of the year (male – top number; female – bottom number) (station number given in parenthesis)

Campaign	Date	Lio Grande (1)	Isola di Crevan (2)	Porto Marghera (7)
2nd	April 1992	14.1 ± 3.4#	3.0 ± 0.7^x	39.6 ± 0.2*
		13.5 ± 2.8*	3.7 ± 0.5^x	20.3 ± 3.6*
3rd	June 1992	5.2 ± 1.1^x	4.6 ± 0.6^x	142 ± 15*
		2.3 ± 0.3^x	2.9 ± 0.7^x	164 ± 46*
Training	October 1992	13.8 ± 4.8*	2.9 ± 0.8*	17.8 ± 7.2*
		22.1 ± 7.0*	4.6 ± 1.1*	11.2 ± 74.2*
4th	May 1993	3.7 ± 0.6^x	N.D.	15.8 ± 4.1*
		N.D.	N.D.	6.9 ± 2.4

Measured in 9000 *g* supernatant or microsomal fractions. Values are means ± SE ($n = 3$ to 6). N.D. not determined; data in same row sharing the same symbol (*, ^{x}, #) do not differ significantly ($p > 0.05$)

seen for selected measurements of CYP1A protein, *viz.* in arbitrary units: 27 6 ± 7.1 (Porto Marghera), 6.4 ± 2.2 (Lio Grande) and 0.1 ± 0.1 (Isola di Crevan) ($p < 0.05$ between all stations; $n = 3$).

The inducibility of CYP1A in the liver of both sexes of goby was confirmed experimentally by exposure to β-naphthoflavone, a classic inducer of the enzyme in vertebrates (Table 20.2). Exposure produced marked increases in both EROD activity and CYP1A protein. The induced EROD activity was similar to the maximal values observed at Porto Marghera (station 7) (Tables 20.1 and 20.2).

Like EROD, the hepatic antioxidant enzyme activities SOD, catalase and putative DT-diaphorase (NADH-dependent resorufin reductase), also varied with sex and season. SOD activity was either similar between sexes, or was higher for males, and showed little consistent variation between stations 1, 2 3 and 7. Seasonal values in males at the different stations varied from 189 to 499 SOD units per g wet weight. In contrast, higher catalase and putative DT-diaphorase activities were seen on occasion in male, but not female, goby at Porto Marghera (station 7) than at Isola di Crevan (station 2) (Table 20.3).

Table 20.2 Induction of hepatic 7-ethoxyresorufin *O*-deethylase (EROD) activity (pmol per min per mg protein) and CYP1A protein (arbitrary units) in goby (*Z. ophiocephalus*) following exposure to the classical cytochrome P451A1-inducer β-naphthoflavone[a]

	EROD		CYP1A protein[b]
Condition	*Male*	*Female*	*Mixed sex*
Corn oil control	2.2 ± 0.5	0.9 ± 0.3	37.8 ± 7.2
β-naphthoflavone	151 ± 27*	113 ± 46*	81 ± 4.1*

[a]single i.p. injection, 20 mg per kg body weight, 30 h exposure
[b]immunoreactive protein (Western blotting) using polyclonal antibody to perch (*P. fluviatilis*) CYP1A
Values are mean ± SE ($n = 3$ to 8); *$p < 0.05$ comparing exposed to control

Table 20.3 Hepatic catalase (mmol per min per g wet wt) and putative DT-diaphorase (NADH-resorufin reductase – nmol per min per g wet wt) activities in male goby (*Z. ophiocephalus*) from Venice Lagoon stations at different times of the year (station number given in parenthesis)

		Catalase		Resorufin reductase	
Campaign	*Date*	*Crevan (2)*	*Porto Marghera (7)*	*Crevan (2)*	*Porto Marghera (7)*
2nd	April 1992	0.6 ± 0.2	2.5 ± 0.4*	19.5 ± 1.1	46.0 ± 5.5*
3rd	June 1992	5.2 ± 0.4	10.2 ± 1.2*	30.0 ± 3.1	50.7 ± 6.3*
Training	October 1992	6.8 ± 0.6	5.4 ± 0.7	2.3 ± 0.4	2.6 ± 0.1

Mean ± SE ($n = 5$ to 7); *$p < 0.05$ comparing Crevan with Porto Marghera

Differences were less evident for Porto Marghera (station 7) compared to Lio Grande (station 1) and Canale di Torcello (station 3).

Cockles and mussels

Cockles were examined in the 1st campaign (December, 1992) only. No consistent differences were found between the only two Lagoon stations from which they were available, *viz.* Lio Grande (station 1) and Canale di Torcello (station 3). The measurements taken (means for the two stations given in parenthesis) on digestive gland microsomes were total cytochrome P450 (19.9–22.6 pmol per mg protein), '418-peak' (7.1–13 arbitrary units), NADPH-cytochrome c reductase activity (8–13.1 nmol per min per mg protein); and putative NADH-dependent DT-diaphorase (dicumarol-inhibitable DCPIP reductase) (1.4–3.0 nmol per min per mg protein).

Results from mussel digestive gland for MFO system components (total cytochrome P450, '418-peak', NADPH-cytochrome c reductase activity), BaP-protein adduct formation, and GST and antioxidant enzyme (catalase, SOD, putative DT-diaphorase) activities are presented in Table 20.4. These were mainly carried out during campaigns 1 to 3. All measurements varied seasonally, and were either similar between stations, or tended to be lower at Canale V.

Table 20.4 Seasonal ranges of mixed-function oxygenase (MFO) system components and activities, and antioxidant enzyme and glutathione *S*-transferase activities in digestive gland of mussel (*M. galloprovincialis*) from Venice Lagoon and Adriatic Sea stations (station number given in parenthesis)

Enzyme activity or content	Stations	
	CNR Platform, Lio Grande Isola di Crevan, Canale di Torcello (0–3)	Punta Salute, Canale V. Emanuele (5,6)
Cytochrome P450[a] '418-peak'	15–64	11–48
(denatured P450)[b] NADPH-cytochrome	3–32	3–33
c (P450) reductase[c]	11–13	N.D.
Catalase[d]	1.6–4.5	1.8–3.0
Superoxide dismutase[e]	340–653	343–659
Putative DT-diaphorase (NADPH-DCPIP		
reductase)[c,f]	13–15	12–13
Benzo[a]pyrene microsomal protein adducts[g]	0.5–4.1	1.3–2.7
Glutathione *S*-transferase[h]	1.5–2.1	1.2–1.8

Seasonal ranges of mean values given for each group of sites; measurements mainly from campaigns 1 to 3; N.D. not determined. [a]pmol mg^{-1} protein; arbitrary units; [c]nmol min^{-1} mg^{-1} protein; [d]mmol min^{-1} g^{-1} wet wt; [e]SOD units g^{-1} wet wt; [f]DCPIP – dichlorophenolindophenol; [g]pmol min^{-1} mg^{-1} protein; [h]mmol min^{-1} g^{-1} wet wt

Emanuele (station 6) and Punta Salute (station 5) than at CNR platform, Lio Grande, Isola di Crevan and Canale di Torcello (stations 0–3). Also, no differences were seen in the isoenzyme composition of GST between the selected stations of CNR platform, Isola di Crevan and Canale V. Emanuele (see Fitzpatrick *et al.*, 1995 for details).

In contrast to the biochemical systems of Table 20.4, differences in the amounts of a CYP1A-like enzyme in mussel digestive gland microsomes were evident between stations (Table 20.5). These studies were carried out during campaign 3 and the Training campaign (September/October, 1992). Levels of *CYP1A*-like mRNA were generally higher in the Lagoon stations than at CNR platform in the Adriatic Sea, and were particularly high at Canale V. Emanuele (station 6) and Canale di Torcello (station 3). Consistent with this, levels of CYP1A-like protein and metabolism of BaP to free polar metabolites were also higher in the Lagoon (Canale V. Emanuele and/or Isola di Crevan) (stations 6 and 2) than at CNR Platform (station 0). Seasonal variability was seen in microsomal BaP metabolism to free polar metabolites, which in September 1992 was higher at Canale V. Emanuele (station 6) than at either Isola di Crevan (station 2) or CNR Platform (station 0), viz. respectively, 1.53 ± 0.68, 0.65 ± 0.08 and 0.71 ± 0.52 pmol per min per mg protein.

DISCUSSION

The major purpose of molecular biomarkers such as CYP1A is to demonstrate that the organism has been impacted by a particular type of pollution, in this case that of organic contaminants such as PAHs, PCBs and dioxins (McCarthy and Shugart, 1990; Stegeman *et al.*, 1992). Such specific biomarkers of pollutant exposure or damage should be used in a comprehensive pollution

Table 20.5 Levels of cytochrome *CYP1A*-like mRNA, CYP1A-like protein and microsomal benzo[a]pyrene (BaP) metabolism (free metabolites – diols, diones, phenols) in digestive gland of mussel (*M. galloprovincialis*) from Venice Lagoon and Adriatic Sea stations (station numbers given in parenthesis)

Site	*CYP1A-like mRNA* (arbitrary units)[a]	*CYP1A-like protein* (arbitary units)[b]	*BaP metabolism* (pmol min^{-1} mg^{-1} protein)
CNR platform (0)	10.5 ± 5.6	6.0 ± 1.3	1.25 ± 0.61
Lio Grande (1)	9.5 ± 1.2	N.D.	N.D.
Isola di Crevan (2)	32.7 ± 128	N.D.	4.05 ± 1.26
Canale di Torcello (3)	58.0 ± 3.4	N.D.	N.D.
Punta Salute (5)	20.2 ± 4.1	N.D.	N.D.
Canale V. Emanuele (6)	$39.5 \pm 0.9*$	10.9 ± 3.8	3.02 ± 0.93

Mean $\pm$ SE ($n = 3$ to 5); $*p < 0.05$ compared with CNR Platform. [a]Northern blotting using trout *CYP1A1* cDNA probe; [b]Western blotting using antibody to perch CYP1A; N.D., not determined monitoring programme comprising also general biomarkers (molecular, cellular, physiological) of animal health or fitness, and measurements of chemical contaminant levels in the environment and biota (Livingstone, 1993). Only with a combination of such measurements can a proper assessment of environmental impact be made. The following discussion, therefore, pertains only to the use of specific biomarkers.

Goby CYP1A

Xenobiotics such as certain planar PAHs, PCBs and dioxins induce the synthesis of CYP1A in vertebrate liver via binding to the cytosolic *Ah*-receptor protein (Nebert *et al.*, 1990). The induction of CYP1A can lead not only to detoxication via metabolism of lipophilic xenobiotics to water-soluble, excretable products, but also to carcinogenesis via metabolism (activation) of PAHs, PCBs, dioxins, etc. to DNA-adducts (Livingstone, 1991a). The *Ah*-receptor has been shown to be present in the liver of several fish species (Hahn *et al.*, 1992) and the induction of CYP1A by PAHs and PCBs, metabolism of PAHs to mutagenic species, and formation of DNA-adducts have been implicated in the etiology of chemically-caused carcinogenesis in fish (Stein *et al.*, 1990; Stegeman and Lech, 1991). Hepatic neoplasms and other lesions in fish have been correlated with both experimental (Fabacher *et al.*, 1991; Schiewe *et al.*, 1991) and field (Myers *et al.*, 1991) exposure to organic pollutants, including PAHs and PCBs. Induction of CYP1A need not necessarily lead to cancer, because, for example, mutagenic metabolites can be detoxified by other pathways, but it nevertheless represents a cause for concern at the environmental level (Livingstone, 1993; Livingstone *et al.*, 1994).

Correlations between CYP1A induction and uptake of PAHs and PCBs in fish liver have been observed in numerous field pollution studies (Livingstone, 1993). Whereas metabolism and elimination of PCBs is slow and tissue bioaccumulation therefore occurs, that of PAHs is fast and tissue levels of the contaminants remain low (Walker and Livingstone, 1992). However, exposure to and uptake of PAHs is evident from levels of PAHs in the environment (sediment, water column) and other biota (mussels, etc.), the presence of metabolites in bile, and the formation of DNA-adducts (Walker and Livingstone, 1992; Livingstone, 1993). In a few cases of very high pollution, induction of CYP1A occurs but the enzyme is inactivated (possibly due to contaminants such as certain PCB congeners, hepatotoxins, cadmium and other metals), resulting in elevation of CYP1A protein but not EROD activity (Livingstone, 1993). Activity of CYP1A has also been observed to vary seasonally, particularly around spawning time, and to a lesser extent with other environmental variables such as temperature (Livingstone, 1993).

The results for goby, including the seasonal variation, are consistent with the observations for many other pollution studies with fish, and indicate that impact by organic contaminants has occurred at Porto Marghera (station 7), resulting in the induction of CYP1A with a potential for activation of contaminants to carcinogenic metabolites. Although only a limited characterisation of the CYP1A enzyme in goby liver has been carried out, the experimental (β-naphthoflavone) exposure study indicates that the induction at Porto Marghera (station 7) is considerable. In contrast, EROD activities at Isola di Crevan (station 2) were low throughout the year, indicating that impact at this site is minimal. Some impact by organic pollution is evident at Lio Grande (station 1), and possibly also at Canale di Torcello (station 3) (the database for this site is limited to campaign 3), but in both cases less than for Porto Marghera (station 7).

Goby antioxidant enzymes

Considerable evidence exists for the occurrence in marine fish and invertebrates of contaminant-stimulated oxyradical generation and oxidative damage (Livingstone *et al.*, 1990; Winston and Di Giulio, 1991; Lemaire *et al.*, 1993), although the extent to which this may contribute to pollution-caused carcinogenesis and other diseases is unknown (Lemaire and Livingstone, 1993; Livingstone *et al.*, 1994). Although the responses are variable, elevated hepatic SOD and catalase activities have been seen in fish experimentally exposed to pollutants, e.g PAH-containing sediments (Livingstone *et al.*, 1993), and from polluted sites in the field (Livingstone *et al.*, 1992a). Such responses are likely to be much less marked than for CYP1A because normal antioxidant enzyme activities must be high to cope with basal oxyradical production, and less specific because oxyradical production may also be enhanced by other contaminants such as metals, and environmental variables such as oxygen tension (Livingstone *et al.*, 1992a). Seasonal variation in antioxidant enzyme activities also occurs (Viarengo *et al.*, 1991). However, despite these problems, the biomarker potential of pro-oxidant and antioxidant responses to pollution has been recognised (Stegeman *et al.*, 1992).

The results for hepatic antioxidant enzyme activities in goby are similar to those of previous studies on marine organisms, showing seasonal variation and some, but not marked, differences in catalase and putative DT-diaphorase activities in only males, between stations. Increases in hepatic catalase, but not SOD, activities were also seen in channel catfish (*Ictalarus punctatus*) exposed to bleached kraft mill effluent (Maher-Mihaich and Di Giulio, 1991). Interpretation is complicated by the fact that responses of these enzymes can be transient (Livingstone *et al.*, 1993) and also that the function of DT-diaphorase in fish (detoxication or toxication) is as yet unknown (Hasspieler and Di Giulio, 1992). However, most significantly, the differences in antioxidant enzyme activities were greatest between Isola di Crevan (station 2) and Porto Marghera (station 7), and therefore consistent with the results for CYP1A, possibly indicating pollutant-caused oxidative stress at the latter site.

Mussel and cockle studies

Hepatic EROD activity is catalysed solely by CYP1A in fish (Stegeman, 1989; Goksøyr and Förlin, 1992) and mammals, but is either not detectable or only present at low activity in invertebrates (Livingstone 1991a, 1994). BaP hydroxylase activity is mainly catalysed by CYP1A, plus some other CYP isoenzymes, in vertebrates (Åstrom and DePierre, 1986; Stegeman, 1989), and is widely detectable in marine invertebrates (Livingstone, 1991a). Apparent induction of the MFO system has been evident in some marine invertebrate species, but responses in most have been absent or low, and certainly much lower than for vertebrates; for example, only a three-fold increase in BaP hydroxylase in the starfish *Asterias rubens* (Den Besten *et al.*, 1993), compared to up to a several hundred-fold increase for hepatic EROD in fish. In field studies with mussels, some success has been achieved with measurements of digestive gland GST, antioxidant enzymes, and the MFO system (BaP hydroxylase activity, '418-peak' content, NADPH-cytochrome c reductase activity), but as yet no single parameter has emerged as a widely used biomarker for organic pollution in molluscs, and a multi-parameter approach has been advocated (Livingstone, 1991b). More recently, however, evidence has been obtained for the existence of a CYP1A-like enzyme in digestive gland of mussel using antibody and cDNA probes to, respectively, the CYP1A enzyme (Porte *et al.*, 1995) and *CYP1A1* mRNA of fish species (Wootton *et al.*, 1995). A chemical aetiology has also been shown in tumour formation in molluscs (Gardner *et al.*, 1992).

Interpretation of the results with mussel and cockle is very much restricted by the above considerations, in particular by the limited understanding of gene regulation of the enzymes, and the paucity of data from other field studies. The marked seasonal differences in the various biochemical measurements are consistent with previous studies on the MFO system and antioxidant enzymes in mussel (Viarengo *et al.*, 1991; Kirchin *et al.*, 1992). Against a background of the chemical contaminant results (Fossato *et al.* Chapter 6), the lack of differences between most of the enzyme measurements and stations would indicate either an absence of an acute impact by pollution, or a long chronic history of it, resulting in some sort of adaptation. A seven-fold increase in '418-peak' (putative denatured cyctochrome P450) was observed in *C. edule* exposed to an oil spill (Moore *et al.*, 1987) which would suggest that acute impact is not responsible for a difference between the two cockle stations of Lio Grande (station 1) and Canale di Torcello (station 3).

The most significant results with mussel were obtained for the CYP1A-like enzyme (protein and mRNA levels; BaP metabolism to free polar metabolites) which shows greater pollutant impact at the Venice Lagoon stations than at the Adriatic Sea station. Higher microsomal BaP metabolism activity was also seen at Canale V. Emanuele (station 6) than at Isola di Crevan (station 2), but the data were limited to one sampling out of two. The information is, therefore, too limited at present to make any definitive conclusions, particularly given the complication of seasonal variability of BaP metabolism (Livingstone, 1985). The results, thus, emphasize the need for further study of the biomarker potential of the CYP1A-like enzyme and its application to investigate the differences between Canale V. Emanuele (station 6) and other Lagoon sites.

CONCLUSIONS

(1) Elevation of cytochrome P4501A (CYP1A) in liver of gobies indicates that they are most impacted by organic pollution (most likely PAHs, PCBs and related compounds) at Porto Marghera (station 7), resulting in an increased potential for the production of mutagenic metabolites, DNA damage and diseases such as cancer. Impact by organic pollution is also evident at Lio Grande (station 1), but less so than at Porto Marghera (station 7).

(2) The results for hepatic CYP1A in goby, and the correlations with tissue organic contaminant levels (Fossato *et al.* Chapter 6), are consistent with its use as a specific biomarker for organic pollution in fish in many other field studies around the world. They indicate a cause for concern at the state of health of goby at Porto Marghera (Livingstone 1993; Livingstone *et al.*, 1994).

(3) Induction of hepatic CYP1A in goby, measured at both the enzyme activity (EROD) and enzyme protein (Western blotting or ELISA methods) levels should be used as a biomarker for exposure to PAHs, PCBs and related compounds for current and future assessment of the state of health of the Venice Lagoon ecosystem.

(4) Studies on a CYP1A-like enzyme in mussels indicate greater impact by organic pollution at most of the Venice Lagoon stations than at the CNR Platform site. Impact is possibly greater at Canale V. Emanuele (station 6) than at Isola di Crevan (station 2) but more analyses are required to confirm this. These results are in general agreement with those for cellular biomarkers (see Lowe and Da Ros, Chapter 21). Further study on the biomarker potential of a CYP1A-like enzyme in mussel is recommended.

(5) No differences in pollutant impact were evident in cockles from Lio Grande (station 1) and Canale di Torcello (station 3).

(6) Antioxidant enzyme activities in goby liver showed some agreement with the results for CYP1A, indicating some oxidative stress in male goby at Porto Marghera (station 7) but more work is required on a wider range of pro-oxidant (e.g. lipid peroxidation, DNA damage) and anti-oxidant (e.g. enzymes, scavengers, stress proteins) measurements before they could be considered for use as biomarkers of pollution exposure (Stegeman *et al.*, 1992; Livingstone *et al.*, 1994).

ACKNOWLEDGEMENTS

This work also formed part of Laboratory Project 5 of the Plymouth Marine Laboratory, a component institute of the UK Natural Environment Research Council (NERC) and was partly carried out during the tenure of a NERC CASE PhD studentship awarded to A.N. Wootton and an EERO post-doctoral fellowship awarded to P. Lemaire.

REFERENCES

Åstrom, A. and DePierre, J. W. (1986). Rat-liver microsomal cytochrome P-450: purification, characterization, multiplicity and induction. *Biochim. Biophys. Acta*, **853**: 1–27

Burke, M. D. and Mayer, R. T. (1974). Ethoxyresorufin: direct fluorimetric assay of a microsomal *O*-dealkylation which is preferentially inducible by 3-methylcholanthrene. *Drug Metab. Disp.*, **2**: 583–8

Den Besten, P. J., Lemaire, P., Livingstone, D. R., Woodin, B., Stegeman, J. J., Herwig, H. J. and Seinen, W. (1993). Time-course and dose-response of the apparent induction of the cytochrome P450 monooxygenase system of pyloric caeca microsomes of the female sea star *Asterias rubens* L. by benzo[a]pyrene and polychlorinated biphenyls. *Aquat. Toxicol.*, **26**: 23–40

Fabacher, D. L., Besser, J. M., Schmitt, C. J., Harshbarger, J. C., Peterman, P. H. and Lebo, J. A. (1991). Contaminated sediments from tributaries of the Great Lakes: chemical characterization and carcinogenic effects in Medaka (*Oryzias latipes*). *Arch. Environ. Contam. Toxicol.*, **20**: 17–34

Fitzpatrick, P. J., Sheehan, D. and Livingstone, D. R. (1995). Studies on isoenzymes of glutathione S-transferase in digestive gland of *Mytilus galloprovincialis* with exposure to pollution. *Marine Environ. Res.*, **39**: 241–4

Gardner, G. R., Pruell, R. J. and Malcolm, A. R. (1992). Chemical induction of tumors in oysters by a mixture of aromatic and chlorinated hydrocarbons, amines and metals. *Marine Environ. Res.*, **34**: 59–63

Goksøyr, A. and Förlin, L. (1992). The cytochrome P-450 system in fish, aquatic toxicology and environmental monitoring. *Aquat. Toxicol.*, **22**: 287–311

Goldstein, B. D. and Witz, G. (1990). Free radicals and carcinogenesis. *Free Rad. Res. Commun.*, **11**: 3–10

Hahn, M. E., Poland, A., Glover, E. and Stegeman, J. J. (1992). The Ah receptor in marine animals: phylogenetic distribution and relationship to cytochrome P4501A inducibility. *Marine Environ. Res.*, **34**: 87–92

Hasspieler, B. M. and Di Giulio, R. T. (1992). DT diaphorase [NAD(P)H: (quinone acceptor)oxidoreductase] facilitates redox cycling of menadione in channel catfish (*Ictalurus punctatus*) cytosol. *Toxicol. Appl. Pharmacol.*, **114**: 156–61

Kirchin, M. A., Wiseman, A. and Livingstone, D. R. (1992). Seasonal and sex variation in the mixed-function oxygenase system of digestive gland microsomes of the common mussel, *Mytilus edulis* L. *Comp. Biochem. Physiol.*, **101C**: 81–91

Lemaire, P., Den Besten, P. J., O'Hara, S. C. M. and Livingstone, D. R. (1993). Comparative metabolism of benzo[a]pyrene by microsomes of hepatopancreas of the shore crab *Carcinus maenas* L. and digestive gland of the common mussel *Mytilus edulis* L. *Polycy. Aromat. Comp.*, **3** (Suppl.): 1133–40

Lemaire, P. and Livingstone, D. R. (1993). Pro-oxidant/antioxidant processes and organic xenobiotic interactions in marine organisms, in particular the flounder *Platichthys flesus* and the mussel *Mytilus edulis*. In *Trends in Comparative Biochemistry and Physiology*. Research Trends, Council of Scientific Research Integration, Trivandrum, India, Vol. 1, 1119–49

Lind, C., Hochstien, P. and Ernster, L. (1982). DT-diaphorase as a quinone reductase: a cellular control device against semiquinone and superoxide radical formation. *Arch. Biochem. Biophys.*, **216**: 175–85

Livingstone, D. R. (1985). Responses of the detoxification/toxification enzyme systems of molluscs to organic pollutants and xenobiotics. *Mar. Pollut. Bull.*, **16**: 158–64

Livingstone, D. R. (1988). Responses of microsomal NADPH-cyctochrome c reductase activity and cytochrome P-450 in digestive glands of *Mytilus edulis* and *Littorina littorea* to environmental and experimental exposure to pollutants. *Mar. Ecol. Prog. Ser.*, **46**: 37–43

Livingstone, D. R., Garcia Martinez, P., Michel, X., Narbonne, J. F., O'Hara, S. C. M., Ribera, D. and Winston, G. W. (1990). Oxyradical production as a pollution-mediated mechanism of toxicity in the common mussel, *Mytilus edulis* L., and other molluscs. *Funct. Ecol.*, **4**: 415–24

Livingstone, D. R. (1991a). Organic xenobiotic metabolism in marine invertebrates. *Adv. Comp. Environ. Physiol.*, **7**: 45–185

Livingstone, D. R. (1991b). Towards a specific index of impact by organic pollution for marine invertebrates. *Comp. Biochem. Physiol.*, **100C**: 151–5

Livingstone, D. R., Archibald, S., Chipman, J. K. and Marsh, J. W. (1992a). Antioxidant enzymes in liver of dab *Limanda limanda* from the North Sea. *Mar. Ecol. Prog. Ser.*, **91**: 97–104

Livingstone, D. R., Lips, F., Garcia Martinez, P. and Pipe, R. K. (1992b). Antioxidant enzymes in the digestive gland of the common mussel *Mytilus edulis*. *Mar. Biol.*, **112**: 265–76

Livingstone, D. R. (1993). Biotechnology and pollution monitoring: use of molecular biomarkers in the aquatic environment. *J. Chem. Technol. Biotechnol.*, **57**: 195–211

Livingstone, D. R. (1994). Recent developments in marine invertebrate organic xenobiotic metabolism. *Toxicol. Ecotoxicol. News*, **1**: 88–94

Livingstone, D. R., Förlin, F. and George, S. (1994). Molecular biomarkers and toxic consequences of impact by organic pollution in aquatic organisms. In Sutcliffe, D.W. (ed.) *Water Quality and Stress Indicators in Marine and Freshwater Systems: Linking levels of organisation*. Freshwater Biological Association, Ambleside, UK. pp. 154–71

Mather-Mihaich, E. and Di Giulio, R. T. (1991). Oxidant, mixed-function oxidase and peroxisomal responses in channel catfish exposed to a bleached kraft mill effluent. *Arch. Environ. Contam. Toxicol.*, **20**: 391–7

McCarthy, J. F. and Shugart, L. R. (1990). *Biomarkers of Environmental Contamination*. Lewis Publishers, Boca Raton, Florida, pp. 457

Myers, M. S., Landhal, J. T., Krahn, M. M. and McCain, B. B. (1991). Relationships between hepatic neoplasms and related lesions and exposure to toxic chemicals in marine fish from the US west coast. *Environ. Health Perspect.*, **90**: 7–15

Narbonne, J. F., Garrigues, P., Ribera, D., Raoux, C., Mathieu, A., Lemaire, P., Salaun, J. P. and Lafaurie, M. (1991). Mixed-function oxygenase enzymes as tools for pollution monitoring: field studies on the French coast of the Mediterranean Sea. *Comp. Biochem. Physiol.*, **100C**: 37–42

Nebert, D. W., Nelson, D. R., Coon, M. J., Estabrook, R. W., Feyereisen, R., Fujii-Kuriyama, Y., Gonzalez, F. J., Guengerich, F. P., Gunsalus, I. C., Johnson, E. F., Loper, J. C., Sato, R., Waterman, M. R. and Waxman, D. J. (1991). The P450 super-family: update of new sequences, gene mapping, and recommended nomenclature. *DNA Cell Biol.*, **10**: 1–14

Nebert, D. W., Petersen, D. D. and Fornace, A. J. Jr. (1990). Cellular responses to oxidative stress: the [*Ah*] gene battery as a paradigm. *Environ. Health Perspect.*, **88**: 13–25

Nims, R. W., Prough, R. A. and Lubet, R. A. (1984). Cytosol-mediated reduction of resorufin: a method for measuring quinone oxidoreductase. *Arch. Biochem. Biophys.*, **229**: 459–65

Pipe, R. K. (1987). Molecular, cellular and physiological effects of oil derived hydrocarbons and molluscs and their use in impact assessment. *Phil. Trans. R. Soc. Lond. B.*, **65**: 3–20

Porte, C., Lemaire, P., Peters, L. D. and Livingstone, D. R. (1995). Partial purification and properties of cytochrome P450 from digestive gland microsomes of the common mussel, *Mytilus edulis* L. *Marine Environ. Res.*, **39**: 27–31

Porte, C., Solé, M., Albaigés, J. and Livingstone, D. R. (1991). Responses of mixed-function oxygenase and antioxidase enzyme system of *Mytilus* sp. to organic pollution. *Comp. Biochem. Physiol.*, **100C**: 183–6

Schiewe, M. H., Weber, D. D., Myers, M. S., Jacques, F. J., Reichert, W. L., Krone, C. A., Malins, D. C., McCain, B. B., Chan, S. L. and Varanasi, U. (1991). Induction of foci of cellular alteration and other hepatic lesions in English sole (*Parophrys vetulus*) exposed to an extract of an urban marine sediment. *Can J. Fish. Aquat. Sci.*, **48**: 1750–60

Sheehan, D., Crimmins, K. M. and Burnell, G. M. (1991). Evidence for glutathione-S-transferase activity in *Mytilus edulis* as an index of chemical pollution in marine estuaries. In *Bioindicators & Environmental Management*. Academic Press, New York, 419–25

Smith, M. T., Evans, G. G., Thor, H. and Orrenius, S. (1985) Quinone-induced oxidative injury to cells and tissues. In Sies, H. (ed.). *Oxidative Stress*, Academic Press, New York, 91–113

Stegeman, J .J. (1989). Cytochrome P450 forms in fish: catalytic, immunological and sequence similarities. *Xenobiotica*, **19**: 1093–110

Stegeman, J. J. (1992). Nomenclature for hydrocarbon-inducible cytochrome P450 in fish. *Marine Environ. Res.*, **34**: 133–8

Stegeman, J. J., Brouwer, M., Di Giulio, R. T., Förlin, L., Fowler, B., Sanders, B. M. and Van Veld, P. A. (1992). Molecular responses to environmental contamination: enzyme and protein synthesis as indicators of chemical exposure and effect. In Huggett, R. J., Kimerle, R. A., Mehrle, P. M. Jr. and Bergman, H. L. (eds) *Biomarkers. Biochemical, Physiological, and Histological Markers of Anthropogenic Stress.* Lewis Publishers, Boca Raton, Florida, 235–335

Stegeman, J. J. and Lech, J. J. (1991). Cytochrome P-450 monooxygenase systems in aquatic species: carcinogen metabolism and biomarkers for carcinogen and pollutant exposure. *Environ. Health Perspect.* **90**: 101–9

Stein, J. E., Reichert, W. L., Nishimoto, M. and Varanasi, U. (1990). Overview of studies on liver carcinogenesis in English sole from Puget Sound: evidence for a xenobiotic chemical etiology II: biochemical studies. *Sci. Total Environ.*, **94**: 51–69

Towbin, H., Staehelin T. and Gordon, J. (1979). Electrophoretic transfer of proteins from polyacrylamide gels to nitrocellulose sheets: procedure and some applications. *Proc. Natl. Acad. Sci. USA*, **76**: 4350–5

Trush, M. A. and Kensler, T. W. (1991). An overview of the relationship between oxidative stress and chemical carcinogenesis. *Free Rad. Biol. Med.*, **10**: 201–9

Viarengo, A., Canesi, L., Pertica, M. and Livingstone, D. R. (1991). Seasonal variations in the antioxidant defense systems and lipid peroxidation of the digestive gland of mussels. *Comp. Biochem. Physiol.*, **100C**: 187–90

Walker, C. H. and Livingstone, D. R. (1992). *Persistent Pollutants in Marine Ecosystems*. Pergamon Press, Oxford, pp. 272

Winston, G. W. and Di Giulio, R. T. (1991). Prooxidant and antioxidant mechanisms in aquatic organisms. *Aquat. Toxicol.*, **19**: 137–61

Wootton, A. N., Herring, C., Spry, J. A., Wiseman, A., Livingstone, D. R. and Goldfarb, P. S. (1995). Evidence for the existence of cytochrome P450 gene families (*CYP1A1, 3A, 4A1, 11A1*) and modulation of gene expression (*CYP1A1*) in the mussel *Mytilus* sp. *Marine Environ. Res.*, **39**: 21–5

Yawetz, A., Manelis, R. and Fishelson, L. (1992). The effects of Arochlor 1254 and petrochemical pollutants on cytochrome P450 from the digestive gland microsomes of four species of Mediterranean molluscs. *Comp. Biochem. Physiol.*, **103C**: 607–14

CHAPTER 21

CELLULAR BIOMARKERS OF CONTAMINANT EXPOSURE AND EFFECT IN MUSSEL (MYTILUS GALLOPROVINCIALIS) AND GOBY (ZOSTERISSOR OPHIOCEPHALUS) FROM THE VENICE LAGOON

D.M. Lowe and L. DaRos

SUMMARY

A range of biomarkers of contaminant exposure and effect were investigated in sections of digestive and reproductive tissues of mussels (*Mytilus galloprovincialis*) and liver tissues of the fish (*Zosterissor ophiocephalis* and *Platichthys flesus*). In addition, an *in vitro* assay of contaminant-induced lyso-somal damage was undertaken on molluscan blood cells.

The results of the *in vitro* assay indicated that mussels from Salute (station 5) and CVE (station 6) were severely impacted, whereas mussels from the Platform (station 0), Alberoni (station 8) and Chioggia (station 9) were only slightly affected.

Consistent with previous studies on environmental impact in fish, levels of lipids were found to be elevated in livers of both fish species from station 7 (Marghera). Levels of lipids in mussel digestive tissues from Lagoon sites were consistently low in comparison with other species of *Mytilus*. Mussels from the Platform (station 0) exhibited higher levels of lipids throughout the study than those at all other sites.

Degeneration of reproductive tissues, quantified by stereological analysis of tissue sections, was found to be very high at all sites throughout the study, giving rise to concern as to the future of mussel production in the Lagoon.

INTRODUCTION

Contaminants in the environment initially impact at the molecular level resulting in higher order tissue changes. Cellular pathology is, therefore, a very powerful tool in that it can investigate impact effects at a very low level of organisation. This provides an opportunity to identify contaminant-induced early onset changes which may ultimately result in organ failure in an individual, before the higher order diseases that may have deleterious effects on the population become apparent.

MATERIALS AND METHODS

For the studies on gamete degeneration and digestive tubule atrophy, samples of *M. galloprovincialis* (size range 30–50 mm) were collected from the populations in the Lagoon as well as at the Platform (station 0) (Figure 21.1). The mantle, which contains the bulk of the reproductive tissues, and the digestive gland were excised from each animal and fixed in Baker's formal calcium (+2.5% NaCl) and stored in a refrigerator. A subsample of each mantle was then removed and dehydrated through an ascending alcohol series, 'cleared' in Histosol and embedded in Paraplast wax. Sections (5 μm) were cut and stained using the Papanicalou technique (Culling, 1963).

For investigations into neutral lipids, as well as total activity of the lysosomal marker enzymes N-acetyl β-D-hexosaminidase (NAH) and β-glucuronidase (β gluc) in mussels, and for high density (HDL) and low density (LDL) lipoproteins in fish, frozen sections of digestive gland and liver, respectively, were prepared as follows. The digestive gland (mussel) or liver (fish) were excised and a small sub-sample frozen in hexane at $-70°C$. Sections (10 μm) were cut in a Bright cryostat at a cabinet temperature of $-26°C$ using dry ice to cool the knife.

Levels of both HDL and LDL were determined immunocytochemically on fresh frozen sections of fish liver using fluorescently-conjugated antibodies and quantified by area thresholding with confocal laser scanning microscopy.

Total activity of β-gluc and NAH was demonstrated using naphthol AS-BI glucoronide and glucosaminide, respectively, and simultaneous coupling with the diazonium salt, Fast Violet LB (Lowe and Clarke, 1989) and measured using microdensitometry on frozen sections of mussel digestive gland. Neutral lipids were demonstrated using the Oil red O reaction (Bancroft, 1967). Contaminant-induced lysosomal damage was shown by exposing molluscan blood cells to the cationic probe, neutral red dissolved in physiological saline, and establishing the time course for the loss of the probe from the lysosomal compartment into the cytosol (Lowe and Pipe, 1994).

RESULTS AND DISCUSSION

Digestive tubule atrophy in mussels

Digestion in mussels occurs in digestive tubule epithelial cells of the hepato-pancreas and is lysosomally mediated. The digestive tubules are arranged in clusters connected to individual ducts that receive food from the stomach at different times throughout the feeding cycle. Because of this structure, indi-vidual tubules are not necessarily all in the same phase of digestive activity at any one time and a section of hepatopancreas shows clusters of tubules in one of four stages, *viz.* resting, digesting, excreting and reconstituting. One of the consequences of exposure to contaminants is that the cycle of cell regener-ation and digestion breaks down, the cells shrink (atrophy) and the digestive

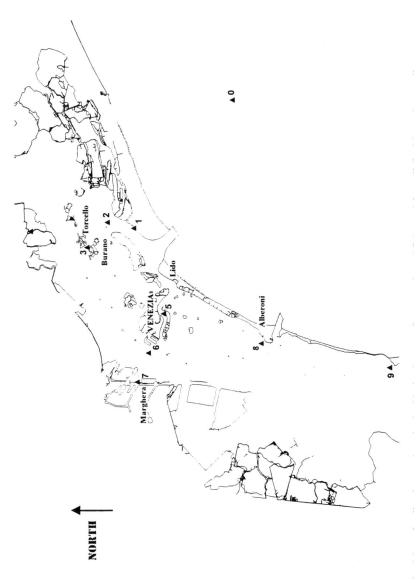

Figure 21.1 Map of collection stations in the Venice Lagoon for mussels and fish used for cellular biomarker studies

epithelium becomes thin and ultimately devoid of lipids (Lowe, 1988; Lowe *et al.*, 1981). Prior to the epithelial thinning, the cells become engorged with neutral lipid which has the effect of disrupting digestive cell membranes (Lowe and Clarke, 1989). Impairment of digestive function would have serious consequences for animal health and reproductive potential and in the long term could influence the balance of the fauna within the Lagoon.

During the course of these investigations, both digestive epithelium atrophy as well as lipid accumulation were observed. Whilst atrophy was apparent in mussels from all sites (with the exception of Alberoni (station 8) and Chioggia (station 9) during the 4th campaign), it was not of a magnitude that might be considered pathological or indeed contaminant-induced. Results obtained during the 1st, 2nd and 4th campaigns indicated that this condition was not present at the Platform site (station 0); however, it was apparent in that population during the 3rd campaign, when lipid levels were also reduced (see below).

Lipidosis in digestive tubules of mussels and hepatocytes of fish

Lipid forms an important storage component for mussels, where it is utilised during gametogenesis and also as an energy source when food is in short supply in the water column. However, a consequence of exposure to low-level contaminants is a breakdown in the capacity to metabolise lipid, resulting in large reserves accumulating in lysosomes in the digestive cells. The digestive cells of unstressed mussels normally contain discrete lipid droplets which are distributed throughout the cytoplasm. By contrast, mussels exposed to contaminants exhibit large amounts of lipid that can rupture the cell membranes resulting in cell death (Lowe, 1988). Alterations in the abundance of lipid in the digestive cells, however, not only represent a sensitive biomarker of impaired metabolic function but also an indicator of accumulated hydrocarbons, in that they are lipophilic. As such, alterations in the lipid levels may reflect on their contaminant burdens and their suitability for human consumption, which is an important consideration for the economics of the Lagoon.

The amount of lipid observed in mussels from all sites throughout these studies was low when compared to that observed in *M. edulis*. Nevertheless, as with *M. edulis*, there was evidence of a strong seasonal influence whereby lipids increased during the early spring (2nd campaign) and late summer months (4th campaign). Platform (station 0) mussels exhibited consistently higher levels of lipids throughout the study than all other sites, which probably reflects a more stable food supply. Results obtained during the 3rd campaign indicated a general reduction in lipids at all sites, with the exception of Torcello (station 3) where they remained unchanged from levels observed during the 2nd campaign. In some respects a reduction in digestive gland lipids at that time of the year (June) is surprising in that one would expect that nutrients would be high in the water column. However, during that period mantle nutrient reserves were very high and the first signs of germ cell development were apparent, which would have placed a heavy demand on lipids.

Some species of fish store lipid in their liver cells (hepatocytes) which remains as discrete membrane-bound pools or droplets. However, as is the case with mussel digestive cells, exposure to contaminants can lead to abnormal lipid accumulation and cell damage (Lowe *et al.*, 1992).

During the 4th campaign, there was a significant increase in the density of lipids in gobies from Marghera (station 7) compared with the gobies from both Alberoni (station 8) ($p < 0.001$) and Lio Grande (station 1) ($p < 0.01$), and in gobies from Lio Grande (station 1) compared with the gobies from Alberoni (station 8) ($p < 0.05$). Similarly, lipids were elevated in flounder (*Platichthys flesus*) from Marghera (station 7), compared to Lio Grande (station 1). These observations are consistent with previous subjective estimates of lipid levels for gobies obtained during the other three campaigns and observations on other fish species exposed to lipophilic contaminants (Lowe *et al.*, 1992), such as those found in the Venice Lagoon.

Mussel digestive cell lysosomal enzyme activity

The two lysosomal marker enzymes examined in these studies were N-acetyl β-D-hexosaminidase (NAH) and β-glucuronidase (β-gluc). NAH is primarily involved in the digestion of microorganisms, whereas β-gluc is involved in the digestion of plant material. The levels of enzyme activity and membrane stability of the lysosomal compartment within a variety of cell types have been demonstrated to alter significantly with diverse stressing factors, including contaminant exposure, anoxia and temperature (Moore, 1976; Lowe, 1988). If digestive function is impaired, then the mussels will neither grow or reproduce optimally, thereby affecting the populations within the Lagoon. Besides, the nature of the particular enzyme activity at a given site reflects the dominant dietary component in the water column for that site, thereby supporting observations on nutrient loading in that body of water.

NAH activity (Figure 21.2) showed a seasonal pattern with very low levels during December (1st campaign), increasing throughout early spring (2nd campaign) to attain a peak in May/June (3rd and 4th campaigns). The results indicate that peak activity occurred in the Platform (station 0) mussels in May and persisted at comparatively high level through June. In contrast, all other populations exhibited peak activity in June. As stated earlier, peak activity for the Platform (station 0) was in May; however, it was significantly lower than at Lio Grande (station 1), Crevan (station 2), and Salute (station 5). During the 1st campaign, when enzyme activity was minimal at Lagoon mussel sites, the Platform (station 0) mussels exhibited the highest level of activity.

The level of β-gluc activity (Figure 21.3) was inversely correlated with NAH in Platform (station 0) mussels throughout the duration of these studies.

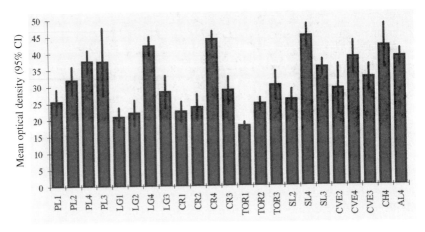

Figure 21.2 Total activity of NAH in mussel digestive cells for all campaigns, plotted in seasonal (not campaign) order

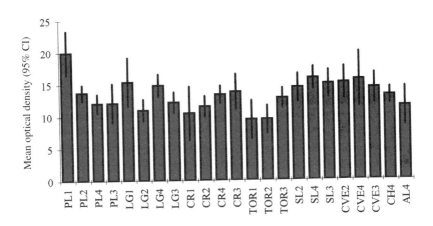

Figure 21.3 Total activity of β-glucuronidase in mussel digestive cells for all campaigns, plotted in seasonal (not campaign) order

The relationship between β-gluc and NAH in Lagoon mussels, however, was less well defined, although both activities were maximal during the spring and early summer months. As compared to NAH, the levels of β-gluc were more stable throughout the study, showing less seasonal variability.

Based on knowledge of the role of these two digestive enzymes in mussels, one possible explanation for the differences in levels of activity may be the nature of available nutrients in the water column. The apparent stability in the levels of β-gluc throughout may reflect a steady source of plant material in the water column in the Lagoon. By contrast, NAH showed a highly significant increase in activity during late spring which may be related to increasing levels of microorganisms in the water column, due possibly to the impact of tourism. The lack of an inverse relationship between the levels of these two enzymes in the Lagoon mussel may also reflect a more mixed diet (Lowe *et al.*, 1999).

Neutral red retention in mussel blood cells

Whilst recognising the contribution made by gross pathology and histopathology, toxicologists and pharmacologists studying the effects of drugs and compounds also use *in vitro* assays. An *in vitro* assay was therefore developed specifically for this programme that could be deployed at any station without sacrificing the mussels, and the sample analysed with the minimum of training and equipment. Such an approach makes it possible to undertake 'spot checks' on mussel health and undertake remedial action, where necessary, which is an important consideration for the management of any body of water.

Contaminant-induced lysosomal membrane damage was determined in mussel blood cells using the retention time of a cationic probe (neutral red) as a marker of membrane dysfunction. The results of the assay (Figure 21.4) for all four campaigns show a clear reduction in lysosomal membrane retention capacity at all of the sites, as compared to the Platform (station 0), and additionally to Alberoni (station 8) and Chioggia (station 9) in the 4th campaign. Retention times for mussels from Salute (station 5) and CVE (station 6) were very low, indicating a high level of cell damage. When the results of this assay were tested against the total body burdens of selected contaminants (Fossato *et al.*, Chapter 25) using multi-stepwise linear regression analysis, it was found that 83% of the biological variability in the cellular response was strongly correlated with the presence of the organochlorine compounds, mercury and cobalt ($p < 0.004$) (Lowe *et al.*, 1995).

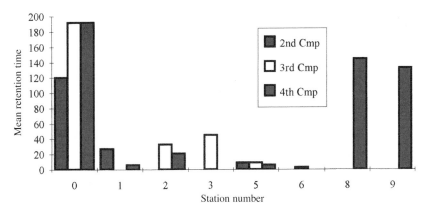

Figure 21.4 Neutral red retention in mussel blood cell lysosomes

Gamete degeneration in mussels

Gamete degeneration is a natural phenomenon which occurs in all populations of mussels and generally becomes apparent after the vitellogenic stage of egg development. The level of atresion (degeneration) in unstressed mussels is about 5% of the total germ cell mass (Lowe, from unpublished results determined stereologically). By contrast, mussels exposed experimentally to contaminants exhibited levels of atresia as high as 30% of the total germ cell mass in 70% of the samples during the latter stages of the reproductive cycle (Lowe and Pipe, 1986, 1987). The natural mussel beds within the Lagoon are a local source of food, in addition to the mussel parks at Chioggia (station 9) and Alberoni (station 8). Any reduction in the reproductive potential of the mussels could, therefore, have serious consequences for the local economy.

There was no difference in the stage of reproductive tissue development between sites during any of the campaigns throughout these studies but the incidence of degenerating gametes (Figure 21.5) was consistently high in mussels. In addition, the amount of degenerating material present was not within what could be considered the norm for *Mytilus edulis*. Unusually, even when the storage reserves were very high and gamete development was in its early stages, there was evidence of gamete atresion in some animals which was not site-specific. It may well be that *M. galloprovincialis* has a different reproductive strategy from that of *M. edulis*, which involves a considerable degree of resorption. For comparative purposes, the incidence of gamete degeneration in the Lagoon mussels in Figure 21.5 was compared with data obtained from studies on mussels from the Oslo Fjord that were exposed to contaminants. The fact that atresion was apparent during the very earliest stages of gamete development suggests that the reason is more likely to be

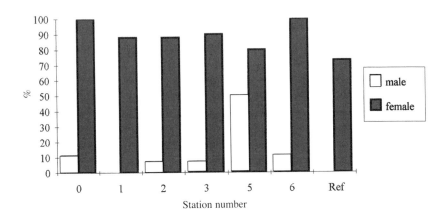

Figure 21.5 Gamete degeneration in mussels

related to some aspect of stress, one possible cause being the levels of copper and zinc in the Lagoon. Maung and Tyler (1982) reported from laboratory studies, using metal levels lower than those at, for example, Crevan (station 3) that copper suppressed vitellogenesis while zinc inhibited oocyte development and caused severe lysis of germinal material.

Low and high density lipoprotein in fish liver tissue

Low density lipoprotein (LDL) is involved in cell repair mechanisms, as a source of cholesterol, and is either produced in the liver or imported via food. Any significant increase in the presence of this protein following exposure to contaminants, in combination with an increase in the LDL scavenging high density lipoprotein (HDL), is taken to be indicative of an imbalance in its regulation, or of a requirement for cell repair following damage by some agent. The failure of cellular repair mechanisms affects both normal cell turnover and repair of contaminant-induced damage and as such is important to the survival of the species.

The results demonstrated considerable seasonal variability in levels of LDL which could not be directly attributed to contaminant exposure. Results obtained after the 2nd and 4th campaigns were tested for significance using t-tests. This indicated high levels in goby livers (also flounder livers during the 4th campaign) from Marghera (station 7). However, the results from the 3rd campaign showed that gobies from Lio Grande (station 1) had significantly more LDL than those from either Crevan (station 2) or Marghera (station 7) ($p < 0.01$ and 0.1, respectively). In addition, fish from Torcello (station 3) had significantly more LDL than those at Crevan (station 2) ($p < 0.1$). HDL levels in fish from Lio Grande (station 1) were significantly higher than in those from Torcello (station 3) and Marghera (station 7) ($p < 0.001$ and 0.01, respectively). Fish from Crevan (station 2) had higher amounts of HDL than those from Torcello (station 3) ($p < 0.01$).

Levels of LDL are also linked to diet. Therefore, no direct correlation between increased levels and contamination can be drawn without further experimentation. However, any increase in its presence, especially when associated with an increase in HDL, should be considered as a cause for concern in that exogenous LDL can also carry certain types of contaminants.

CONCLUSIONS

The cellular pathology investigated a range of biomarkers of contaminant exposure and effect covering various aspects of major cellular processes in molluscs and fish. The results indicate, as might be expected from the sediment, water and tissue chemistry, that mussels and fish at the 'positive control sites' of Salute (station 5) and Marghera/CVE (stations 6/7) are the most impacted.

The fact that the mussels collected at the mussel parks of Alberoni (station 8) and Chioggia (station 9), considered as 'negative control sites', also showed signs of being affected suggests that the contaminant problems are not to be considered as restricted entirely to the northern Lagoon. Indeed, further studies should be undertaken to ascertain their suitability as culture areas.

Whilst differences were apparent in samples taken from the 'core' sites of the Platform (station 0), Lio Grande (station 1), Crevan (station 2) and Torcello (station 3), some parameters were not of a magnitude that signals the Lagoon as being in great danger. There is, however, a disturbing trend that should be addressed in future research planning for the area. The abnormalities observed in the reproductive cycle of the mussels from all sites, including those at the Platform (station 0), suggest that there might be a possible decline in stocks in the coming years.

The differences observed for some of the indices between Platform mussels and those of the 'core' lagoon sites may be due more to a varied diet in the Adriatic Sea than to contaminant impacts. As such, the Platform should not be employed as a reference site for future studies.

REFERENCES

Bancroft, D. (1967). *An Introduction to Histochemical Technique*. Butterworths, London

Culling, C. F. A. (1963). *Handbook of Histopathological Techniques*. 2nd ed. Butterworths, London,

Lowe, D. M., Moore, M. N. and Clarke, K. R. (1981). Effects of oil on digestive cells in mussels: quantitative alterations in cellular and lysosomal structure. *Aquatic Toxicol.*, **1**: 213–26

Lowe, D. M. (1988). Alterations in cellular structure of *Mytilus edulis* resulting from exposure to environmental contaminants under field and experimental conditions. *Mar. Ecol. Prog. Ser.*, **46**: 91–100

Lowe, D. M. and Fossato, V. U. (1999). The influence of environmental contaminants on lysosomal activity in the digestive cells of mussels (*Mytilus galloprovincialis*) from the Venice Lagoon. *Aquatic Toxicol.* (in press)

Lowe, D. M., Fossato, V. and Depledge, M. H. (1995). Contaminant-induced lyposomal membrane damage in blood cells of mussels, *Mytilis galloprovincialis*, from the Venice Lagoon: an *in vitro* study. *Mar. Ecol. Prog.*, **129**, 189–96

Lowe, D. M. and Clarke, K. R. (1989). Contaminant-induced changes in the structure of the digestive epithelium of *Mytilus edulis*. *Aquatic. Toxicol.*, **15**: 345–58

Lowe, D. M., Moore M. N. and Evans, B. M. (1992). Contaminant impact on interactions of molecular probes with lysosomes in living hepatocytes from Dab, *Limanda limanda*. *Mar. Ecol. Prog. Ser.*, **91**: 135–40

Lowe, D. M. and Pipe, R. K. (1986). Hydrocarbon exposure in mussel: a quantitative study of the responses in the reproductive and nutrient storage cell systems. *Aquatic Toxicol.*, **8**: 265–72

Lowe, D. M. and Pipe, R. K. (1987). Mortality and quantitative aspects of storage cell utilization in mussel, *Mytilus edulis*, following exposure to diesel oil hydrocarbons.

Mar. Environ. Res., **22**: 243–51

Lowe, D. M. and Pipe, R. K. (1994). Contaminant-induced lysosomal membrane damage in marine mussel digestive cell: an *in vitro* study. *Aqatic. Toxicol.*, **30**: 357–65

Maung-Myint, U. and Tyler, P. A. (1982). Effects of temperature, nutritive and metal stressors on the reproductive biology of *Mytilus edulis*. *Mar. Biol.*, **51**: 558–63

Moore, M. N. (1976). Cytochemical demonstration of latency of lysosomal hydrolases in digestive cells of the common mussel, *Mytilus edulis*, and changes induced by thermal stress. *Cell Tissue Res.*, **175**: 279–87

CHAPTER 22

CONTAMINANT EFFECTS IN THE VENICE LAGOON: PHYSIOLOGICAL RESPONSES OF MUSSELS (MYTILUS GALLOPROVINCIALIS)

J. Widdows and C. Nasci

SUMMARY

Mussels (*Mytilus galloprovincialis*) were sampled from six sites in the Venice Lagoon in May 1993 for measurement of their physiological responses (clearance rate, absorption efficiency and respiration rate) under standard laboratory conditions. Scope for growth (SFG) calculated for a 'standardised food/ration level' (0.4 mg particulate organic matter/l) reflected the underlying pollution-induced stress caused by toxicants accumulated in body tissues. Mussels from Chioggia had the highest SFG (16 J $g^{-1} h^{-1}$) and those at Lio Grande, Crevan and Alberoni had slightly lower SFG (9–10 J $g^{-1} h^{-1}$) rates. Mussels from the inner part of the Lagoon (Salute and Canale Vittorio Emanuele) had significantly reduced SFG values (2 and −4 J $g^{-1} h^{-1}$) which indicated severe pollution-induced stress. There were significant negative correlations between SFG and the tissue concentrations of petroleum hydrocarbons, PCBs, DDT and HCH. SFG was also calculated on the basis of measured POM levels prevailing at each site and this demonstrated that the inhibitory effects of toxicants on the growth of mussels living in the more polluted inner sites are at least partially offset by the positive effects of eutrophication and the higher food/ration levels at these sites.

INTRODUCTION

The primary objective of this study was to assess the degree of pollution impact on mussels (*Mytilus galloprovincialis*) living at various sites in the Venice Lagoon. Mussels are widely used for monitoring spatial and temporal changes in chemical contamination of the marine environment, due to their ability to pump large volumes of water and bioconcentrate metals and organic contaminants in their tissues with minimal metabolic transformation. In order to establish whether the accumulated chemical contaminants induce deleterious effects, the tissue residue chemistry data should be combined with measurements of biological impact.

Growth provides one of the most sensitive measurements of stress in an organism, since growth represents an integration of major physiological responses and, specifically, the balance between processes of energy acquisition (feeding and digestion) and energy expenditure (metabolism and excretion).

Each of these physiological responses can be converted into measurements of energy flow ($J\,g^{-1}\,h^{-1}$) and alterations in the amount of energy available for growth and reproduction (termed scope for growth, SFG) can be quantified by means of the energy budget. Therefore, SFG provides an instantaneous measurement of the energy status of an animal, which can range from maximum positive values under optimal conditions, declining to negative values when the animal is severely stressed and utilising body reserves. SFG has proved to be a sensitive, rapid and cost-effective method of detecting and quantifying pollution impact in the marine environment (Widdows and Johnson, 1988; Widdows *et al.*, 1990; Widdows and Donkin, 1992; Widdows *et al.*, 1995).

The aim of the present study is to use SFG to quantify the degree to which mussels in various parts of the Venice Lagoon are stressed by chemical contaminants and to examine the extent to which this negative impact may be partially offset by the positive effects of increased food availability resulting from eutrophication.

MATERIALS AND METHODS

Mussels (*Mytilus galloprovincialis*) were collected from six sites in the Venice Lagoon (Figure 22.1) in May 1993 and returned to the Istituto di Biologia del Mare, Venice, for measurement of physiological responses.

The mussels were either obtained from wooden pilings marking the shipping channels (e.g. Canale Vittorio Emanuele (CVE), Salute, Crevan and Lio Grande) or from mussel cultivation sites (e.g. Alberoni and Chioggia). At the same time as sampling mussels, a 40 l water sample was taken from each site for measurements of the concentration of suspended matter. Measurements of physiological responses were made on mussels sampled from two sites per day. Sixteen individuals of a standard body size (46 mm shell length) were selected from each group, cleaned of epibionts and labelled. The physiological responses (clearance or feeding rate, food absorption efficiency and respiration rate) were measured under standard laboratory conditions (filtered offshore seawater, 18°C, 34 PSU) after a 2-h period of recovery in seawater.

Clearance rate, or the volume of water cleared of suspended particles per hour, was measured in a 'closed system'. Individual mussels were placed in tanks containing 2.5 l of filtered offshore seawater which was mixed by gentle aeration. After a period of 30 min to allow the mussels to open their shell valves and resume pumping, algal cells (*Tetraselmis suiecica*) were added to each tank to give an initial concentration of 15,000 cells / ml. Twenty ml aliquots were sampled from each tank at 30 min. intervals over a period of 1 h, and the decline in cell concentration was monitored using a Coulter counter (Model TAII). Control tanks, without mussels, showed no significant decline

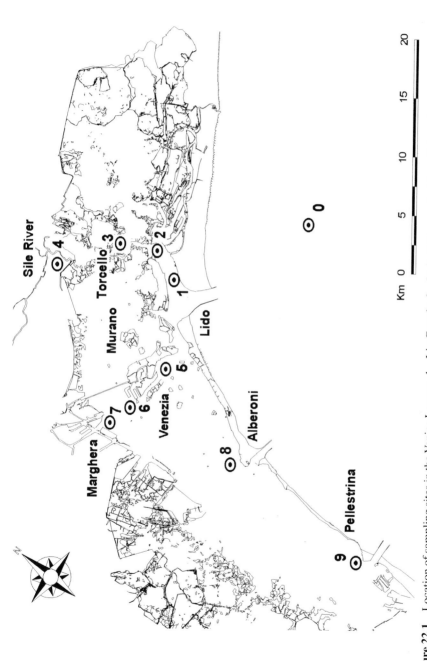

Figure 22.1 Location of sampling sites in the Venice Lagoon. 1 – Lio Grande, 2 – Isola di Crevan, 5 – Punta Salute, 6 – Canale Vittorio Emanuele, 8 – Alberoni, 9 – Chioggia

in cell concentration over 1 h. The clearance rate (CR, volume of water cleared of particles h^{-1}) by individual mussels was then calculated using the following equation:

$$CR \text{ (l/h)} = 2.5l \, (\log_e C_1 - \log_e C_2) \, / \text{ time interval (h)}$$

where C_1 and C_2 are the cell concentrations at the beginning and end of each time increment (i.e. 0.5 h).
The clearance rate was based on the average rate over the one hour period.

Food absorption efficiency was measured by comparing the proportion of organic matter in the seston and the faeces of mussels. The amount of seston in the seawater collected from each site was estimated by filtering known volumes of water through washed, ashed and pre-weighed GF/C filters. Salts were washed out of the filters with 3×10 ml of distilled water and the filters dried at 90°C. They were then weighed before ashing in a furnace at 450°C for 4 h and re-weighed. Faecal pellets produced during the 2-h recovery period were collected by means of a pipette, filtered onto pre-weighed GF/C filters and processed in the same manner as for the filters with seston (i.e. washed, dried, weighed, ashed and re-weighed). The composition of this faecal material will reflect that of the ambient seston. The ash-free dry weight : dry weight ratios of the food (seston) and the faeces were then used to calculate the absorption efficiency by the following equation:

$$\text{Absorption efficiency} = (F - E) \, / \, [(1 - E) \, F]$$

where F = ash-free dry weight : dry weight ratio for the food, and E = ash-free dry weight : dry weight ratio for the faeces.

Respiration rate was determined by placing individual mussels in glass respirometers (500 ml) containing air-saturated seawater which was stirred by a magnetic stirrer bar beneath a perforated glass plate supporting the mussel. The respirometer was sealed and the decline in oxygen concentration was measured by means of a radiometer oxygen electrode (E5046) connected to a Strathkelvin oxygen meter.

After physiological measurement the mussel tissues were removed from their shells and dried at 90°C to obtain their dry tissue weight.

The individual clearance rates (l h^{-1}) and oxygen consumption rates (μmol O$_2$h^{-1}) were converted to mass specific rates for a 'standard mussel' of 1 g dry weight using the standard weight exponent (b = 0.67). Each physiological rate was then converted to energy equivalents (J g^{-1} h^{-1}) in order to calculate the energy budget and the scope for growth (SFG), which represents the difference between the energy gained (absorbed) from the food and the energy lost via metabolic energy expenditure (for details see Widdows, 1993).

RESULTS AND DISCUSSION

The physiological responses of mussels from the six sites in the Venice Lagoon are presented in Table 22.1. Clearance rate is the component of the energy budget which is typically the most responsive to pollutants. In the present study, mussels from 6-CVE, 5-Salute and 8-Alberoni had significantly lower clearance rates ($p < 0.05$) than mussels from 1-Lio Grande and 9-Chioggia (i.e. 25%, 57% and 66%, respectively, of the clearance rates recorded at Chioggia). The rates of oxygen consumption were high and showed some variations amongst sites. The variations primarily reflect the recent spawning (<1 month) and the low tissue weight of the Crevan and Lio Grande mussels compared to those from CVE and Alberoni which had higher tissue weights and body reserves.

Scope for growth (SFG) provides an integrated stress response which can be calculated in two ways, (i) using a 'standardized' food/ration level (i.e. 0.4 mg particulate organic matter / l), and (ii) using the prevailing ration levels in the field. Clearly, ration is a major factor determining growth of an animal and the physiological energetics approach allows the effects of this parameter to be removed or examined separately.

SFG based on a 'standardized' ration level of 0.4 mg POM / l and an absorption efficiency of 0.45, is used to show underlying pollution-induced stress, or the reduction in growth potential caused by contaminants accumulated within the tissues of mussels when all other natural and potential environmental stressors are held constant. The 'standard' ration level of 0.4 mg POM / l is typical of northern temperate oceanic waters (e.g. North Atlantic) and has been used in many previous pollution studies (Widdows and Donkin, 1992; Widdows *et al.*, 1987, 1995). Therefore, it allows a comparison among different field studies, thus placing each study in a wider geographical and ecological context. On the other hand, SFG based on field ambient POM levels, and associated absorption efficiencies, reflects the growth potential of mussels under the prevailing seston (i.e. food/ration) levels at each site at the time of sampling and it, therefore, provides an indication of the overall balance between eutrophic (positive) effects and pollutant (negative) effects.

SFG, based on a standardized ratio, shows that mussels from Chioggia, Lio Grande and Crevan all had rates which were not significantly different (i.e. 10 to 16 J g^{-1}h^{-1}; Figure 22.2). Mussels from Alberoni had SFG values of 8.5 J g^{-1}h^{-1} which were significantly lower than those of the mussels at Chioggia ($p < 0.05$). The mussels from Salute and CVE were more severely stressed and had markedly reduced SFG values (1.71 and −4.34 J g^{-1}h^{-1}, respectively) compared to all other sites. The 'standardised' SFG values recorded in the Venice Lagoon study appear to be comparable to the range of SFG values measured in previous mussel studies elsewhere (Widdows and Donkin, 1992). The maximum SFG values recorded under 'optimal conditions' in pristine environments with minimal contamination are 20–25 J g^{-1}h^{-1}. The SFG values between 10–16 J g^{-1}h^{-1} are typical of areas influenced by moderate

Table 22.1 Components of the energy budget of mussels (*M. galloprovincialis*) sampled from sites in the Venice Lagoon in May 1993. (Mean ± 95% C.I.; Physiological responses n = 15; Seston and particulate organic matter n = 5)

Sampling site	Mean CR $1\,g^{-1}h^{-1}$	O_2 uptake ($\mu mol\,O_2\,g^{-1}h^{-1}$)	Scope for growth ($J\,g^{-1}h^{-1}$) [based on standard ratio of 0.4 mg POM/l and 0.45 absorption efficiency]	Absorption efficiency	Seston conc. ($mg\,l^{-1}$)	Particulate organic matter ($mg\,POM/l$)	Scope for growth ($J\,g^{-1}h^{-1}$) [based on POM conc. and absorption efficiency at site]
6-Canale Vittorio Emanuele	1.75 ± 0.44	25.41 ± 1.36	-4.34 ± 2.1	0.16 ± 0.06	27.55 ± 2.29	8.15 ± 0.24	41 ± 13
5-Punta Salute	3.90 ± 0.96	31.72 ± 1.62	1.71 ± 4.3	0.26 ± 0.06	18.99 ± 1.06	5.00 ± 0.45	102 ± 28
8-Alberoni	4.51 ± 0.50	22.24 ± 1.68	8.55 ± 2.3	0.66 ± 0.02	3.52 ± 0.52	0.98 ± 0.17	57 ± 8
2-Isola di Crevan	5.92 ± 1.02	32.56 ± 2.60	9.67 ± 3.8	0.50 ± 0.07	9.54 ± 0.64	3.02 ± 0.20	192 ± 35
1-Lio Grande	6.59 ± 1.36	37.10 ± 1.84	10.36 ± 5.6	0.68 ± 0.06	9.04 ± 0.66	2.61 ± 0.15	251 ± 54
9-Chioggia	6.88 ± 1.04	27.51 ± 1.50	15.95 ± 4.4	0.63 ± 0.04	3.63 ± 0.52	1.24 ± 0.15	111 ± 19

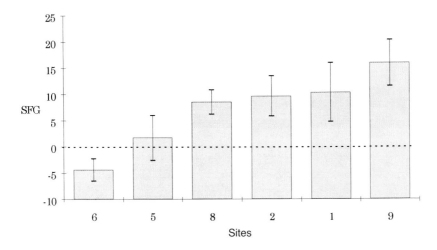

Figure 22.2 Scope for growth (SFG; J g^{-1}h^{-1}) of mussels (*M. galloprovincialis*) collected from 6 sites in the Venice Lagoon. Mean ± 95% C.I. $n = 16$. Sites: 6 – Canale Vittorio Emanuele; 5 – Punta Salute; 8 – Alberoni; 2 – Isola di Crevan; 1 – Lio Grande; 9 – Chioggia

levels of pollution (e.g. coastal regions with urban influence), whereas SFG values < 5 J g^{-1}h^{-1} are more typical of industrialised and urbanised areas with significant levels of pollution. Negative SFG values are indicative of severe stress where the animals are in negative energy balance and have to utilise body reserves in order to survive.

There are statistically significant negative correlations ($r = -0.81$; $p < 0.05$) between SFG and the size of the 'unresolved complex mixture' (UCM), reflecting petroleum hydrocarbon contamination, and the concentrations of HCH ($r = -0.96$; $p < 0.05$), DDT ($r = -0.81$; $p < 0.05$) and PCBs ($r = -0.84$; $p < 0.05$) in the mussel tissues (Figure 22.3; chemistry data provided by Fossato *et al.*, Chapter 25). The apparent lack of a significant correlation between SFG and the sum of 16 PAHs in the tissues ($r = -0.65$) is probably due to: (i) the inclusion of higher molecular weight PAHs which do not exert toxic effects on the SFG of *Mytilus*, and (ii) changes in the relative composition of the 16 PAHs at each site. Previous studies (Widdows and Donkin, 1992) have demonstrated that < 50% of these 16 compounds contribute to the established line representing the Quantitative Structure–Activity Relationship (QSAR) for hydrocarbon toxicity. Ideally, the extraction and analysis should include the toxic lower molecular weight PAHs, because they are known to be widespread contaminants in the coastal environment, are toxicologically significant and thus provide a significant contribution to the 'total toxic load' accumulated by bivalves. Consequently, a detailed quantitative toxicological interpretation of the tissue hydrocarbon data is not possible. The PCB and DDT concentrations in the mussels at CVE (6) and Salute (5) are relatively

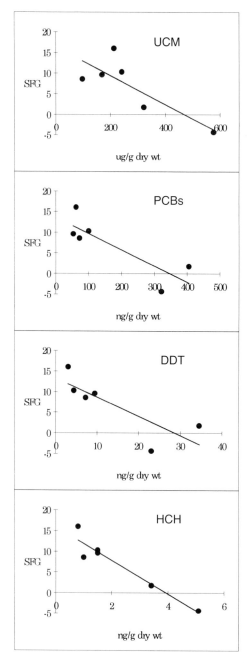

Figure 22.3 Relationships between scope for growth (SFG J $g^{-1}h^{-1}$) and major organic contaminants accumulated in the tissues of mussels (*M. galloprovincialis*). UCM – unresolved complex mixture of hydrocarbons; PCBs – polychlorinated biphenyls; DDT; HCH – hexachlorocyclohexane

high (i.e. > 300 and 20 ng dry weight, respectively) and equivalent levels in the North Sea study (Widdows *et al.*, 1995) were also associated with very low SFG values (i.e. < 2 J g^{-1} h^{-1} at the 'standard ration').

SFG based on the measured POM concentration in the seston at each site demonstrates that mussels living in the Venice Lagoon are capable of potentially high growth rates, primarily due to the high ration levels and the generally eutrophic conditions. Mussels at the inner Venice Lagoon sites (e.g. CVE and Salute) experience high seston and POM levels (> 5 mg POM/l in May) and consequently have relatively high SFG values (41 and 102 J g^{-1} h^{-1}, respectively). This is in spite of the pollution-induced reductions in clearance rates and the lower absorption efficiencies which are associated with increased food availability. The results therefore demonstrate that the inhibitory effects of toxicants on the growth of mussels living in the most polluted inner sites are at least partially offset by the positive effects of eutrophication at these sites. Mussels living at Crevan and Lio Grande experienced intermediate seston and POM levels (2.6 to 3 mg POM/l in May) but had the highest SFG values (192 and 251 J g^{-1} h^{-1}).

At the Alberoni site, mussels appear to experience the lowest ration levels (1 mg POM/l) which, when combined with a slightly lower clearance rate (compared to Chioggia), results in a markedly lower SFG. Mussels at Alberoni are cultivated and grown in nylon mesh bags and those sampled appeared to be tightly packed within the mesh bags, this may have resulted in a measurable reduction in clearance rate and SFG.

CONCLUSIONS

Mussels sampled from the most contaminated sites (6-CVE and 5-Salute) showed a significant reduction in clearance rate and scope for growth (based on a standardized ration level) compared to those living in the outer regions of the Venice Lagoon. The low SFG values (i.e. < 5 J g^{-1} h^{-1}) were indicative of severely stressed mussels and poor water quality at these two sites. There was a significant negative correlation between SFG and concentrations of petro-leum hydrocarbons, PCBs, DDT and HCH accumulated in the mussel tissues. However, the study also showed that these pollution-induced stress effects were partially offset by the higher ration levels at the innermost contaminated sites, which thus enabled the mussels to survive and grow in this region.

This present study, together with other worldwide environmental monitor-ing programmes in both temperate regions (e.g. North Sea coastline, Oslofjord, Shetland Islands, New Zealand) and tropical regions (e.g. Bermuda, India, Thailand), has demonstrated the application of SFG as a sensitive biomarker which represents a powerful and cost-effective method of assessing environmental pollution with rapid feedback of information to environmental managers. The SFG technique not only enables a rapid detection and quanti-fication of the biological effects of pollution, but also provides a means of identifying the cause(s), through toxicological interpretation of tissue residue chemistry, using an ever-increasing toxicological database.

The combined measurement of SFG and contaminant levels in mussels could, therefore, provide a valuable contribution to environmental monitoring programmes designed to assess the 'health' of the Venice Lagoon ecosystem.

REFERENCES

Widdows, J. (1993). Marine and estuarine invertebrate toxicity tests. In Calow, P. (ed.) *Handbook of Ecotoxicology*. Blackwell Scientific, Oxford, Vol. 1, Chap. 9, 145–66

Widdows, J., Burns, K. A., Menon, N. R., Page, D. and Soria, S. (1990). Measurement of physiological energetics (scope for growth) and chemical contaminants in mussel (*Arca zebra*) transplanted along a contamination gradient in Bermuda. *J. Exp. Mar. Biol. and Ecol.,* **138**: 99–117

Widdows, J. and Donkin, P. (1992). Mussels and Environmental Contaminants: Bioaccumulation and Physiological Aspects. In Gosling, E. (ed.) *The Mussel Mytilus*. Elsevier Press: Amsterdam, Chap. 8, 383–424

Widdows, J., Donkin, P., Brinsley, M. D., Evans, S. V., Salkeld, P. N., Franklin, A., Law, R. J. and Waldock, M. J. (1995). Scope for growth and contaminant levels in North Sea mussels, *Mytilis edulis. Mar. Ecol. Prog. Series,* **127**: 131–48

Widdows, J., Donkin, P., Salkeld, P. N. and Evans, S. V. (1987). Measurement of scope for growth and tissue hydrocarbon concentrations of mussels (*Mytilus edulis*) at sites in the vicinity of the Sullom Voe oil terminal: A case study. In Kuiper, J. and Van den Brink, W. J (eds). *Fate and Effects of Oil in Marine Ecosystems*. Martinus Nijhoff: Dordrecht, 269–77

Widdows, J. and Johnson, D. (1988). Physiological energetics of *Mytilus edulis*: scope for growth. *Mar. Ecol. Prog. Series,* **46**: 113–21

CHAPTER 23

POSSIBLE LINKS BETWEEN IMPAIRED IMMUNE FUNCTION AND CONTAMINANT LEVELS IN MUSSELS (MYTILUS GALLOPROVINCIALIS) SAMPLED FROM THE VENICE LAGOON

R.K. Pipe, J.A. Coles and M.E. Thomas

SUMMARY

Mussels, *Mytilus galloprovincialis*, were sampled from sites within the Venice Lagoon and a reference Platform site in the north Adriatic Sea during December 1991, April 1992 and June 1992. The immunocompetence of the mussels was assessed using a range of asssays which included total and differential cell counts, phagocytosis, degradative enzyme levels and release of reactive oxygen metabolites. Chlorinated hydrocarbons, including lindane, DDT and PCBs, together with trace metal levels, were measured in digestive gland tissues from the mussels. The contaminant levels and immune responses showed seasonal fluctuations. However, the results did demonstrate significant differences in a number of immunocompetence assays which, in some instances, showed good correlations with the levels of contaminants measured in the tissues. The results confirm the requirement for using a range of assays for monitoring the effects of pollution.

INTRODUCTION

The main function of the immune system for all metazoan organisms is to sustain host defence mechanisms and maintain homeostasis within the organism. In mammals it has been demonstrated that exposure to contaminants in the environment can lead to modification of the normal host defence mechanisms through stimulation or impairment of the immune response or, alternatively, may lead to an abnormal immune response. It has been established that the immune system is particularly sensitive for detecting cellular toxicity resulting from chemical exposure, probably due to the rapid proliferation of cells during an immune response combined with the complex intercellular signalling mechanisms which are present.

The immune response can be split into two functional divisions, innate (non-anticipatory) and acquired (anticipatory). Vertebrates, including fish, are capable of both immune responses, with the anticipatory response being effected by lymphocytes. Metazoan invertebrates have the capacity for non-anticipatory immune responses but the question of anticipatory responses remains somewhat controversial (Klein, 1989, Kelly *et al.*, 1992), with present evidence indicating that only vertebrates and, possibly ancestral deutero-stomes, possess the anticipatory immune response which is characterized by its specificity, inducibility and memory (Marchalonis and Schluter, 1990). Given the complex nature of the immune response, demonstrated even with the invertebrates (Pipe, 1990a,b, 1992; Pipe *et al.*, 1993, 1997), it is very impor-tant when trying to evaluate immunological effects resulting from environ-mental chemicals to use a range of selected assays (Pipe *et al.*, 1992, 1995a).

The purpose of the present study was to identify possible correlations between the immune response of mussels, *Mytilus galloprovincialis*, and chemical contaminants from a range of sites within the Venice Lagoon. The immune response of the mussels was assessed using a range of assays which included total and differential cell counts, phagocytosis, degradative enzyme levels and respiratory burst activity.

MATERIALS AND METHODS

Mussels were collected for immunotoxicology studies during three campaigns in December, March and June 1991–92. During the first campaign 4 sites were studied; 0 – Reference Platform, 1 – Lio Grande, 2 – Isola Crevan and 3 – Canale di Torcello. Two additional sites, 5 – Punta Salute and 6 – Can. V. Emanuele (CVE) were studied during the second and third campaigns (Figure 23.1).

Mussels (50–75 mm shell length) were collected from the sites and trans-ferred directly to the laboratory for analyses. Samples of between 20 and 40 mussels were used to estimate overall immune function using a suite of assays which varied slightly during the different campaigns, as described in detail by Pipe *et al.* (1995b). Haemolymph was extracted from the posterior adductor muscle using a 2.5 ml syringe. Total blood cell numbers were counted with a haemocytometer from haemolymph samples extracted into Baker's formol calcium. Differential cell counts, identifying eosinophilic and basophilic blood cell populations, were carried out on Wrights-stained cytospin preparations of Baker's fixed haemocytes. Phagocytosis was estimated by counting the numbers of blood cells containing particles following a 40-min incubation in suspensions of carbon (first campaign) or Congo red-stained zymosan (second and third campaign). Respiratory burst activity was determined as release of superoxide radicals and was quantified as the percentage of blood cells, prepared as monolayers, capable of reducing nitro-blue tetrazolium to insoluble formazan dye. Cytochemical determinations for phenol oxidase (using L-dopa as substrate), peroxidase (using diaminobenzidine as substrate) and arylsulfatase (naphthol-AS-BI-sulfate as substrate) were carried out on Baker's fixed cytospin preparations.

Digestive gland tissues from 6–10 mussels were dissected out and analysed for organic and metal contaminants. Chlorinated hydrocarbons, including lindane, DDT and PCBs were measured by fluorescence HPLC and gas chromatography following lyophilisation, hexane extraction and alumina/

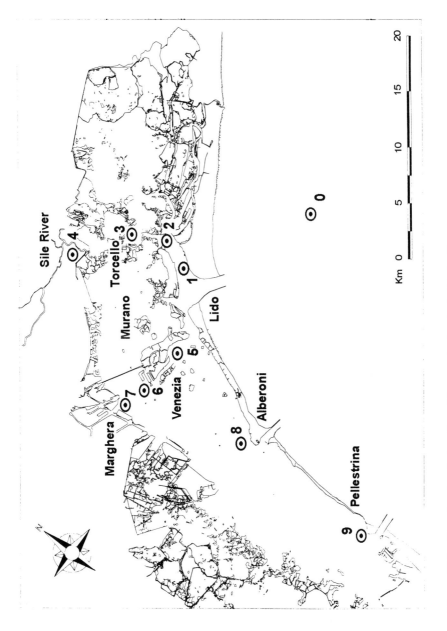

Figure 23.1 Map showing the location of the sample sites in the Venice Lagoon

silica gel chromatography (Fossato *et al.*, 1988). Metal analysis was carried out using atomic absorption spectrophotometry, following overnight digestion with ultrapure HNO_3 (Fossato *et al.*, 1989).

RESULTS

The results showed considerable differences for some of the assays carried out, both in terms of the sites and the time of year. The results for total blood cell numbers in *M. galloprovincialis* demonstrated clear seasonal variability. Mussels collected at the Platform (station 0) and Lio Grande (station 1) sites had significantly lower ($p < 0.024$) numbers of blood cells in June compared with April and December. Mussels from Salute (station 5) and CVE (station 6) had significantly higher ($p < 0.032$) blood cell counts in April compared with June. Inter-site comparisons of the blood cell counts also demonstrated significant differences for each month sampled. In December, the numbers of circulating blood cells in mussels from Crevan (station 2) and Torcello (station 3) were significantly lower ($p < 0.029$) than those from the Platform (station 0) or Lio Grande (station 1). The only significant difference in blood cell numbers in June was at Torcello (station 3) where the levels were significantly higher ($p < 0.05$) than those at Salute (station 5). In April the blood cell numbers showed greater variability but correlations between contaminants and the blood cell counts were evident (Figure 23.2). The haemocyte numbers increased with increasing levels of PCB, DDT and cadmium.

The differential blood cell counts also demonstrated variability between sites and time of year. In only two instances were there more eosinophils than basophils; both of these were in June, at Lio Grande (station 1) and Torcello (station 3), where the proportion of eosinophils exceeded 60%. There was no correlation between total cell numbers and cell type, indicating that high total cell numbers were not a result of proliferation of a single cell population.

The results from the phagocytosis studies showed considerable differences for comparisons between sites for April and June. In April, the blood cells from Lio Grande (station 1) mussels showed a significantly higher ($p < 0.004$) percentage phagocytosis than all other sites except Torcello (station 3). The tissue metal levels at Salute (station 5), CVE (station 6) and Crevan (station 2) were also higher than at Lio Grande (station 1), but this was also true for Torcello (station 3) but not for the Platform (station 0). In addition the percentage phagocytosis at Salute (station 5) was significantly higher ($p < 0.026$) than at CVE (station 6) and the Platform (station 0) despite the fact that the total contaminant burden was considerably higher at this site. Clearly, the differences observed in phagocytosis in April cannot be simply interpreted in terms of contaminant levels. In mussels sampled in June the percentage phagocytosis at Lio Grande (station 1) and the Platform (station 0) was significantly higher ($p < 0.038$) than that recorded for all the other sites (except for the Platform (station 0) versus Torcello (station 3)). The mussels sampled in

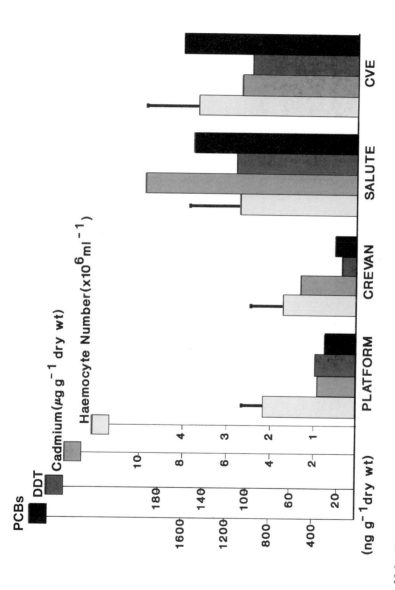

Figure 23.2 The number of circulating haemocytes in mussels, *M. galloprovincialis*, sampled from a range of sites in the Venice Lagoon and Adriatic Sea during April, demonstrating a correlation between increasing numbers and higher levels of specific contaminants. Values are means of total numbers $\times 10^6 \pm 1$ SE

June from Crevan (station 2), CVE (station 6) and Salute (station 5) showed, not only the lowest levels of phagocytosis, but also the greatest overall contaminant burden, indicating a possible correlation (Figure 23.3) between these factors at this time of the year.

The reduction of nitro-blue tetrazolium showed significant seasonal effects with very low values in December compared with April and June. The highest levels of NBT reduction were observed in June at the Platform (station 0) site and in April at Crevan (station 2) and Torcello (station 3); values at these sites more than doubled the mean of all the remaining observations. Comparisons between the sites showed very clearly the significantly higher ($p < 0.018$) levels of NBT reduction observed in mussel blood cells from the Platform (station 0) in June and Crevan (station 2) and Torcello (station 3) in April when compared with all other sites. These high levels of NBT reduction correlate well (Figure 23.4) with the relatively low levels of organic contaminants (PCBs, DDT, and lindane) which were found in tissues from the Platform (station 0) in June and Crevan (station 2) and Torcello (station 3) in April.

The results for percentage haemocytes showing cytochemically determined enzyme activities for phenol oxidase and peroxidase showed considerable variability. The differences appear to be linked to seasonal changes, as has been demonstrated previously (Coles and Pipe, 1994), with no evidence for direct correlations between contaminant levels and percentage positive cells. The levels of activity for the lysosomal enzyme, arylsulfatase, were very low at all sites at the periods of the year when the investigations were done. There did not appear to be any correlation between the arylsulfatase activity and contaminant levels.

CONCLUSIONS

In field situations animals may be exposed to relatively low levels of contaminants on a chronic basis without demonstrating acute effects. Interpretation of field data requires considerable care as there are many factors which will contribute to a measured physiological response; this is particularly important when comparing a range of different sites which will be subjected to varying local conditions including temperature, salinity and nutrition, as well as possible contaminants. It is, therefore, very important to use a range of assays for determining potentially deleterious effects resulting from contaminant exposure (Pipe and Coles, 1995). The results from the present study indicate differences in immune response in mussels from the different sites. These differences fluctuated according to the time of the year but in some instances showed good correlation with the levels of contaminants measured in the tissues.

The total number of circulating haemocytes in mussels fluctuates greatly but has been shown to increase significantly under stressed conditions (Renwrantz, 1990). In *Crassostrea virginica* it increased following exposure

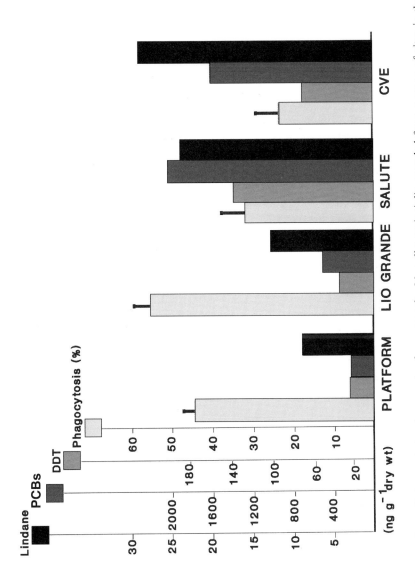

Figure 23.3 The percentage of phagocytosing haemocytes from mussels, *M. galloprovincialis*, sampled from a range of sites in the Venice Lagoon and Adriatic Sea during June, showing a correlation between decreasing phagocytosis and increasing levels of specific contaminants

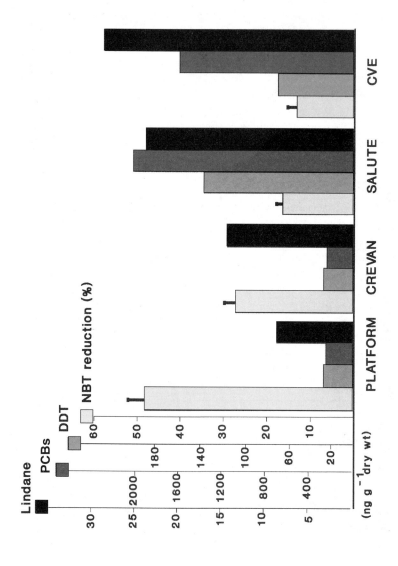

Figure 23.4 The percentage of haemocytes from mussels, *M. galloprovincialis*, sampled from a range of sites in the Venice Lagoon and Adriatic Sea during June, showing reduction of nitroblue tetrazolium related to increasing levels of specific contaminants

to cadmium (Cheng, 1988a) and in *M. edulis*, following exposure to cadmium (Coles *et al.*, 1995) and fluoranthene (Coles *et al.*, 1994). In some bivalve species a reduction in total haemocyte numbers has been noted following exposure to copper (Suresh and Mohandas, 1990).

Phagocytosis is the principal mechanism of bacterial clearance in molluscs; the process can be divided into recognition, attachment, uptake and killing of potential pathogens. Most studies on the effects of exposure to contaminants have addressed the overall process, measuring either clearance of injected bacteria from the haemolymph or uptake of bacteria into haemocytes. Laboratory experiments have demonstrated that phagocytic indices for molluscs can become enhanced following short-term exposure to low level contaminants (Anderson, 1981; Cheng and Sullivan, 1984; Cheng, 1988b) but tend to decline on long-term exposure to higher concentrations (Fries and Tripp, 1980; Anderson, 1981; Anderson *et al.*, 1981). Results from field studies have demonstrated that haemocytes from oysters living in a PAH-contaminated site showed reduced levels of phagocytosis compared with those from a clean site (Sami *et al.*, 1992). The present studies on the Venice Lagoon indicate the possibility of a correlation between high metal levels and low rates of phagocytosis, but there are clear anomalies to this being a general rule; other factors obviously contribute to the overall levels of phagocytosis.

The reduction of NBT provides a measure of the release of reactive oxygen metabolites and is, in part, a reflection of the ability of the haemocytes to kill invading pathogens. Studies on release of oxygen radicals from blood cells in relation to contaminant levels have shown a suppression of chemi-luminescence in *C. virginica* following exposure to a range of metal and organic pollutants (Larsen *et al.*, 1989; Anderson *et al.*, 1992). Results from the present study support these findings and indicate a correlation between high levels of oxygen radical release and low levels of organic contaminants.

Synthesising all the data, it would appear that mussels sampled from the sites of CVE (station 6) and Salute (station 5) show potentially reduced capability in terms of immune response. This can lead to an increased suscepti-bility to diseases and can have deleterious effects on the populations as a whole.

ACKNOWLEDGEMENTS

This work forms part of the UNESCO Venice Lagoon Ecosystem Project and was funded in part by the Commission of the European Community Research Programme on the Fisheries Sector (FAR) contract, AQ 2 419 and UK Dept. of the Environment contract 7/7/386.

REFERENCES

Anderson, R. S. (1981). Effects of carcinogenic and noncarcinogenic environmental pollutants on immunological functions in a marine invertebrate. In Dawe, C. J. (ed.).

Phyletic Approaches to Cancer. Japan Sci. Soc. Press, Tokyo, 319–31

Anderson, R. S., Giam, C. S., Ray, L. E. and Tripp, M. R. (1981). Effects of environmental pollutants on immunological competency of the clam *Mercenaria mercenaria*: impaired bacterial clearance. *Aquatic Toxicol.*, **1**: 187–95

Anderson, R. S., Oliver, L. M. and Jacobs, D. (1992). Immunotoxicity of cadmium for the eastern oyster (*Crassostrea virginica* [Gmelin, 1791]): Effects on hemocyte chemiluminescence. *J. Shellfish Res.*,**11**: 31–5

Cheng, T. C. (1988a). *In vivo* effects of heavy metals on cellular defense mechanisms of *Crassostrea virginica*: total and differential cell counts. *J. Invertebr. Pathol.* **51**: 207–14

Cheng, T. C. (1988b). *In vivo* effects of heavy metals on cellular defense mechanisms of *Crassostrea virginica*: phagocytic and endocytotic indices. *J. Invertebr. Pathol.*, **51**: 215–20

Cheng, T. C. and Sullivan, J. T. (1984). Effects of heavy metals on phagocytosis by molluscan haemocytes. *Mar. Environ. Res.*, **14**: 305–15

Coles, J. A., Farley, S. R. and Pipe, R. K. (1995). Alteration of the immune response of the common marine mussel, *Mytilus edulis* resulting from exposure to cadmium. *Dis. Aquat. Org.*, **22**: 59–65

Coles, J. A., Farley, S. R. and Pipe, R. K. (1994). The effects of fluoranthene on the immunocompetence of the common marine mussel, *Mytilus edulis. Aquat. Toxicol.*, **30**: 367–79

Fossato, V. U., Van Vleet, E. S. and Dolci, F. (1988). Chlorinated hydrocarbons in sediments of Venetian canals. *Archo. Oceanogr. Limnol.*, **21**: 151–61

Fossato, V. U., Campesan, G., Raboledda, L. C. and Stocco, G. (1989). Trends in chlorinated hydrocarbons and heavy metals in organisms from the Gulf of Venice. *Archo. Oceanogr. Limnol.*, **21**: 179–90

Fries, C. R. and Tripp, M. R. (1980). Depression of phagocytosis in *Merceneria* following chemical stress. *Dev. Comp. Immunol.*, **4**: 233–44

Kelly, K. L., Cooper, E. L. and Raftos, D. A. (1992). *In vitro* allogeneic cytotoxicity in the solitary urochordate *Styela clava. J. Exp. Zoology,* **262**: 202–8

Klein, J. (1989). Are invertebrates capable of anticipatory immune responses? *Scand. J. Immunol.*, **29**: 499–505

Larson, K. G., Roberson, B. S. and Hetrick, F. M. (1989). Effect of environmental pollution on the chemiluminescence of hemocytes from the American oyster *Crassostrea virginica. Dis. Aquat. Org.*, **6**: 131–6

Marchalonis, J. J. and Schluter, S. F. (1990). Origins of immunoglobulins and immune recognition molecules. *BioScience*, **40**: 758–68

Pipe, R. K. (1990a). Hydrolytic enzymes associated with the granular haemocytes of the marine mussel *Mytilus edulis. Histochem. J.*, **22**: 595–603

Pipe, R. K. (1990b). Differential binding of lectins to haemocytes of the mussel *Mytilus edulis. Cell & Tissue Res.*, **261**: 261–8

Pipe, R. K. (1992). The generation of reactive oxygen metabolites by the haemocytes of the mussel *Mytilus edulis. Dev. Comp. Immunol.*, **16**: 111–22

Pipe, R. K. and Coles, J. A. (1995). Environmental contaminants influencing immune function in marine bivalve molluscs. *Fish and Shellfish Immunol.* **5**: 581–95

Pipe, R. K., Holden, J. A. and Farley, S. R. (1992). The development of immunocompetence assays in mussels for use with a microtiter plate reader. *ICES, C.M. 1992/E:13*, pp. 10

Pipe, R. K., Porte, C. and Livingstone, D. R. (1993). Antioxidant enzymes associated with the blood cells and haemolymph of the mussel *Mytilus edulis. Fish & Shellfish Immunol.*, **3**: 221–33

Pipe, R. K., Farley, S. R. and Coles, J. A. (1997). The separation and characterisation of haemocytes from the mussel *Mytilus edulis. Cell & Tiss. Res.*, **289**: 537–45

Pipe, R. K., Coles, J. A. and Farley, S. R. (1995a). Assays for measuring immune response in the mussel *Mytilus edulis*. In Stolen, J.S., Fletcher, T.C., Smith, S.A., Zelikoff, J.T., Kaattari, S.L., Anderson, R.S., Söderhall, K. and Weeks-Perkins, B.A. *Techniques in Fish Immunology-4. Immunology and Pathology of Aquatic Invertebrates*. SOS Publications, Fair Haven, NJ

Pipe, R. K., Coles, J. A., Thomas, M. E., Fossato, V. U. and Pulsford, A. L. (1995b). Evidence for environmentally derived immunomodulation in mussels from the Venice Lagoon. *Aquatic Toxicol.*, **32**: 59–73

Renwrantz, L. (1990). Internal defence system of *Mytilus edulis*. In Sefano, G. B (ed.) *Neurobiology of Mytilus edulis*. Manchester University Press, Manchester, 256–75

Sami S., Faisal, M. and Huggett, R. J. (1992). Alterations in cytometric characteristics of haemocytes from the American oyster *Crassostrea virginica* exposed to a polycyclic aromatic hydrocarbon (PAH) contaminated environment. *Mar. Biol.*, **113**: 247–52

Suresh, K. and Mohandas, A. (1990). Effect of sublethal concentrations of copper on hemocyte number in bivalves. *J. Invertebr. Pathol.*, **55**: 325–31

CHAPTER 24

IMMUNOMODULATORY EFFECTS OF CONTAMINANT EXPOSURE IN SEDIMENT-DWELLING GOBY (ZOSTERISSOR OPHIOCEPHALUS)

A.L. Pulsford, M.E. Thomas, J.A. Coles, S. Lemaire-Gony and R.K. Pipe

SUMMARY

The sediment-dwelling goby, *Zosterissor ophiocephalus*, was collected from four sites in the Venice Lagoon: Crevan (station 2 reference site for fish), Lio Grande (station 1), Torcello (station 3) and Porto Marghera (station 7, industrial site) during April, June and October 1992. A range of tissue and cellular parameters were measured on the immune system of 8–10 fish from each site. Gobies from the polluted site at Porto Marghera (station 7) showed statistically significant ($p \leq 0.05$) depression of spleen somatic indices, increased levels of phagocytic activity in spleen adherent cells and increased proportions of phagocytes in the peripheral blood. Application of a range of simple assays to measure immunocompetence of the goby *Z. ophiocephalus* demonstrated significant differences in the function of the immune system and response of phagocytic cells which correlated with exposure to high levels of pesticides, PAHs, PCBs and metals at the most polluted site at Porto Marghera (station 7).

INTRODUCTION

Fish exposed to contaminated sediments or poor water quality may exhibit impaired immune function which may result in immunosuppression and an increased incidence of infectious or neoplastic diseases (Anderson, 1990). Pollutants, such as heavy metals (Zelikoff, 1993), pesticides (Dunier *et al.*, 1991), chlorinated dioxins, polycyclic aromatic hydrocarbons (PAHs) and various sources of radiation, drugs and sewage (Secombes *et al.*, 1991, 1992; Tahir *et al.*, 1993), have all been shown to have an effect on the immune system of fish. Fish living on sediments contaminated with high levels of heavy metals, PCBs and PAHs, resulting from accumulated industrial discharges, may also be subjected to chronic stress, which may modulate the responses of the cells of the immune system (Pulsford *et al.*, 1992a, 1994, 1995; Secombes *et al.*, 1991, 1992; Tahir *et al.*, 1993; Weeks and Warriner, 1984; Weeks *et al.*, 1986, 1987, 1990a,b).

Laboratory exposure experiments, both *in vivo* and *in vitro*, have also shown a range of effects on the immune system of fish (Pulsford *et al.*, 1994, 1995; Lemaire-Gony *et al.*, 1995) and invertebrates (Coles *et al.*, 1994, 1995; Pipe *et al.*, 1995), and have underscored the requirement to carry out a wide range of assays to measure immunomodulation. Cells for immunocompetence assays were taken from the spleen and kidney and it has been shown that macrophages from these organs are affected by stress (Peters *et al.*, 1991) and can accumulate particulate metals from sediments (Pulsford *et al.*, 1992b), which could interfere with their function. The present study applied a range of simple assays to cells from the immune system of the goby *Zosterissor ophiocephalus* collected from four sites, Lio Grande (station 1), Crevan (station 2), Torcello (station 3) and Porto Marghera (station 7), from the Venice Lagoon during April, June and October 1992. The aim was to correlate possible contaminant stress to modulation of the immune system.

MATERIALS AND METHODS

Gobies, *Z. ophiocephalus*, were collected from four sites; Crevan (station 2, reference site for fish), Porto Marghera (station 7, industrial site), Torcello (station 3) and Lio Grande (station 1) during April, June and October 1992 (Figure 24.1). Eight to ten fish were collected from each site and transported to the laboratory in tanks of aerated seawater at 20 °C, and either sampled immediately or within a few hours of collection.

The fish were killed by a blow to the head and weights of the spleen and kidney recorded immediately following dissection in order to calculate the somatic index. Blood smears were prepared from heart blood, fixed in methanol and stained with May–Grunwald Giemsa, and differential blood cell counts recorded. Percentage phagocytosis was recorded from adherent cells isolated from the spleen and kidney. The cells were allowed to attach to glass slides for 1 h and to phagocytose a yeast suspension for a further hour. These were fixed in methanol and stained with May–Grunwald Giemsa and counted (100 per slide). The number of cells which had ingested yeast and the number of yeast particles per phagocytic cell were recorded. The phagocytic index was calculated as the average number of yeast particles ingested by the phagocytic cells. All data were collected without prior knowledge of the contaminant status at the various sites.

Analytical chemistry for the organics was performed on liver samples following lyophilisation, hexane extraction and alumina/silica gel chromatography. The hydrocarbons, PCBs and DDTs in liver samples were measured by fluorescence HPLC and gas chromatography (Fossato *et al.*, 1988, 1989). Metals were measured by standard procedures (Zatta *et al.*, 1992).

Data were examined using a Student's *t* test or analysis of variance (ANOVA) with an *F* test. Levels of confidence less than or equal to 0.05 were considered statistically significant. Data were expressed throughout as the mean ± one standard error of the mean.

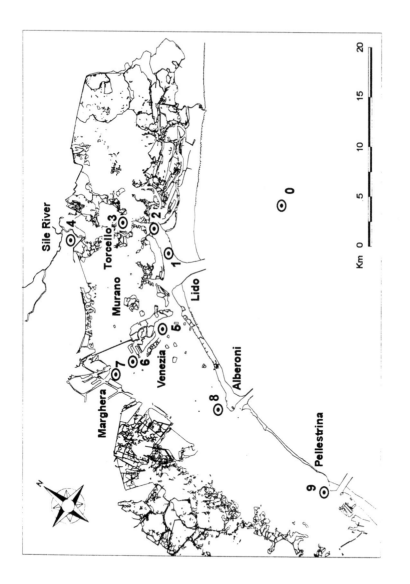

Figure 24.1 Map showing the location of the sample stations in the Venice Lagoon for fish immunology studies

RESULTS

The significant results recorded from the study on the immune system of the goby *Z. ophiocephalus* were reduction in the spleen somatic index, changes in differential blood cell counts and a reduction in total phagocytic capacity of spleen adherent cells in fish collected from Porto Marghera (station 7) in comparison with Crevan (station 2), Torcello (station 3) and Lio Grande (station 1).

A significant reduction ($p \leq 0.05$) in the spleen somatic index was apparent at Porto Marghera (station 7) in comparison with the other sites sampled in June (Figure 24.2), April and October. The kidney somatic index was also lower in fish from Porto Marghera (station 7) but this was not significant.

Differential blood cell counts in June (Figure 24.3) and October showed a significantly ($p \leq 0.05$) increased proportion of phagocytes in fish from Porto Marghera (station 7) with a reduction in lymphocytes and thrombocytes, although these differences were not apparent in April.

A highly significant ($p \leq 0.01$) reduction in phagocytic capacity, in both total percentage phagocytosis and the number of particles ingested by individual cells, of isolated spleen adherent cells was recorded in fish from Porto Marghera (station 7) compared with Crevan (station 2), Torcello (station 3) and Lio Grande (station 1) in June (Figure 24.4) and April. A similar pattern of results was only apparent in kidney phagocytic cells in June.

Contaminant levels obtained from the analysis of livers from gobies collected from the various sites in April, June and October 1992 are summarised in Table 24.1, and include metals, PAHs, PCBs and DDT. Fish collected from Porto Marghera (station 7) showed elevated levels of PAHs in April and October, and PCBs at all times of the year. The notable differences in metals between Porto Marghera (station 7) and other sites in April and October were the levels of nickel and chromium.

Levels of DDT were much higher in June (Figures 24.1–3) at all the sites, more so at Porto Marghera (station 7), presumably reflecting agricultural activity in the surrounding regions.

DISCUSSION

The significant results recorded from this study on the immune system of the goby, *Z. ophiocephalus*, included a consistent reduction in the spleen somatic index, increased numbers of phagocytes in peripheral blood and a reduction in the overall phagocytic capacity of adherent spleen cells, in fish collected from Porto Marghera (station 7) in comparison with Crevan (station 2), Torcello (station 3) and Lio Grande (station 1). Pavoni *et al.* (1990) found the sediments and overlying waters of Porto Marghera (station 7) to be the most contaminated part of the Lagoon in all the measurements taken, which included nutrient concentrations as well as chemical, physical and biological parameters.

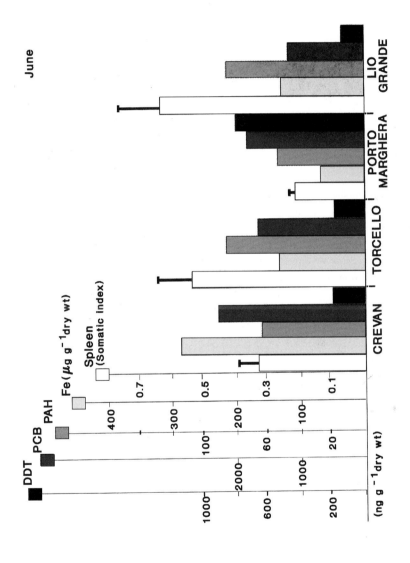

Figure 24.2 Spleen somatic indices from gobies, *Z. ophiocephalus*, sampled from four stations in the Venice Lagoon in June and contaminant levels recorded from liver analyses of the same fish

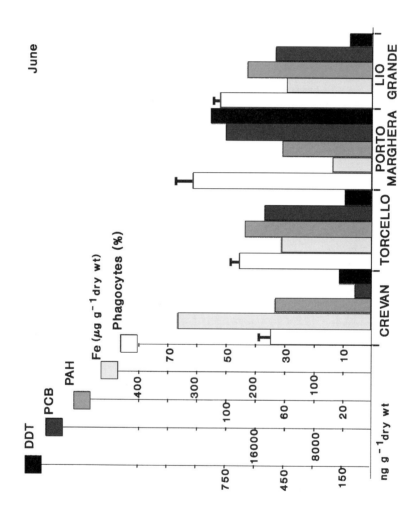

Figure 24.3 Differential blood cell counts from gobies, *Z. ophiocephalus* sampled from four stations in the Venice Lagoon in June 1992 and contaminant levels recorded from liver analyses of the same fish

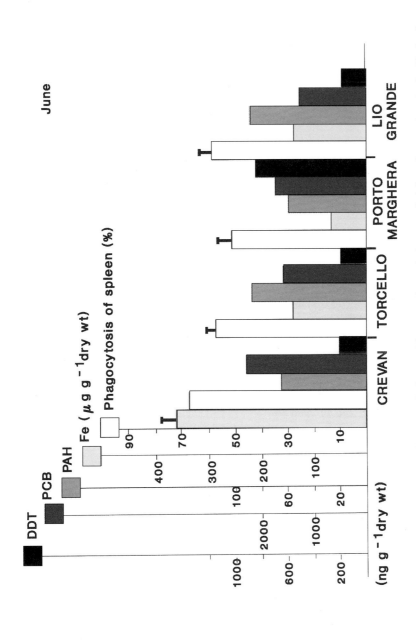

Figure 24.4 Spleen macrophage percentage phagocytosis from gobies, *Z. ophiocephalus*, sampled from four stations in the Venice Lagoon in June 1992 and contaminant levels from liver analyses of the same fish

Table 24.1 Selected contaminants in liver samples in gobies, *Z. ophiocephalus*, collected from the Venice Lagoon in April, June and October 1992. Trace metals in μg/g dry weight; polyaromatic hydrocarbons in ng/g dry weight; PCBs Aroclor 1260

Trace metals	April			June				October		
	Crev	L.Gra	P.Mar	Crev	L.Gra	P.Mar	Torc	Crev	L.Gra	P.Mar
Iron	185	178	49	336	129	67	144	147	134	67
Chromium	0.37	0.00	3.76	1.22	<0.03	<0.03	0.03	0.16	2.23	4.48
Nickel	0	0.50	4.17	1.64	0.69	0.38	0.19	1.61	0.60	5.22
PAH {16C	51	197	233	64	87	57	86	94	126	193
PCB	1993	2727	8863	2250	1240	18110	1567	1360	1740	7200
{DDT	1.0	0.9	2.2	194.7	147.5	827.0	192.0	184	217	461

The high level of DDT found in the livers of fish collected in June was correlated with a reduction in spleen somatic index, percentage phagocytosis, phagocytic index and a significant increase in the proportion of phagocytes in the peripheral blood in June. Suppression of neutrophil phagocytic activity, serum lysozyme and leucopenia has also been demonstrated in carp (Dunier and Siwicki, 1992) following exposure to organophosphorus insecticides.

The significant reduction in somatic index recorded from fish at Porto Marghera (station 7) in comparison with the other sites sampled in June, April and October could be an indication of stress in fish from this polluted site. Laboratory experiments have also shown that a significant reduction in spleen somatic index occurs in acutely stressed fish (Pulsford *et al.*, 1995). The spleen, a major storage area for red blood cells, is known to contract under the influence of catecholamines in stressed fish (Randall and Perry, 1992), forcing stored erythrocytes into the circulation (Fange, 1992). For fish living in the polluted and potentially anoxic conditions at Porto Marghera (station 7) this may contribute to increased oxygen levels in arterial blood.

The consistent reduction in spleen weight was accompanied by lower iron levels in the liver of fish collected from Porto Marghera (station 7), a condition which could be associated with anaemia. The spleen functions not only as a storage organ for erythrocytes in fish, but also as a site for recycling aged erythrocytes (Fange, 1992); exposure to contaminants may increase the rate of destruction of red cells. The haemoglobin is transformed into bile pigments and iron which is stored as haemosiderin or ferritin in the lymphomyeloid tissues and liver and may be reused in erythropoiesis. Pollutant-induced anaemia could, therefore, affect both spleen weight and liver iron levels, as laboratory experiments have already demonstrated effects on haemoglobin biosynthesis and anaemia in fish, following exposure to heavy metals (reviewed by Zelikoff, 1993).

Significant increases in the proportion of phagocytes in the peripheral blood were recorded from the fish from Porto Marghera (station 7) in June and

October, with a reduced proportion of lymphocytes and thrombocytes, although these differences were not apparent in April. Similar haematological changes have been recorded in fish exposed to contaminated sediments (Pulsford *et al.*, 1992a), sewage sludge (Secombes *et al.*, 1991) and those subjected to stress (Ellsaesser and Clem, 1986; Pulsford *et al.*, 1994). Increases in phagocytes in the peripheral blood have also been linked to disease (Bucke *et al.*, 1989; Pulsford *et al.*, 1992a).

Although haematological differences were recorded in fish collected from the various Lagoon sites, the possibility that these may be due to seasonal differences cannot be excluded. Other studies on fish have shown seasonal variations in many haematological and serological parameters (Folmar, 1993), although the proximity of all the sampling sites to each other makes it unlikely that seasonal differences between sites would occur. Coastal lagoons, such as the Venice Lagoon, are characterised by strong seasonal variations of physical, chemical and biological parameters (Calvo *et al.*, 1991) and large seasonal variations in nutrients (Sfriso *et al.*, 1988). This may affect the food available to fish, which may subsequently affect immunocompetence, as nutritional status has been shown to affect immune function and susceptibility to disease in fish (Blazer, 1992).

A significant reduction in overall phagocytic capacity of isolated spleen adherent cells was recorded in fish from Porto Marghera (station 7) compared with Crevan (station 2), Torcello (station 3) and Lio Grande (station 1) in June and April. A similar pattern of results was only apparent in kidney phagocytic cells in June. The findings of reduced phagocytic activity in macrophages from gobies collected from known contaminated sites in the Venice Lagoon are in agreement with Weeks *et al.* (1990b) who found significantly reduced phagocytic activity, both in percentage phagocytosis and phagocytic index, in fish collected from a polluted compared with an uncontaminated river. They also found a reduction in chemotactic activity but a simultaneous stimulation of pinocytotic activity. This concurs with our laboratory exposure experiments (Pulsford *et al.*, 1994) where some aspects of the immune response are stimulated and others suppressed following exposure to contaminants. This emphasises the requirement to conduct a range of assays in order to achieve a balanced evaluation of the effects of environmental stressors on the immune system.

The immune system functions within the complex physiology of the fish, and is regulated by the nervous and endocrine systems (Fange, 1992). Interpretation of the effects on the immune system of pollutant exposure should, therefore, take account of the possible indirect effects of stressors on the immune system. The environmental influences are also complex, in terms of the mixture of contaminants to which the fish are exposed and physical effects such as temperature changes can add further stress to fish already living under conditions of poor water quality. However, this study has demonstrated that the application of a range of simple assays to measure the immunocompetence of the goby *Z. ophiocephalus* has produced statistically significant differences in the function of the immune system and response of phagocytic cells which correlated with exposure to high levels of pesticides, PAHs, PCBs and metals at the most polluted site at Porto Marghera (station 7).

ACKNOWLEDGEMENTS

This work forms part of the UNESCO Venice Lagoon Ecosystem Project and was funded in part by the Commission of the European Community Research Programme on the Fisheries Sector (FAR) contract, AQ2419 and UK Department of the Environment contract 7/7/386.

REFERENCES

Anderson, D. P. (1990). Immunological indicators: effects of environmental stress on immune protection and disease outbreaks. *Amer. Fish. Soc. Symp.*, **8**: 38–50

Blazer, V. S. (1992). Nutrition and disease resistance in fish. *Ann. Rev. Fish Dis.*, **2**: 309–23

Bucke, D., Dixon, P. F. and Feist, S. W. (1989). The measurement of disease susceptibility in dab *Limanda limanda* L. following long-term exposure to contaminated sediments: preliminary studies. *Mar. Envir. Res.*, **28**: 363–7

Calvo, C., Donazzolo, F., Guidi, F. and Orio, A. A. (1991). Heavy metal pollution studies by resuspension experiments in the Venice lagoon. *Water Res.*, **25**(10): 1295–302

Coles, J. A., Farley, S. R. and Pipe, R. K. (1994). The effects of fluoranthene on the immunocompetence of the common marine mussel, *Mytilus edulis*. *Aquat. Toxicol.*, **30**: 367–79

Coles, J. A., Farley, S. R. and Pipe, R. K. (1995). Alteration of the immune response of the common marine mussel, *Mytilus edulis,* resulting from exposure to cadmium. *Dis. Aquat. Org.*, **22**: 59–65

Dunier, M. and Siwicki, A. K. (1992). Effect of organophosphorus insecticides used against ectoparasites on immune response of carp (*Cyprinus carpio*). In Michel, C. and Alderman, D. J. (eds). *Chemotherapy in Aquaculture: from Theory to Reality.* Office International des epizooties Paris, France

Ellsaesser, C. F. and Clem, L. W. (1986). Haematological and immunological changes in channel catfish stressed by handling and transport. *J. Fish Biol.*, **28**: 511–21

Fange, R. (1992). Fish Blood Cells. In Hoar, W. S., Randall, D. J. and Farrell, A. P. (eds). *Fish Physiology, The Cardiovascular System,* Academic Press Inc. London, Vol. XII, Part B, Chap. 1, 1–54

Folmar, L. C. (1993). Effects of chemical contaminants on blood biochemistry of teleost fish: a bibliography and synopsis of selected effects. *Envir. Toxicol. & Chem.*, **12**: 337–75

Fossato, V. U., Van Vleet, E. S. and Dolci, F. (1988). Chlorinated hydrocarbons in sediments of Venetian Canals. *Archo Oceanogr. Limnol.*, **21**: 151–61

Fossato, V. U., Campesan, G. Raboledda, L. C.and Stocco, G. (1989). Trends in chlorinated hydrocarbons and heavy metals in organisms from the Gulf of Venice. *Archo Oceanogr. Limnol.*, **21**: 179–90

Lemaire-Gony, S., Lemaire, P. and Pulsford, A. L. (1995). Effects of cadmium and benzo(a)pyrene on the immune system, gill ATPase and EROD activity of European sea bass *Dicentrarchus labrax*. *Aquatic Toxicol.*, **31**: 297–313

Pavoni, B., Donazzolo, R. and Orio, A. A. (1990). Sediments as a source of pollutants in the Venice Lagoon. *MAP Tech. Rep.*, **45**: 146–60

Peters, G., Nubgen, A., Raabe, A. and Mock, A. (1991). Social stress induces structural and functional alterations of phagocytes in rainbow trout *Oncorhynchus mykiss*. *Fish and Shellfish Immun.*, **1**: 17–21

Pipe, R. K., Coles, J. A., Thomas, M. E., Fossato, V. U. and Pulsford, A. L. (1995). Evidence for environmentally derived immunomodulation in mussels from the Venice lagoon. *Aquatic Toxicol.*, **32**: 59–73

Pulsford, A. L., Lemaire-Gony, S. and Farley, S. R. (1992a). Effects of environmental stress on dab *Limanda limanda* immune system. *Marine Environmental Quality Committee, ICES, C.M. 1992/E14*

Pulsford, A. L., Ryan, K. R. and Nott, J. (1992b). Metals and melanomacrophages in flounder *Platichthys flesus*, spleen and kidney. *J. Mar. Biol. Ass. U.K.*, **72**: 483–98

Pulsford, A. L., Thomas, M. E., Lemaire-Gony, S., Holden, J., Fossato, V. U. and Pipe, R. K. (1995). Studies on the immune system of the goby, *Zosterissor ophiocephalus*, from the Venice lagoon. *Mar. Poll. Bull.*, **30**: 586–91

Pulsford, A., Lemaire-Gony, S., Farley, S. R., Tomlinson, M., Collingwood, N. and Glynn, P. J. (1994). Effects of acute stress on the immune system of the dab *Limanda limanda*. *Comp. Biochem. Physiol. B*, **109**: 129–39

Randall, D. J. and Perry, S. F. (1992). Catecholamines. In Hoar, W. S., Randall, D. J. and Farrell, A. P. (eds). *Fish Physiology. The Cardiovascular System*, Academic Press Inc. London, Vol. XII, Part B, Chap. 4, 255–300

Secombes, C. J., Fletcher, T. C., O'Flynn, J. A., Costello, M. J., Stagg, R. and Houlihan, D. F. (1991). Immunocompetence as a measure of the biological effects of sewage sludge pollution in fish. *Comp. Biochem. Physiol.*, **100**c (1/2): 133–6

Secombes, C. J., Fletcher, T. C., White, M. J., Costello, M. J., Stagg, R. and Houlihan, D. F. (1992). Effect of sewage sludge on immune responses in the dab *Limanda limanda* L. *Aquat. Toxicol.*, **23**: 217–30

Sfriso, A., Pavoni, A., Marcomini, A. and Orio, A. A. (1988). Annual variations of nutrients in the lagoon of Venice. *Mar. Pollut. Bull.*, **19**(2): 54–60

Tahir, A., Fletcher, T. C., Houlihan, F. and Secombes, C. J. (1993). Effects of short-term exposure to oil-contaminated sediments on the immune response of dab, *Limanda limanda*. *Aquat. Toxicol.*, **27**: 71–82

Weeks, B. A. and Warinner, J. E. (1984). Effects of toxic chemicals on macrophage phagocytosis in two estuarine fishes. *Mar. Envir. Res.*, **14**: 327–35

Weeks, B. A., Keisler, A. S., Warinner, J. E. and Matthews, E. S. (1987). Preliminary evaluation of macrophage pinocytosis as a technique to monitor fish health. *Mar. Envir. Res.*, **22**: 205–13

Weeks, B. A., Warinner, J. E., Mason, P. L. and McGinnis, D. S. (1986). Influence of toxic chemicals on the chemotactic response of fish macrophages. *J. Fish Biol.*, **28**: 653–8

Weeks, B. A., Huggett, R. J. and Hargis, W. J. (1990a). Integrated chemical, pathological and immunological studies to assess environmental contamination. In Sandhu, S. S., Lower, W. R., Serres, F. J., De Suk, W. A. and Tice, R. R. (eds). *In situ evaluation of biological hazards of environmental pollutants*. Plenum Press, New York, 233–40

Weeks, B. A., Huggett, J. E., Warinner, J. E. and Matthews, E. S. (1990b). Macrophage responses of estuarine fish as bioindicators of toxic contamination. In McCarthy, J. F. and Shugart, L. R. (eds). *Biomarkers of Environmental Contamination*. L.R. Lewis Publishers CRC Press. Florida, 193–201

Zatta, P., Gobbo, S., Rocco, P., Perazzolo, M. and Favarato, M. (1992). Evaluation of heavy metal pollution in the Venetian lagoon by using *Mytilus galloprovincialis* as biological indicator. *Sci. Tot. Environ.*, **119**: 29–41

Zelikoff, J. T. (1993). Metal pollution-induced immunomodulation in fish. *Ann. Rev. of Fish Diseases*, **9**: 305–25

CHAPTER 25

PERSISTENT CHEMICAL POLLUTANTS IN MUSSELS AND GOBIES FROM THE VENICE LAGOON

V. U. Fossato, G. Campesan, L. Craboledda, F. Dolci
and G. Stocco

SUMMARY

Hydrocarbon, chlorinated hydrocarbon and trace metal contents in tissues of mussels (*Mytilus galloprovincialis*) and gobies (*Zosterissor ophiocephalus*) from different sites in the Venice Lagoon and adjacent Adriatic Sea were measured during five seasonal surveys between December 1991 and May 1993. The results are discussed in terms of differences in tissue types, seasons and stations, and possible risks to human health. Comparisons based on mussel data from a few stations sampled between 1976 and 1991–93 indicate a general decrease of HCH and DDT, steady levels for PCBs and most of the trace metals, and a significant increase of petroleum hydrocarbons and lead at some sites.

INTRODUCTION

It is generally recognised that there is no single or simple method of assessing marine environmental pollution and that it must ultimately be a combination of physicochemical (cause) and biological (effect) measurements. Since the 1970s indicator organisms have proved to be useful tools in identifying spatial and temporal variations of chemical contamination in coastal waters, as well as in coupling cause and effect measurements (Widdows and Donkin, 1992).

Several papers have reported the accumulation of petroleum hydrocarbons (Fossato and Siviero, 1974, 1975; Fossato and Dolci, 1976), chlorinated hydrocarbons (Fossato and Craboledda, 1979, 1980; Fossato *et al.*, 1979; Fossato, 1982; Nasci and Fossato, 1982) and heavy metals (Campesan *et al.*, 1981, Perdicaro, 1984; Zatta *et al.*, 1990) in molluscs and fish from the Venice Lagoon. The response in reproduction, growth and death of Lagoon organisms on exposure to chemical pollution was studied at population (Dalla Venezia and Fossato, 1977; Dalla Venezia *et al.*, 1982) and at molecular and cellular (Nasci *et al.*, 1985, 1989) levels. The present study was undertaken to investigate the co-occurrence of several types of persistent chemical contaminants in gobies and mussels. The current state of health of these organisms was also investigated by analysing a large spectrum of biochemical, toxicological and immunological parameters in the most sensitive organs, i.e. reproductive tissues and digestive gland in molluscs and liver in fish (see Lowe and DaRos, Chapter 21). As a single indicator organism, in general, does not reflect the impact of all pollutants in the water body and in the bottom sediment, two indicator species were selected for the monitoring.

The sessile filter-feeding mussel, *Mytilus galloprovincialis*, is found widespread in Adriatic lagoons, both as wild and cultured populations (annual production in the Venice Lagoon is about 2500 tons). It grows on pilings and rocks in the upper water layers and assimilates pollutants from solution and through ingestion of food and other suspended particulate matter. The mussel is the most popular indicator organism of chemical pollution in coastal zones throughout the world and can be used for realistic inter-regional comparisons (Goldberg, 1975; Fowler, 1990).

The goby, *Zosterissor ophiocephalus*, is a non-migratory bottom fish which burrows into sediments where the maximum concentrations of chemical pollutants are found (Van Vleet *et al.*, 1988). It accumulates pollutants from ingestion of food and absorption from solution through gills. The goby has a wide distribution in the Adriatic lagoons and is used for monitoring organo-chlorine residues and trace metals in lagoon waters (Campesan *et al.*, 1980; Fossato, 1982).

Samples in the Lagoon and the adjacent Adriatic Sea were collected from sites representing a wide range of environmental conditions and pollution loads (Figure 25.1).

MATERIALS AND METHODS

Sampling was done during five field surveys between December 1991 and May 1993. Mussels were sampled from six stations: Platform (station 1), Lio Grande (station 2), Crevan (station 3), Torcello (station 4), Salute (station 5), and Canale Vittorio Emanuele (CVN, station 6). Gobies were obtained from three stations: Lio Grande (station 2), Crevan (station 3), and Porto Marghera (station 7).

Composite samples from 10–15 specimens of mussels and 8–10 gobies of similar size were prepared for chemical analyses. The soft parts and digestive gland of mussels and livers and skinless fillets of gobies were dissected out and pooled separately for each category. The tissues were freeze-dried, blended, and their wet/dry weight ratios recorded.

Hydrocarbons and chlorinated hydrocarbons were Soxhlet-extracted for 8 h with *n*-hexane. CB-29 was added as an internal standard to check the recovery rate. The extract was evaporated at 50 °C to constant weight for determination of extractable organic matter (EOM). It was then dissolved in 1 ml of *n*-hexane and fractionated by chromatography on an alumina/silica gel column. Individual column eluates were analysed by FID and ECD gas chromatography (Carlo Erba 4160 GC) using a 30 m × 0.32 mm i.d. SE-54 fused silica column and hydrogen as carrier gas. Quantification was based on peak height/

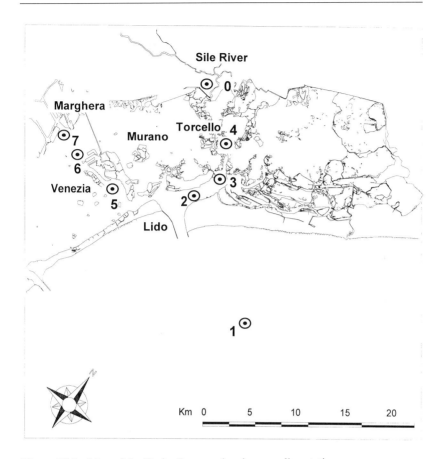

Figure 25.1 Map of the Venice Lagoon showing sampling stations

area measurement and comparisons with reference standards (*n*-alkanes from *n*-C$_{15}$ to *n*-C$_{30}$, Kuwait crude oil, Aroclor 1254 and 1260, DDTs, HCH). Identification and quantification of PCB congeners were also carried out for the congeners No. 52-101-110-(118+149)-138-153-180 (nomenclature according to Ballschmiter and Zell, 1980). Recovery was 94% for CB-29.

Polyaromatic hydrocarbons (PAH) analyses were performed on a Hewlett-Packard 1090A liquid chromatograph equipped with a reverse-phase column (200 × 2.1 mm ODS Hypersil 5 μm) and a Hewlett-Packard 1046A fluorescence detector. Identification and quantification of PAHs were done with a mixture of 12 external standards: fluorene (Fl), phenanthrene (Phe), fluoranthene (Ft), pyrene (Py), benz[a]anthracene (BaA), chrysene (Chy), benzo[b]fluoranthene (BbFt), benzo[k]fluoranthene (BkFt), benzo[a]pyrene (BaP), dibenz[a,h]anthracene (DBA), benzo[ghi]perylene (Bpe) and indeno[1,2,3-cd]pyrene (InP). Blanks were run at regular intervals and these showed that no significant contamination was introduced during the analytical procedure.

For heavy metal analyses, aliquots of about 0.5 g of freeze-dried tissue were digested with 5 ml of concentrated HNO_3 in a Teflon bomb at 120 °C for 4 h. The solution was then made up to 50 ml with distilled water. Measurements of heavy metals were carried out in a Perkin-Elmer model 372 atomic absorption spectrophotometer. Copper, iron, zinc and manganese were determined directly in an air–acetylene flame and cadmium, lead, nickel, cobalt and chromium, in a graphite furnace. Mercury was reduced with $NaBH_4$ and measured by a flameless procedure. Deuterium background corrector was used for Cd, Fe, Pb and Zn. Blank corrections were applied for each set of analyses.

RESULTS AND DISCUSSION

The results of field surveys on mussels and gobies are summarised in Tables 25.1 to 25.4 where average concentrations and standard deviations of organic pollutants and trace metals, expressed on a wet weight basis, are given. Wet to dry weight ratio of tissues and their lipid contents (EOM) are also reported in order to facilitate comparisons with other studies and with historical data from the Venice Lagoon. Note, that the standard deviation includes analytical uncertainty and seasonal variability, on the assumption that any consistent individual variability related to sex, age and size of the specimens has been eliminated by preparing composite samples.

The quality control of analytical data was done by repeated analyses of hydrocarbons, chlorinated hydrocarbons and trace metals in homogenate samples and/or reference materials distributed in the framework of the MURST Project Venetian Lagoon System and by participating in the Worldwide and Regional Inter-comparison for Determination of Organochlorine Compounds and Petroleum Hydrocarbons in Sediment Sample, IAEA-357. The results indicate that a good level of recovery (> 85%) and precision (SD < 10%) may be obtained when the concentrations of the analyte are relatively high, notwith-standing the complex nature of parameters such as aliphatic hydrocarbons (UCM) and PCBs, for which poor precision is generally obtained in the inter-laboratory tests (Farrington *et al.*, 1988; Villeneuve and Mee, 1992).

Levels of petroleum hydrocarbons, chlorinated hydrocarbons and trace metals in mussels and gobies of the study area appear to be dependent on the type of tissues, time of collection and site. Despite some concordance, distri-bution patterns of different classes of pollutants are strikingly different and hence need separate discussion.

Hydrocarbons

An unresolved complex mixture (UCM) in capillary gas chromatography made up 96–98% of the aliphatic fraction, the resolved part being mainly

Table 25.1 Aliphatic, polyaromatic and chlorinated hydrocarbons in *Mytilus galloprovincialis*. Mean ± SD in ng/g wet weight

Station	Wet wt/ Dry wt	EOM %	nC15–30	UCM 1000	PAH 12 C	HCH	DDT	PCB Ar 1254	PCB 8 C
Soft parts									
Platform	5.1 ± 1.0	1.3 ± 0.5	718 ± 573	58 ± 43	17 ± 9	0.7 ± 0.6	1.4 ± 0.6	13.8 ± 5.9	8.1 ± 3.8
Lio Grande	5.5 ± 0.8	1.2 ± 0.5	752 ± 516	65 ± 39	31 ± 17	0.7 ± 0.7	1.9 ± 1.0	25.1 ± 10.8	15.6 ± 8.1
Crevan	5.6 ± 1.1	1.2 ± 0.6	782 ± 492	55 ± 40	27 ± 8	0.7 ± 0.7	2.2 ± 1.1	30.1 ± 21.4	17.6 ± 13.4
Torcello	6.1 ± 1.2	1.0 ± 0.6	584 ± 231	66 ± 59	24 ± 3	0.2 ± 0.2	1.8 ± 0.7	25.7 ± 19.5	14.4 ± 12.9
Salute	5.1 ± 0.9	1.3 ± 0.8	1000 ± 639	151 ± 101	139 ± 111	1.0 ± 1.0	6.6 ± 4.6	141 ± 124	89 ± 76
CVE	4.3 ± 0.6	1.6 ± 0.6	2311 ± 1296	133 ± 62	74 ± 26	2.0 ± 1.5	5.6 ± 2.4	121 ± 57	75 ± 34
Digestive glands									
Platform	3.9 ± 0.5	4.8 ± 3.1	1086 ± 497	98 ± 50	36 ± 21	1.1 ± 0.9	5.9 ± 3.4	52 ± 18	29 ± 10
Lio Grande	4.5 ± 0.8	2.4 ± 0.7	1135 ± 357	135 ± 71	43 ± 19	0.9 ± 1.3	5.6 ± 2.6	67 ± 35	38 ± 16
Crevan	4.2 ± 0.6	2.0 ± 0.8	983 ± 381	127 ± 86	40 ± 25	1.1 ± 1.5	5.5 ± 3.9	55 ± 26	31 ± 14
Torcello	4.6 ± 0.8	2.0 ± 1.5	1201 ± 582	131 ± 96	21 ± 14	1.6 ± 2.6	4.5 ± 4.5	55 ± 39	29 ± 14
Salute	3.7 ± 0.4	3.8 ± 0.6	3530 ± 3662	732 ± 669	208 ± 208	2.8 ± 3.4	34.6 ± 6.6	511 ± 130	272 ± 67
CVE	3.6 ± 0.3	4.0 ± 0.4	1569 ± 356	242 ± 133	120 ± 104	3.5 ± 3.8	21.4 ± 7.1	438 ± 76	221 ± 29

EOM = extractable organic matter; *n*C 15–30 = total *n*-alkanes with 15 to 30 carbon atoms; UCM = unresolved complex mixture of aliphatic hydrocarbons; PAH 12C = polyaromatic hydrocarbons, sum of 12 individual compounds; HCH = hexachlorocyclohexane, sum of α and γ isomers; DDT = sum of pp'DDT, pp'DDD, pp'DDE; PCB 1254 = total polychlorinated biphenyls expressed as aroclor 1254; PCB 8C = polychlorinated biphenyls, sum of 8 individual congeners

Table 25.2 Aliphatic, polyaromatic and chlorinated hydrocarbons in *Zosterissor ophiocephalus*

Station	Wet wt/ Dry wt	EOM %	nC15–30	UCM 1000	PAH 12 C	HCH	DDT	PCB Ar 1254	PCB 8 C
Fillets									
Lio Grande	5.0 ± 0.2	5.0 ± 0.2	596 ± 348	34 ± 7	14.3 ± 17.2	0.5 ± 0.2	1.5 ± 0.6	11.0 ± 2.7	6.3 ± 1.6
Crevan	5.0 ± 0.1	5.0 ± 0.1	713 ± 341	46 ± 30	11.4 ± 8.1	0.5 ± 0.3	1.1 ± 0.3	12.6 ± 3.4	7.0 ± 1.3
P. Marghera	4.7 ± 0.4	0.3 ± 0.1	711 ± 353	50 ± 11	14.6 ± 9.7	1.1 ± 0.9	2.0 ± 0.4	47.7 ± 23.0	27.8 ± 12.0
Livers									
Lio Grande	2.2 ± 0.6	34.3 ± 16.2	4040 ± 1428	446 ± 364	63 ± 12	9.5 ± 14.1	92 ± 25	865 ± 153	503 ± 130
Crevan	2.3 ± 0.7	31.4 ± 16.4	3198 ± 1932	379 ± 287	35 ± 21	10.5 ± 15.8	85 ± 32	846 ± 179	475 ± 77
P. Marghera	2.3 ± 0.5	30.4 ± 16.6	4965 ± 2720	674 ± 395	76 ± 35	27.4 ± 50.3	238 ± 149	4738 ± 3321	2612 ± 1532

Mean ± SD in ng/g wet weight, abbreviations as in Table 25.1

Table 25.3 Trace metals in *Mytilus galloprovincialis*. Mean ± SD in $\mu g/g$ wet weight

Station	Wet wt/ Dry wt	Cu	Zn	Fe	Mn	Cr	Cd	Pb	Co	Ni	Hg
Soft parts											
Platform	5.1 ± 1.0	1.75 ± 1.14	24 ± 10	18 ± 6	1.67 ± 0.33	1.07 ± 1.16	0.48 ± 0.18	2.60 ± 1.34	0.15 ± 0.04	0.58 ± 0.49	0.04 ± 0.02
Lio Grande	5.5 ± 0.8	1.37 ± 0.23	25 ± 5	18 ± 3	1.25 ± 0.53	0.30 ± 0.36	0.29 ± 0.17	1.71 ± 0.89	0.21 ± 0.05	0.41 ± 0.30	0.06 ± 0.02
Crevan	5.6 ± 1.1	1.46 ± 0.50	25 ± 6	41 ± 12	1.88 ± 0.47	0.54 ± 0.81	0.48 ± 0.27	0.95 ± 0.21	0.22 ± 0.07	0.39 ± 0.39	0.11 ± 0.03
Torcello	6.1 ± 1.2	1.01 ± 0.20	18 ± 5	37 ± 14	1.27 ± 0.31	0.30 ± 0.24	0.58 ± 0.31	0.56 ± 0.34	0.16 ± 0.05	0.34 ± 0.16	0.10 ± 0.02
Salute	5.1 ± 0.9	2.62 ± 1.02	32 ± 17	22 ± 10	1.86 ± 0.57	0.66 ± 0.83	0.56 ± 0.38	3.19 ± 0.81	0.29 ± 0.13	0.66 ± 0.70	0.06 ± 0.01
CVE	4.3 ± 0.6	1.94 ± 0.99	26 ± 7	30 ± 23	2.11 ± 0.69	0.24 ± 0.09	0.54 ± 0.08	0.52 ± 0.24	0.28 ± 0.20	0.46 ± 0.19	0.05 ± 0.03
Digestive glands											
Platform	3.9 ± 0.5	2.83 ± 0.67	19 ± 6	108 ± 68	5.61 ± 2.78	0.85 ± 0.53	0.37 ± 0.21	1.94 ± 1.63	1.08 ± 0.27	2.05 ± 0.94	0.13 ± 0.09
Lio Grande	4.5 ± 0.8	3.57 ± 1.10	18 ± 5	171 ± 79	4.86 ± 2.82	0.74 ± 0.67	0.36 ± 0.21	1.52 ± 0.91	0.93 ± 0.26	2.43 ± 1.39	0.14 ± 0.05
Crevan	4.2 ± 0.6	3.10 ± 1.53	22 ± 6	261 ± 85	7.26 ± 0.93	0.60 ± 0.20	0.37 ± 0.19	2.19 ± 0.71	1.28 ± 0.36	2.77 ± 0.82	0.21 ± 0.05
Torcello	4.6 ± 0.8	3.41 ± 0.55	22 ± 7	253 ± 141	5.10 ± 1.61	0.70 ± 0.47	0.68 ± 0.40	1.51 ± 0.38	0.91 ± 0.14	2.14 ± 0.54	0.26 ± 0.05
Salute	3.7 ± 0.4	8.57 ± 3.91	27 ± 9	209 ± 136	7.63 ± 3.01	0.99 ± 0.41	1.00 ± 0.99	4.12 ± 0.99	1.90 ± 0.70	2.87 ± 1.21	0.20 ± 0.15
CVE	3.6 ± 0.3	3.06 ± 0.33	21 ± 7	146 ± 73	5.39 ± 2.00	2.10 ± 2.69	0.98 ± 0.23	3.76 ± 0.87	2.67 ± 1.31	4.61 ± 4.07	0.08 ± 0.04

Table 25.4 Trace metals in *Zosterissor ophiocephalus*. Mean ± SD in μg/g wet weight

Station	Wet wt/ Dry wt	Cu	Zn	Fe	Mn	Cr	Cd	Pb	Co	Ni	Hg
Fillets											
Lio Grande	5.0 ± 0.2	0.58 ± 0.19	8 ± 2	7 ± 3	0.72 ± 0.07	0.63 ± 0.07	< 0.04	0.56 ± 0.48	< 0.10	0.50 ± 0.10	0.12 ± 0.05
Crevan	5.0 ± 0.1	< 0.20	8 ± 2	5 ± 4	0.62 ± 0.11	1.09 ± 0.32	< 0.04	< 0.50	< 0.10	0.98 ± 025	0.18 ± 0.05
P. Marghera	4.7 ± 0.4	0.53 ± 0.58	6 ± 2	9 ± 8	< 0.50	1.30 ± 1.16	< 0.04	< 0.50	< 0.10	1.08 ± 1.06	0.15 ± 0.07
Livers											
Lio Grande	2.2 ± 0.6	2.36 ± 0.34	19 ± 6	68 ± 5	1.73 ± 0.81	< 0.20	< 0.04	2.33 ± 2.55	< 0.10	0.29 ± 0.09	0.10 ± 0.04
Crevan	2.3 ± 0.7	3.01 ± 0.28	23 ± 6	102 ± 38	1.72 ± 0.11	0.68 ± 0.52	< 0.04	4.08 ± 4.51	< 0.10	0.63 ± 0.35	0.18 ± 0.07
P. Marghera	2.3 ± 0.5	4.23 ± 1.73	23 ± 10	47 ± 28	2.11 ± 2.01	1.43 ± 1.02	0.10 ± 0.10	2.52 ± 3.48	0.13 ± 0.05	1.35 ± 1.15	0.15 ± 0.08

composed of *n*-alkanes (*n*-C15 to *n*-C$_{30}$), isoprenoids (pristane and phytane) and C$_{21}$ *n*-alkenes with 1, 4, 5 and 6 double bonds, probably derived from a dietary source (Blumer *et al.*, 1971; Youngblood *et al.*, 1971; Rowland and Volkman, 1982).

Hydrocarbons in the marine environment may result from direct petroleum input, fossil fuel combustion and biogenic sources. Our analyses of the aliphatic hydrocarbon fraction of water and suspended particulate matter (Fossato *et al.*, Chapter 6) and mussels and gobies show that hydrocarbons are accumulated and stored in the tissues of organisms without major apparent changes. Based on differences in composition and complexity, it is possible to distinguish between biogenic and pollutant hydrocarbons, although the relative contribution is often difficult to estimate. A high percentage of unresolved components and a continuous abundance of *n*-alkanes with an odd/even ratio of ≈ 1, point to either a recent small spill or to continuous exposure to low levels of petroleum. Biogenic hydrocarbons are characterised by less complexity, a predominance of odd carbon alkanes, a high pristane/ phytane ratio and ubiquitous presence of some individual alkenes. In the present study, hydrocarbons extracted from mussels and fish resemble those derived from direct petroleum inputs and fossil fuel combustion (dominance of parent PAHs over alkylated homologues), especially at the most polluted stations (Salute, CVE and Porto Marghera, station 5, 6, 7).

Hydrocarbon UCM concentrations in soft parts of mussels ranged from 13.1 μg/g wet weight (2.05 mg/g lipid weight) at Platform to 260 μg/g wet weight (32.9 mg/g lipid weight) at Salute. These concentrations are similar to those recorded in mussels from polluted areas in the Mediterranean Sea and elsewhere in the world (UNEP/IOC, 1988; Widdows and Donkin, 1992). Similar or higher levels of hydrocarbons (39.4–1490 μg/g wet weight; 0.94–47.6 mg/g lipid weight) have also been found in the digestive gland of mussels.

Since three sites (Platform, Salute and CVE) in the present study were the same as in earlier surveys, a temporal comparison became possible. The results are summarised in Table 25.5 which shows that aliphatic hydrocarbon concentrations in soft parts of mussels have increased substantially between 1974/76 and 1991/93, most likely as a result of the increase in boat traffic in the Lagoon near Venice and Porto Marghera and at the site where the platform is located.

In gobies, hydrocarbon concentrations ranged from 26.9 to 80 μg/g wet weight in muscles and from 198 to 1260 μg/g wet weight in liver, consistent with the distribution of lipids in the tissues. However, when the data are expressed on a lipid weight basis, muscles showed higher hydrocarbon contents than liver (9.1–19.8 mg/g vs 0.4–5.7 mg/g lipid weight). Fish liver is known to be a temporary storage site for chemical pollutants, since it serves mainly as an avenue for uptake and transport.

Table 25.5 Temporal comparisons of hydrocarbon and chlorinated hydrocarbon content in mussels. Mean ± SD in ng/g wet weight

Platform	1974–1976 (a,b,c)	1977–1980 (d)	1986–1988 (e)	1991–1993 (f)
AH*10^3	6.5 ± 2.3	–	–	58 ± 39
αHCH	1.1 ± 1.3	0.5 ± 0.5	0.2 ± 0.1	0.4 ± 0.5
γHCH	1.3 ± 0.7	0.7 ± 0.6	0.2 ± 0.1	0.3 ± 0.2
ΣHCH	2.4 ± 1.7	1.2 ± 1.0	0.4 ± 0.1	0.6 ± 0.5
pp'DDE	2.2 ± 1.4	4.5 ± 2.3	1.4 ± 0.6	0.8 ± 0.3
pp'DDD	2.2 ± 1.9	2.5 ± 1.3	0.9 ± 0.4	0.4 ± 0.1
pp'DDT	5.1 ± 3.9	6.0 ± 3.6	0.5 ± 0.2	0.2 ± 0.1
ΣDDT	9.5 ± 5.7	13.0 ± 7.0	2.8 ± 0.9	1.4 ± 0.6
ΣPCB	56 ± 24	41 ± 27	22 ± 6	13.8 ± 5.2
Salute	1974–1976(a,b,c)	1977–1980(d)	1986–1988	1991–1993(f)
AH*10^3	45 ± 19	109 ± 50		151 ± 82
αHCH	1.4 ± 0.6	1.0 ± 0.6		0.7 ± 0.7
γHCH	1.6 ± 0.5	4.0 ± 1.9		0.3 ± 0.2
ΣHCH	2.9 ± 1.0	5.0 ± 2.3		1.0 ± 0.8
pp'DDE	1.9 ± 1.1	12.3 ± 4.8		4.7 ± 2.9
pp'DDD	2.9 ± 1.3	9.2 ± 6.9		1.4 ± 0.7
pp'DDT	6.4 ± 2.7	9.6 ± 6.3		0.4 ± 0.2
ΣDDT	11.2 ± 4.7	31 ± 14		6.6 ± 3.7
ΣPCB	130 ± 52	268 ± 99		141 ± 101
CVE	1974–1976 (a,b,c)	1977–1980 (d)	1986–1988	1991–1993(f)
AH*10^3	33 ± 14	–		133 ± 54
αHCH	1.2 ± 0.3			1.6 ± 1.2
γHCH	1.5 ± 0.5			0.5 ± 0.3
ΣHCH	2.5 ± 0.9	3.0 ± 1.5		2.0 ± 1.3
pp'DDE	1.8 ± 0.8			3.3 ± 2.5
pp'DDD	3.0 ± 1.9			1.0 ± 0.5
pp'DDT	7.3 ± 4.4			0.3 ± 0.1
ΣDDT	12.2 ± 6.4	21 ± 10		5.6 ± 2.1
ΣPCB	107 ± 25	100 ± 35		121 ± 49

(a) Fossato and Siviero (1975); (b) Fossato and Dolci (1976); (c) Fossato and Craboledda (1979); (d) Fossato and Craboledda (1980); (e) Fossato *et al.* (1989); (f) Fossato *et al.* present study

The presence of petroleum hydrocarbons in marine organisms at the concentrations found in mussels and gobies does not represent a health problem for humans but may lead to tainting of edible species. An objectionable oily taste in mussels from the most polluted sites of the Venice Lagoon was noticed in earlier surveys (Fossato and Dolci, 1977) and was confirmed in the present study, although the extent of this problem remains unknown.

Twelve individual PAHs among the aromatic hydrocarbons that we analysed are listed as priority pollutants by the US Environmental Protection Agency. These PAHs are least volatile and hence are recovered sufficiently well during analyses. Moreover, they were almost fully resolved chromato-graphically and gave consistently measurable peaks in tissue extracts. Levels of PAHs in mussels are usually high relative to those in gobies inhabiting the same environment. This is probably due to differences in their ability to metabolise aromatic hydrocarbons; molluscs have little capacity for PAH metabolism compared to fish (Lee *et al.*, 1972).

No correlation was found between PAH levels and lipid contents of the tissues, hence even when concentrations are expressed on lipid weight basis, variability of PAH still remains. PAHs ranged from 9.4 to 265 ng/g wet weight (0.54 to 11.9 μg/g lipid weight) in soft parts of mussels and from 9 to 437 ng/g wet weight (0.44 to 10.6 μg/g lipid weight) in digestive gland. The major contributors are usually some three, four and five aromatic condensed ring compounds (phenanthrene, fluoranthene, pyrene, benzo(a)anthracene, benzo(ghi)perylene), although notable differences exist between tissues.

In gobies, PAHs ranged from 3.9 to 34.3 ng/g fillet wet weight (1.3 to 11.8 μg/g lipid weight) and from 17.7 to 113 ng/g liver wet weight (0.08 to 0.65 μg/g lipid weight). The PAH composition in fillet and liver of gobies showed a relatively high percentage of three and four aromatic condensed ring com-pounds than the polyaromatic hydrocarbons of higher molecular weight.

To our knowledge, PAH contents in organisms of the Venice Lagoon have never been reported, with the exception of benzo(a)pyrene and perylene measured in mussels during 1975 and 1976 using thin-layer chromatography and fluorescence spectroscopy (Fossato *et al.*, 1979). Unfortunately, differ-ences in the analytical methods between earlier and present studies do not allow a comparison, hence no inferences of the trends in BaP levels can be drawn.

The spatial distribution of both aliphatic and polyaromatic hydrocarbons was similar: the most contaminated site appears to be the central Lagoon (stations 5, 6, 7), followed by the northern lagoon (stations 2, 3, 4) and the Gulf of Venice (station 1).

Chlorinated pesticides and polychlorinated biphenyls

This survey shows that HCH and DDT contamination of mussels and gobies is widespread, in spite of the ban on agricultural and civil use of chlorinated pesticides since 1974–75. Measurable amounts (0.1 ng/g wet weight) of α and γ isomers of HCH were found in 90% of samples analysed, and DDT residues were present in all species at all seasons.

HCH and DDT values showed considerable seasonality and dependence on species and tissue types which are related to their lipid content; the highest values were measured in the digestive gland of mussels (8.9 ng/g wet weight for HCH; 39.1 ng/g wet weight for DDT) and liver of gobies (103 ng/g wet weight for HCH; 435 ng/g wet weight for DDT) in May/June, a period of active lipid accumulation after reproduction.

The results also showed that most of pp' DDT (58% to 95%) had been degraded to pp' DDD and pp' DDE, the latter generally being the most abundant degradation product. The amount of DDT degraded versus the total amount may be an index of DDT ageing in the environment and leads to the belief that the chlorinated pesticides detected at present were, to a large extent, introduced into the Lagoon years ago. This is also consistent with the decrease of the amount of ΣDDT between 1976 and 1991–93 in mussels (Table 25.5). A slight decrease in ΣDDT content in gobies from the area opposite to Porto Marghera was also found between 1977–80 (15 ± 12 ng/g dry weight for DDT at S. Giorgio in Alga – Fossato, 1982) and 1991–93 (9.6 ± 2.2 ng/g dry weight for DDT at Porto Marghera).

PCBs are synthetic compounds used in a variety of industrial and technical applications. Having been recognised as environmental priority pollutants, their use in Italy was restricted from 1982 to non-dispersive systems (DPR No. 82/915). Among the chlorinated hydrocarbons, PCBs were predominant in all species at all stations of the Venice Lagoon, as in industrialised areas (UNEP/FAO/WHO/IAEA, 1990).

The entire data collected in earlier chromatographic analyses were obtained with packed-glass columns which gave total PCB contents instead of individual component spectra. Hence, the quantification of PCBs was carried out using both procedures so that a comparison could be made. The PCB chromatographic pattern matches that of aroclor 1254 in mussels while it is similar to that of aroclor 1260 in gobies.

From the data presented in Tables 25.1 and 25.2, it seems that PCBs accumulated in mussel and goby tissues are related to their lipid (EOM) content and that the differences between species and between tissues greatly diminish when the concentrations are expressed on lipid weight basis rather than on wet weight basis. On the other hand, differences in PCB levels between stations still remain significant ($p < 0.05$): (Salute > CVE > Lio Grande ≈ Crevan ≈ Torcello > Platform, based on mussel data; Porto Marghera > Lio Grande ≈ Crevan, based on goby data).

An extended monitoring of chlorinated hydrocarbons (HCH, DDT, PCB) in mussels from the Venice Lagoon and adjacent Adriatic Sea was carried out during 1976 (Fossato and Craboledda, 1979). Further sampling and analysis of mussels from Platform, Salute and CVE was done in 1977–80 and later in 1986–88 at Platform (Fossato *et al.*, 1989). A comparison of ΣHCH, ΣDDT and ΣPCB content in mussels between 1976 and 1991–93 (Table 25.5) indicates that HCH and DDT concentrations decreased significantly at all sites and that PCB concentrations declined at Platform but remained essentially unchanged at Salute and CVE.

A review of the existing database on the state of chlorinated hydrocarbon pollution of the Mediterranean Sea was compiled by UNEP/FAO/WHO/IAEA (1990). A comparison of the present data with this shows that current levels of HCH, DDT and PCBs in mussels and gobies of the Venice Lagoon are similar to those reported from a majority of other industrial and urban areas of the Mediterranean Sea and the rest of the world; however, it is necessary to be cautious while comparing data from different regions/sources when inter-calibration exercises show a large dispersion of data and/or very uneven sampling between areas.

Trace metals

Trace metals in seawater exist partly in solution and partly in suspension, adsorbed to organic and inorganic particulate matter (Fossato *et al.*, Chapter 6). They enter the tissues of the organisms from solution and through ingestion of food and other particulate material. Bivalves accumulate trace elements from the three possible routes, while teleosts assimilate elements from both food and solution (Phillips, 1980).

Trace metal concentrations in mussels and gobies are summarised in Tables 25.3 and 25.4. The concentrations are at $\mu g/g$ wet weight level, i.e., about 10^3–10^4 times higher than the concentrations in water, depending on the species, tissue and metal. According to Italian regulations, the Hg and Pb contents must not exceed 0.7 $\mu g/g$ and 2 $\mu g/g$ wet weight, respectively, in mussels and fish destined for human consumption. All the samples analysed in the present study had a Hg content (0.02–0.41 $\mu g/g$) below this limit but 28% of the mussel samples had Pb contents higher than this limit, with the highest values being normally found at Salute and Platform.

Distribution patterns for the various elements are basically different, reflecting varying levels of local inputs and/or different biochemical processes. Some of them correlated significantly ($p < 0.05$) with each other: Cu/Mn, Hg/Fe and Cr/Ni, in both soft parts and digestive glands of mussels, Hg/Zn and Cr/Ni, in both fillets and livers of gobies. Differences in the concentrations of metals in the soft parts of mussels between stations were evident: Cu, Zn, Mn, Co and Ni contents were higher in the mussels at stations near Venice (CVE and Salute), while Fe and Hg were higher in the mussels in the northern part of the Lagoon (Crevan and Torcello). Unusually, high concentrations of Cr and Pb were found in the mussels at Platform. Digestive glands of mussels generally had higher metal contents than soft parts, especially in the case of Fe, Co and Ni, which were 2 to 9 times higher. Despite some differences, the distribution pattern of the elements in digestive glands resembles that of soft parts; Cu, Zn, Mn and Cd were higher at Salute, Cr, Co and Ni at CVE, and Fe, Pb and Hg at Crevan and Torcello in the northern part of the Lagoon.

On the whole, trace element contents were lower in gobies than in mussel tissues; in fact concentrations of Cu, Cr, Cd, Pb and Co were below the detection limits of the analytical methods, respectively, in 36% of fillet and 24% of liver samples. Trace metals generally accumulated more in livers than in fillets of goby, with statistically significant ($p < 0.05$) differences for Cu,

Zn, Fe, Mn and Pb. As for the distribution between the stations, trace metal levels in goby were relatively uniform, because the highest values were generally associated with large standard deviations. The data base is insufficient to discern significant differences between stations.

Table 25.6 shows a comparison of trace metal contents in mussels between 1977 and 1993 at three stations (Platform, Salute and CVE). At Platform, Cu, Fe, Cr and Pb concentrations were different between years but only the increase of Pb and the decrease of Fe were statistically significant ($p < 0.01$). Comparisons between 1979–80 and 1991–93 for five metals at Salute and CVE show significantly higher Pb levels at Salute and CVE in the present study, while no such clear variations are seen with data for other elements. Zatta *et al.* (1992) in 1988 surveyed the trace metal contents in mussels at 44 stations in the whole Lagoon, which included Lio Grande, Crevan and Torcello stations of the present study. A comparison of the present data with those of Zatta *et al.* (1992) showed good agreement for Cu, Zn, Cr and Hg, whereas concentrations of Fe and Mn were significantly higher than in the 1988 survey. This comparison also confirms the relatively high concentrations of Hg in mussels from the northern part of the Lagoon.

Table 25.6 Temporal comparison of trace metal contents in mussels. Mean and SD in $\mu g/g$ wet weight

Platform	1977–1979(a)	1979–1980(b)	1986–1988(c)	1991–1993(d)
Cu			0.99 ± 0.26	1.75 ± 1.14
Zn			17 ± 3	24 ± 10
Fe			47 ± 16	18 ± 6
Mn		1.80 ± 0.82	1.93 ± 0.31	1.67 ± 0.33
Cr		0.30 ± 0.08	0.28 ± 0.09	1.07 ± 1.11
Cd	0.33 ± 0.08	0.18 ± 0.05	0.23 ± 0.09	0.48 ± 0.18
Pb	1.23 ± 0.27	1.11 ± 0.27	0.45 ± 0.22	2.60 ± 1.34
Ni			0.69 ± 0.31	0.58 ± 0.49
Hg	0.05 ± 0.02	0.06 ± 0.05	0.03 ± 0.01	0.04 ± 0.02
Salute		1979–1980(b)		1991–1993(d)
Mn		2.80 ± 0.38		1.86 ± 0.57
Cr		0.23 ± 0.03		0.66 ± 0.83
Cd		0.64 ± 0.18		0.56 ± 0.38
Pb		1.83 ± 0.41		3.19 ± 0.81
Hg		0.09 ± 0.03		0.06 ± 0.01
CVE		1979–1980(b)		1991–1993(d)
Mn		2.06 ± 0.30		2.11 ± 0.69
Cr		0.17 ± 0.03		0.24 ± 0.09
Cd		0.89 ± 0.33		0.54 ± 0.08
Pb		1.26 ± 0.48		0.52 ± 0.24
Hg		0.07 ± 0.04		0.05 ± 0.03

a: Viviani *et al.* (1983), b: Campesan *et al.* (1981), c: Fossato *et al.* (1989), d: Present study

Information on trace metal content in fish from the Lagoon is sparse. Data are available for Hg in *Z. ophiocephalus* measured in 1977–80 at Porto Marghera (Campesan *et al.*, 1980): Hg levels do not appear to have changed either in fillets (0.20 ± 0.13 $\mu g/g$ *vs* 0.15 ± 0.07 $\mu g/g$) or livers (0.13 ± 0.09 $\mu g/g$ *vs* 0.15 ± 0.08 $\mu g/g$) between 1977 and 1993.

CONCLUSIONS

Levels of hydrocarbons, chlorinated hydrocarbons and trace metals in mussels and gobies of the Venice Lagoon appear to be dependent on species, tissues, time and space.

Seasonal variability of lipophilic pollutants and their distribution in the tissues are related to the lipid content of tissues; despite some concordance, the distribution patterns of some trace metals are more complex and erratic.

The spatial distribution of aliphatic, polyaromatic and chlorinated hydrocarbons is similar: the most contaminated area appears to be the central Lagoon, followed by the northern Lagoon and the Gulf of Venice. Concentrations of trace elements in the organisms are quite different between stations, which indicates varying local inputs and/or different biochemical processes. Cu, Zn, Mn, Cd, Cr, Co and Ni in mussels were higher in the central Lagoon between Venice and Porto Marghera, while Fe, Pb and Hg were higher in the northern part of the Lagoon.

A comparison of mussel data from a few stations sampled between 1976 and 1991–93 showed different trends for different classes of compounds, possibly related to their use and release into the Lagoon. In particular, the chlorinated pesticides, HCH and DDTs, decreased at all sites whereas PCB concentrations remained essentially the same. At the same time, aliphatic hydrocarbons and Pb significantly increased at some sites, probably as a result of increased boat traffic and combustion of leaded gasoline.

The continued use of mussels is recommended to extend the long-term time series data already generated and to monitor the chemical pollutants in the water body and the food web. On the whole, current levels of hydrocarbons, chlorinated hydrocarbons and trace metals in mussels and gobies of the Venice Lagoon are similar to those reported from a majority of other industrial and urban areas of the Mediterranean Sea and the rest of the world and are, when measurable, mostly within established limits. It is apparent that, for the most part, neither molluscs and fish stocks nor man are at risk from levels of persistent chemical pollutants in the Lagoon; however, studies on sublethal responses of Lagoon organisms to chemical exposure might prove interesting.

REFERENCES

Ballschmiter, K. and Zell. M. (1980). Analysis of polychlorinated biphenyls (PCB) by glass capillary gas chromatography. *Fresenius Z. analyt. chem.*, **302**: 20–31

Blumer, M., Guillard, R. R. L. and Chase, T. (1971). Hydrocarbons of marine phytoplankton. *Mar. Biol.*, **8**: 183–9

Campesan, G., Fossato, V. U. and Stocco, G. (1981). Metalli pesanti nei mitili, *Mytilus* sp., della Laguna Veneta. *Ist. Ven. Sci., Rapporti e Studi*, **8**: 141–52

Campesan, G., Capelli, R., Pagotto, G., Stocco, G. and Zanicchi, G. (1980). Heavy metals in organisms from the Lagoon of Venice. *V^es Journées Etud. Pollutions, Cagliari, CIESM*, 317–22

Dalla Venezia, L. and Fossato, V. U. (1977). Characteristics of suspension of Kuwait oil and Corexit 7664 and their short- and long-time effects on *Tisbe bulbisetosa* (Copepoda harpacticoida). *Mar. Biol.*, **42**: 233–7

Dalla Venezia, L., Fossato, V. U. and Scarfì, S. (1982). First observations on physiological and behavioural response of *Mytilus galloprovincialis* to PCB Aroclor 1254 pollution. *VIes Journées Etud. Pollutions, Cannes, CIESM*, 669–75

Farrington, J. M., Davis, A. C., Frew, N. M. and Knap, A. (1988). ICES/IOC intercomparison exercise on the determination of petroleum hydrocarbons in biological tissues (mussel homogenate). *Mar. Pollut. Bull.*, **19**: 372–80

Fossato, V. U. and Siviero, E. (1974). Oil pollution monitoring in the Lagoon of Venice using the mussel *Mytilus galloprovincialis*. *Mar. Biol.*, **25**: 1–6

Fossato, V. U. and Siviero, E. (1975). Idrocarburi alifatici in mitili prelevati da una stazione del Golfo di Venezia, scelta quale riferimento nella valutazione del grado di inquinamento della Laguna. *Accademia Nazionale dei Lincei Rendiconti, Serie VIII*, **58**: 641–6

Fossato, V. U. and Dolci, F. (1976). Inquinamento da idrocarburi nel bacino settentrionale della Laguna Veneta. *Archo Oceanogr. Limnol.*, **19**: 169–78

Fossato, V. U. and Dolci, F. (1977). Inquinamento da idrocarburi nel bacino centrale e meridionale della Laguna Veneta. *Archo Oceanogr. Limnol.*, **19**: 47–54

Fossato, V. U. and Craboledda, L. (1979). Chlorinated hydrocarbons in mussels, *Mytilus* sp., from the Laguna Veneta. *Archo Oceanogr. Limnol.*, **19**: 169–78

Fossato, V. U. and Craboledda, L. (1980). Chlorinated hydrocarbons in organisms from the Italian coast of the Northern Adriatic Sea. *V^es Journées Etud. Pollutions, Cagliari, CIESM*, 169–74

Fossato, V. U. (1982). Etude des hydrocarbures chlorés dans l'environnement de la Lagune de Venise. *VIes Journées Etud. Pollutions, Cannes, CIESM*, 465–8

Fossato, V. U., Campesan, G., Craboledda, L. and Stocco, G. (1989). Trends in chlorinated hydrocarbons and heavy metals in organisms from the Gulf of Venice. *Archo Oceanogr. Limnol.*, **21**: 179–90

Fossato, V. U., Nasci, C. and Dolci, F. (1979). 3,4-benzopyrene and perylene in mussels, *Mytilus* sp., from the Laguna Veneta, north-east Italy. *Mar. Environ. Res.*, **2**: 47–53

Fowler, S. W. (1990). Critical review of selected heavy metal and chlorinated hydrocarbon concentrations in the marine environment. *Mar. Environ. Res.*, **29**: 1–64

Goldberg, E. D. (1975). The mussel Watch – A first step in global marine monitoring. *Mar. Pollut. Bull.*, **6**: 111

Lee, R. F., Sauerheber, R. and Dobbs, G. H. (1972). Uptake, metabolism and discharge of polycyclic aromatic hydrocarbons by marine fish. *Mar. Biol.*, **17**: 201–8

Nasci, C., Campesan, G. and Fossato, V. U. (1985). Indici fisiologici e biochimici di stress in *Zosterissor ophiocephalus* della laguna di Venezia. *Oebelia*, **11**: 883–5

Nasci, C., Campesan, G., Fossato, V. U., Dolci, F. and Menetto, A. (1989). Hydrocarbon content and microsomal BPH and reductase activity in mussel, *Mytilus* sp., from the Venice area, north-east Italy. *Mar. Environ. Res.*, **28**: 109–12

Nasci, C. and Fossato, V. U. (1982). Studies on physiology of mussels and their ability in accumulating hydrocarbons and chlorinated hydrocarbons. *Environ. Technol. Letters*, **3**: 273–80

Perdicaro, R. (1984). Study of cadmium, chromium, copper, iron, lead, mercury and nickel bioaccumulation in three mollusc species, *Crassostrea gigas, Ostrea edulis* and *Venerupis decussata*, breeding in an experimental park situated in Venice Lagoon. *Riv. Idrobiol.*, **23**: 129–43

Phillips, D. J. H. (1980). *Quantitative Aquatic Biological Indicators*. Applied Science Publishers Ltd., London, pp. 488

Rowland, S. J. and Volkman, J. K. (1982). Biogenic and pollutant aliphatic hydrocarbons in *Mytilus edulis* from the North Sea. *Mar. Environ. Res.*, **7**: 117–30

UNEP/FAO/WHO/IAEA (1990). *Assessment of the state of pollution of the Mediterranean Sea by organohalogen compounds.* MAP Technical Report Series No. 39 UNEP, Athens, pp. 224

UNEP/IOC (1988). *Assessment of the state of pollution of the Mediterranean Sea by petroleum hydrocarbons.* MAP Technical Report Series No. 19 UNEP, Athens, pp. 130

Van Vleet, E. S., Fossato, V. U., Shervin, M. R., Lovett, H. B. and Dolci, F. (1988). Distribution of coprostanol and chlorinated hydrocarbons in sediments from canals and coastal waters of Venice, Italy. *Org. Geochem.* **13**: 757–63

Villeneuve, J. P. and Mee, L. D. (1992). World-wide and regional intercomparison for the determination of organochlorine compounds and petroleum hydrocarbons in sediment sample IAEA-357. *IAEA Rep. No. 51*, p. 53

Viviani, R., Crisetig, G., Cortesi, P., Poletti, R. and Serrazanetti, G. P. (1983). Heavy metals (Hg, Pb, Cd) in selected species of marine animals from the North and Middle Adriatic Sea. *Thalassia Jugosl.*, **19**: 383–91

Widdows, J. and Donkin, P. (1992). Mussels and environmental contaminants: bioaccumulation and physiological aspects. *The Mussel Mytilus: Ecology, Physiology, Genetics and Culture.* In Gosling, E. (ed.). Elsevier, Amsterdam, 383–424

Youngblood, W. W., Blumer, M., Guillard, R. R. L. and Fiore, F. (1971). Saturated and unsaturated hydrocarbons in marine benthic algae. *Mar. Biol.*, **8**: 190–201

Zatta, P., Gobbo, S., Rocco, P., Perazzolo, M. and Favarato, M. (1992). Evaluation of heavy metal pollution in the Venetian Lagoon by using *Mytilus galloprovincialis* as biological indicator. *Sci Total Environ.*, **119**: 29–41

Section 5

Modelling

CHAPTER 26

MODELLING TROPHIC EVOLUTION OF THE VENICE LAGOON

G. Bendoricchio

SUMMARY

The modelling of trophic dynamics carried out in the framework of the UNESCO-MURST project on the Venice Lagoon is summarized. A system approach (Figure 26.2) was considered in order to account for the complexity of the trophic structure, simplifications needed for modelling purposes. The modelling approach is useful and needed to quantify behaviour of the system.

The mathematical model for seasonal changes of *Ulva* biomass in such a eutrophic environment is discussed. A main finding is the necessity to take into account physical disturbance due to winds and tides in simulating the sharp and sudden changes in *Ulva* biomass. The biological processes simulated with the model can predict the expected biomass evolution driven by light, temperature and nutrients. This evolution can be quite different from observations because of advective transport of floating mats of algae due to water movements. Planned investigations need to be carried out to verify this plausible hypothesis.

A shift in the composition of the primary producer community, from a phanerogam-dominated environment to an *Ulva*-dominated one, was observed. The presence of *Zostera* sp. is usually considered an indicator of good quality environment. A simple but satisfactory model built for *Zostera noltii* (phanerogam typical of lagoon systems) can be integrated into a structural dynamic approach for such an evolutionary system.

A SYSTEM OVERVIEW OF THE VENICE LAGOON

An ideal and comprehensive model of a natural system would include, as far as possible, all state variables, forcing functions and feedbacks of the system. Since natural systems are inherently complex, such models also tend to be very complex. Such complex models are almost impossible to construct and, when made, unusable for practical purposes of research and management of the system. Nevertheless, a rough conceptual sketch of the general system is still needed to evaluate the degree of simplification and to quantify the loss of information as a consequence of restriction of the model to only a subsystem of the general system.

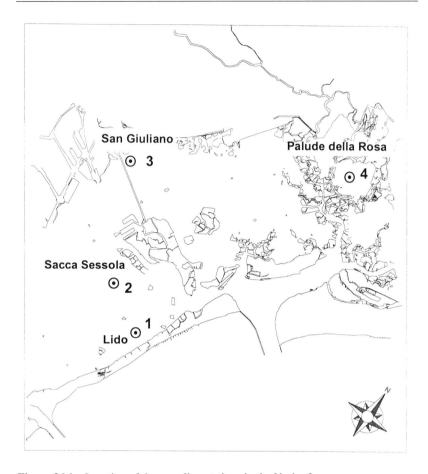

Figure 26.1 Location of the sampling stations in the Venice Lagoon

As shown in Figure 26.1, the Venice Lagoon system can be considered part of a more general system that includes both the watershed and the Adriatic Sea. The first advects freshwater and pollutants into the Lagoon and the second dilutes them with the tide-induced circulation.

With attention focused only on water quality and trophic dynamics in the Lagoon, the feedbacks between the Lagoon subsystem and the watershed and sea subsystems are negligible. The nutrient export from the Lagoon towards the watershed is almost non-existent compared with the quantity of nutrients entering the Lagoon from the watershed. Similarly, freshwater loss by evaporation from the Lagoon surface contributes little to rainfall on the mainland. The magnitude of seawater exchange during a tidal cycle through the three inlets is determined by astronomical, meteorologic and physical conditions that are unrelated to the state of the Lagoon. The quality of seawater entering

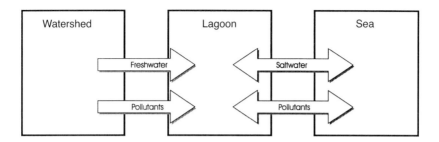

Figure 26.2 Watershed and sea are the subsystems interacting with the Lagoon subsystem through two main forcing actions: the input of freshwater and pollutants from the watershed and the input induced by the circulation of seawater. Feedbacks, such as transfer of pollutants from the Lagoon into the sea are of minor importance

the Lagoon is somewhat related to natural cycles occurring in the Lagoon because part of polluted water leaving the Lagoon subsystem during an ebb tide returns with the following flood tide. This fact generates an appreciable effect of Lagoon water quality on the seawater. In spite of this, under the restrictive hypothesis of well mixed water in the sea, and considering the large differences between water quality of the inner Lagoon and the sea, this effect can be neglected for at least some important indices (eg. salinity, nutrients, dissolved oxygen, etc.).

These simplifications of the general system enable us to consider the Lagoon subsystem model separately from others. The inputs from the sea and watershed can be regarded as forcing functions without feedback effects. The Lagoon becomes our new system and Figure 26.3 shows a possible selections of compartments and flows describing its trophic dynamics. Though the trophic relations might appear complex, the diagram is by no means the most complicated. In fact, each of these compartments can be described in more detail as autonomous subsystems. The degree of complexity and detail depends on the user's selection of system constituents (state variables, forcing functions and parameters). Each compartment refers to some state variables and the arrows connecting them show the direction of flows of matter in the network. Arrows entering the system indicate the functions forcing it in terms of nutrients, toxicants, water and energy, and those leaving the system describe permanent output of the same quantities from the system. The biomass can be selected as a trophic index and quantified in terms of mass or energy, and its flows and accumulation functions are driven by the inputs and outputs of energy and nutrients to and from the system.

A further simplification of this conceptual diagram can be made by considering only the compartments that have high productivity and, over an annual cycle, maximum stored biomass. The *primary producers* compartment is the more productive and can be isolated from the rest of the network. The

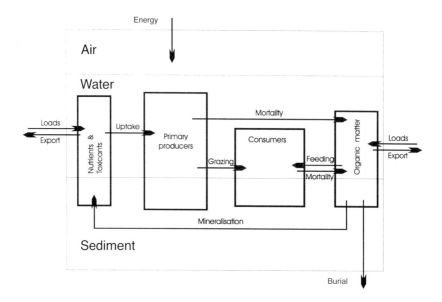

Figure 26.3 A schematic representation of the trophic network of the Lagoon ecosystem

importance of this compartment resides in its high productivity and the dystrophic cycle that such overproduction induces. In the last half of the century the natural evolution of the Venice Lagoon has been altered from its previous equilibrium by input of nutrients of anthropogenic origin. The present situation of the Lagoon cannot ensure structural stability; it can lead to some discontinuities of the network and an irreversible state of the system, unacceptable from the environmental point of view.

Attempts at modelling to solve some problems of the Venice Lagoon should mainly consider the primary producer compartment. Figure 26.4 shows a representation of the components of this compartment. The macroalgal compartment plays an important role in the Lagoon ecosystem and a large amount of the biomass is stored in it. The rapid decomposition of this huge biomass in wide areas of the Lagoon results in depletion of the dissolved oxygen-inducing anaerobic processes that are typical of eutrophic systems. This justifies the assumption that the macroalgae blooms are at the origin of major environmental problems in the Lagoon. Phytoplankton and benthic diatoms play a marginal role (10%) in terms of aquatic biomass. Due to increased erosion, the extent of emerged areas (11 km²) is only marginal compared with shallow water areas (245 km²). As a consequence, the saltmarsh vegetation shows lower annual production compared with the macroalgae compartment.

Even though annual seagrass production in the Venice Lagoon is limited compared with macroalgal production, its presence is an indication of good

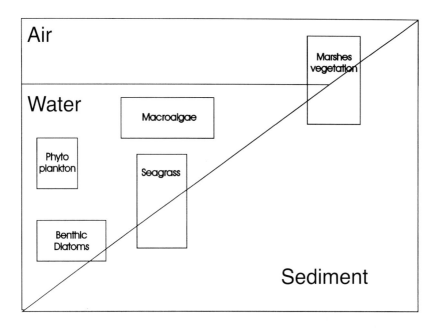

Figure 26.4 Major components of the conceptual diagram for the *primary producers* compartment

environmental conditions, whereas the presence of *Ulva* has a negative impact. Further, competition for natural resources, particularly light, occurs between seagrasses and macroalgae and among different species of macroalgae.

ECOLOGICAL MODELS FOR THE VENICE LAGOON

Ecological models are powerful tools to understand and simulate behaviour of ecological systems. They are usually built with mathematical functions that reproduce the major processes of the ecological system, and their use is rather recent compared with more mature experimental research. Nevertheless, the results of the application of models are very important and are promising for further interesting developments.

The most important reasons for using models of ecological systems are:

(1) Complexity can be simulated better than with physical models such as micro- or mesocosms;
(2) Mathematical description forces the modeller to simplify and give a rational and quantitative description of processes based on experimental results and general knowledge;
(3) Models explain experimental data and indicate further experimental research needed; and

(4)   Models save money and time in simulating past evolution of the eco-
      logical system and in predicting its future behaviour.

Many ecological systems (lakes, rivers, bays, estuaries and terrestrial ecosystems) have been successfully simulated with different techniques and the
models have provided significant information for their ecological restoration.
In the 1980s and 1990s several ecological models have been developed for the
Venice Lagoon (see Coffaro and Sfriso, Chapter 28; Dejak *et al.*, 1987; DHI,
1987; Bendoricchio *et al.*, 1994; Bergamasco *et al.*, 1994; Zharova *et al.*,
1994) with different aims (distribution of nutrients, growth of phytoplankton
and macroalgae) and various scales of time and space. These models stress the
dynamic behaviour of the Lagoon with respect to hydrology, dispersion of
nutrients and biogeochemical cycles, and simulate with different detail the
growth cycle of macroalgae and/or phytoplankton. The models use well-
known basic processes of phytoplankton and macroalgal physiology. Some
models have been calibrated and validated with experimental data. The major
constraints of these models are the limited volume of experimental data
available for calibration and validation and lack of laboratory studies focused
specifically on obtaining quantitative values for some crucial parameters and
processes. Notwithstanding these, the models developed for the Venice
Lagoon are compatible with others developed for similar ecological systems
around the world and can compete scientifically with those published recently
in the international literature.

MODELLING ACTIVITIES IN THE UNESCO PROJECT

In the general framework of biological research carried out for the Venice
Lagoon, the mathematical models of trophic dynamics of the system can help
to clarify the principal roles some species of algae and macrophytes play in
biological cycles of an ecotone such as the Venice Lagoon.

The model reproduces, with mathematical functions, some experimental data
collected in the field. The values of parameters of functions are usually found
in the literature or directly measured in the laboratory. The degree of uncertainty introduced by the simulation is somehow similar to, and of the same
order of magnitude as, that introduced by laboratory experiments. The mathematical and laboratory methods are complementary and integrate acquisition
of field data for the final purposes of understanding the complex evolution of
a natural system.

The modelling work done at the University of Padova is a continuation of
earlier work (Bendoricchio *et al.*, 1993) focused mainly on an *Ulva* growth
cycle model. This work began with a literature review (Coffaro, 1993) of the
processes involved in the macroalgal and macrophyte life cycle and
culminated in development of models for the life cycles of *Ulva* and *Zostera*.
The model for *Ulva* (see Coffaro and Sfriso, Chapter 28) has been validated

with the field data collected in the Palude della Rosa. The model for *Zostera* (see Bocci, Chapter 27) is a simplified version of previous models described in the literature.

FIELD DATA COLLECTION

The field data from Palude della Rosa were collected on two different time scales. The data on the longer time scale (nutrients, temperature, *Ulva* biomass) were collected using the methodology described by Sfriso *et al*. (1993). Data on short-term (every few minutes) variations affecting the *Ulva* cycle were recorded with electrochemical probes.

Figure 26.5 (from Carrer *et al*., 1993) is an example of diel variations in dissolved oxygen at different periods of the *Ulva* growth season recorded automatically. The oxen depletion is strikingly evident and also critical in late June when the standing stock of *Ulva* was low; at this time, decomposition was predominant over production. On the other hand, when biomass was high (5 kg fresh weight/m^2) in May, there was net production of oxygen. These changes demonstrate clearly that oscillations in some parameters are more evident on short time scales and hence have to be simplified before inclusion in the model.

Figure 26.6 shows long-term changes of some parameters at station 3. A comparison of the standing stock of *Ulva* with total nitrogen concentration clearly shows the limiting effect of nitrogen in the 1992 growth season.

The macroalgal growth in 1993 in the sampling area (station 3 on the map) was very low compared with growth in the previous year and at other sites in the Lagoon. The same data also show that total N was limiting in 1992 and non-limiting in 1993, suggesting that some other processes, such as grazing, light limitation and washout, control growth. The validation of the model with this new finding was a good test of its flexibility and reliability.

The field data also demonstrated a large variability in biomass at the sampling site even when sampling frequency was high. This is mainly due to physical disturbance induced by tides and winds, necessitating their introduction into the model as forcing functions due to mechanical energy inputs. A check on field and laboratory data also showed that they are affected by a large error range that has to be considered in comparing results from the model with them.

THE ULVA MODEL

The system simulated with the *Ulva* model is a further simplification of the conceptual network shown in Figure 26.3. It is a zero dimensional model, i.e. it reproduces the dynamics of *Ulva* biomass at one point of the Lagoon characterized by shallow waters, high biomass, and high rates of growth and decomposition. The forcing functions considered in the model are light

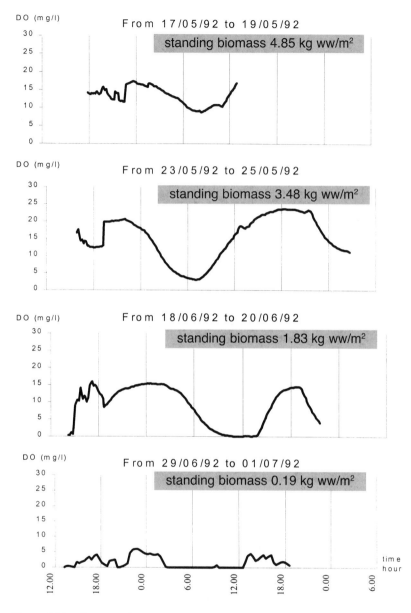

Figure 26.5 Changes of dissolved oxygen concentrations at different moments of the *Ulva* growth season with different biomasses

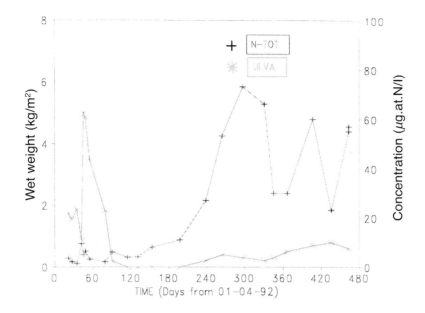

Figure 26.6   Long-term records of *Ulva* biomass and total nitrogen concentration

intensity, water temperature, concentration of dissolved nutrients and wind-induced washout of macroalgae. The state variables are *Ulva* biomass and internal quota of nutrients. The first three forcing functions are related to trophic structure, while the wind is a physical disturbance. In fact, a sufficiently strong wind causes a washout of macroalgae from shallow areas to the canals; this forcing function is valid for all three sites (1, 2 and 3) considered in this project.

Light dependence was taken into account with a functional relationship supported by experimental data (Keith and Murray, 1980; Brocca and Felicini, 1981). An adaptation effect was considered in modelling temperature dependence: the optimum temperature was modified as a function of past temperature changes that algae were subjected to, as suggested by O'Neil *et al.* (1972). Nutrient uptake was simulated with the classic Michaelis–Menten uptake kinetics.

Using the three forcing functions, the model can duplicate the biological cycle of *Ulva* and reproduce well critical features of the phenomenon, such as expected spring biomass peak, summer collapse and dormant transition from one growth season to the next, during winter. The model takes into account only the energy input from light and temperature, both of which are related to solar radiation and not inputs from tides and wind-induced currents.

The canals of the Lagoon and areas close to Lagoon inlets have high turbulence and the mechanical energy input is a few orders of magnitude greater than solar energy input. On the other hand, average water velocity is low in shallow water areas, and the input of mechanical energy becomes comparable with that of solar energy. In such a case, additional wind energy input is critical for energy balance.

Sometimes, due to strong tides and winds, all the *Ulva* is washed out of the system. This would explain the sudden reduction of biomass recorded in sampling areas. Even if strong wind and high tide events are unpredictable, they generate a significant decrease of the biomass not only through washoff but also by increasing turbidity and reducing growth rate. The inclusion of these effects in the model has clearly improved agreement between measured and simulated data, accounting for the sharp and sudden reductions of biomass. Nevertheless, such an effect, because of its random nature, cannot be taken into account when the model is used for forecasting purposes.

The *Ulva* model was used to predict the biomass cycle in the simplified system under the hypothesis of a reduction of nutrients. The results of such a simulation clearly show that nitrogen limits the growth cycle of *Ulva* and that only a reduction of 80% in the nutrient load can ensure a reduction of 70% of the peak standing stock, necessary to arrive at an acceptable trophic state.

The over-simplifications applied to the system, however, produce some modelling problems that limit general validity of the results. The zero dimension feature of the model cannot take into account circulation of water and nutrients in the Lagoon. A possible application of such a model within a bidimensional model of the Lagoon would incorporate circulation effects, but other problems might arise when linking together hydrodynamic, dispersive and trophic models in a single interconnected model and running them separately.

Because of the unpredictability of strong wind events and, more generally, of hydrodynamic input of energy, the model predicts only *average* evolution of biomass due mainly to biological forcing functions: light, temperature and nutrients. Figure 26.7 shows the results of calibration and validation of the *Ulva* model.

The species of macroalgae considered in the model cannot be changed during simulation. This means that the model cannot incorporate natural and continuous structural dynamics of the system, leading from one species of macroalga to another, or even other macrophytes, as a consequence of a change in values of the forcing functions. Thus the model cannot estimate the effect of a reduction of nutrient loads on reduction of biomass. In other words, the predicted biomass reduction due to a scarcity of nutrients could be greater if a change in macroalgal species could be inserted into the model. The state of the art of mathematical modelling in ecology shows few preliminary attempts at simulation of such structural dynamics in a natural system.

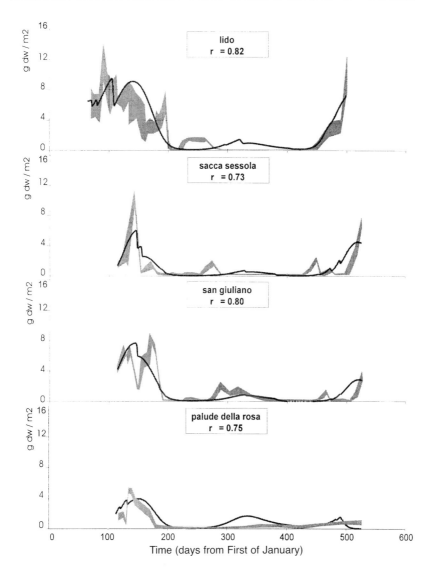

Figure 26.7 Comparison of modelled biomass (line) with observed biomass (mean and standard deviation) (band) during calibration (top 3 panels) and validation (lower panel). Correlation coefficients between observed and calculated biomass are reported on the panels

THE SEAGRASS MODEL

To orient the research in the direction of structural dynamic modelling, a preliminary rough model of seagrass dynamics was built and calibrated with data from Consorzio *Venezia Nuova* (1991). The model is a simplified and

451

rearranged version of previous models (Back, 1993; Verhagen and Nienhuis, 1983). The simplifications were necessary to match this model with the *Ulva* model in the general framework of structural dynamic modelling of the system. It is not yet a suitable tool for this purpose, but it demonstrates that the major processes of the seagrass cycle are reproduced by the model using few forcing functions, i.e. those used in the *Ulva* model.

The forcing functions used were light and water temperature. In accordance with earlier models on seagrasses and experimental studies, nutrient limitation was not considered in the model. The state variables were the root and shoot biomasses.

The results of the model, after arriving at a steady state through simulations over several years, showed good agreement with the few available field data (Figure 26.8).

CONCLUSIONS

The models provide rational and quantitative explanation of the major phenomena related to macroalgal blooms in the Lagoon. What was once qualitatively explained is today quantitatively related to phenomena that force the evolution of macroalgal biomass. The role of light and temperature on the growth process is clear, as is the limitation on growth set by dissolved nitrogen. The role of energy input due to winds and tides has been analysed. This effect appears to be critical for simulation of sudden variations of

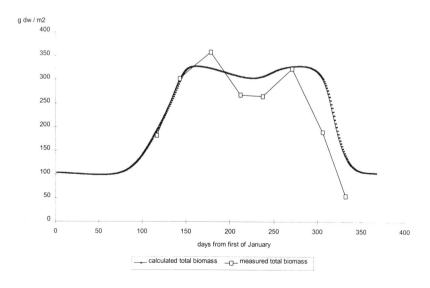

Figure 26.8 Comparison of simulated and measured biomass of *Zostera noltii*

biomass and has been included in the model for the simulation of the past evolution of *Ulva* growth.

There is general agreement that a more holistic approach to the ecology of the Lagoon is urgently needed to understand the complexity of such a system. Changes in environmental conditions can cause significant modifications in the state of the ecosystem, such as variations of structure and internal dynamics. Modifications of the structure of the primary producer community give rise to changes at all trophic levels. In the Venice Lagoon, a shift in primary producer community has occurred in recent years with the substitution, in large areas, of marine macrophytes such as opportunistic macroalgae, such as *Ulva rigida*.

With the final aim of studying competition between a seagrass and a seaweed, a model for simulation of *Zostera noltii* dynamics was separately constructed. As a first step, the interaction between two dominant species of aquatic plants (*Ulva* sp. and *Zostera* sp.) is being pursued. These models point out the complexity of the lagoon ecosystem. This is a key point for all future environmental research activities and this complexity must be taken into consideration when planning and implementing reclamation projects. The only effective way to forecast the impacts of reclamation is to develop a good model of the whole Lagoon and use it for simulation of the effects of the planned actions on the system. The models developed so far are good examples of the types of models needed.

ACKNOWLEDGEMENTS

This work was partially supported by EU – MASTCT – 92-0036 contract.

REFERENCES

Back, H. J. (1993). A dynamic model describing the seasonal variations in growth and the distribution of eelgrass (*Zostera marina* L.), I. Model theory. *Ecol. Model.*, **65**: 31–50

Bendoricchio, G., Coffaro, G. and De Marchi, C. (1993). Modelling the photosynthetic efficiency for *Ulva rigida* growth. *Ecol Model.*, **67**: 221–32

Brocca, M. and Felicini, G. P. (1981). Autoecologia di *Ulva rigida*: 1. Influenza dell'intensità luminosa e della temperatura sulla produzione di ossigeno. *Giorn. Bot. Ital.*, **115**: 285–90

Consorzio, Venezia Nuova (1991). *Rilievi sui popolamenti delle barene ed aree circostanti a vegetazione dei bassi fondi*. Studio, A-3.16: rapporto finale

Coffaro, G. (1993). Model for *Ulva rigida* growth in lagoon of Venice. *Technical Report to UNESCO*, Contract n. SC 218.354.1

Keith, E. A. and Murray, N. M. (1980). Relationship between irradiance and photosynthesis for marine benthic green algae (*Clorophyta*) of different morphologies. *J. Exp. Mar. Bio. Ecol.*, **43**: 183–92

Jørgensen, S. E. (1992). *Integration of Ecosystem Theories: A Pattern*. Kluwer

Academic Pub., pp. 383

O'Neil, R. V., Goldstein, R. A., Shugart, H. H. and Mankin, J. B. (1972) Terrestrial Ecosystem Energy Model. *Eastern Deciduous Forest Biome Memo Report,* 19–72

Sfriso, A., Pavoni, B., Marcomini, A. and Orio, A. A. (1990). Eutrofizzazione e Macroalghe: la Laguna di Venezia come Caso Esemplare. *Inquinamento*, **4**: 62–78

Verhagen, J. H. G. and Nienhuis, P. H. (1983). A simulation model of production, seasonal changes in biomass and distribution of eelgrass (*Zostera marina*) in Lake Grevelingen. *Mar. Ecol.*, **10**: 187–95

CHAPTER 27

A MATHEMATICAL MODEL FOR STUDYING ZOSTERA NOLTII DYNAMICS IN THE VENICE LAGOON

M. Bocci

SUMMARY

Changes in environmental conditions can cause significant modifications in the structure and dynamics of ecosystems. Modifications of the structure of the primary producer community gives rise to changes at all trophic levels. In the Venice Lagoon, such a shift has occurred in recent years with the opportunistic macroalga *Ulva rigida* replacing other macrophytes over large areas.

In this study, a model for simulation of *Zostera noltii* dynamics was developed with the objective of studying competition between a marine phanerogam and a seaweed. The characteristics of the model were calibrated with field data from the Venice Lagoon. The model has been coupled with the *Ulva* model described in Chapter 28 by Coffaro and Sfriso and is undergoing improvement.

INTRODUCTION

Relatively small changes in forcing functions do not affect the ability of a model to provide a good description of the behaviour of an ecosystem but large changes could give rise to problems. As a matter of fact, the system response in this case can be a global modification of its structure (Nielsen, 1992). The aim of a 'structural dynamic model' is to describe deep and fundamental variations in ecosystems, such as changes in species composition or trophic relationships.

In a shallow water system like the Venice Lagoon, a structural dynamic model can take into account the possibility of a shift from the dominance of a 'rooted' species, a marine phanerogam, to a 'non-rooted' species, an opportunistic macroalga, and vice versa, through a wide variety of other species that become important in different ecological situations.

A possible approach to this problem is the design of a coupled model for a 'rooted' and a 'non-rooted' species that explicitly takes into account the interactions between the two. The dynamics of such a coupled model with variations of boundary conditions and forcing functions can facilitate a study of the interaction of two species in a changing ecosystem. In the case of the Venice Lagoon, such a model can focus on the interaction of *Ulva rigida* and *Zostera noltii* which are, respectively, the most typical macroalgae and marine phanerograms in the shallow waters.

A necessary first step in this direction is the identification of two *independent* models, one for *Ulva* and another for *Zostera*. In this work, the structure and calibration of a model for a marine phanerogam is presented; the calibration was done with data on the biomass of *Zostera noltii* from the Venice Lagoon. This model has already been coupled with the *Ulva* model constructed by Coffaro and Sfriso (Chapter 28) but the calibration of the coupled model is still in progress and, hence, no results about competition are presented here.

MARINE PHANEROGAMS IN THE VENICE LAGOON

The main marine phanerogams in the Venice Lagoon are *Zostera noltii* (Hornem), *Zostera marina* (L.) and *Cymodocea nodosa* (Uncria). Although their presence is well known, not many data are available about their precise location. Recently, Caniglia *et al.* (1992) undertook an extensive sampling covering the whole Lagoon: the target of this work was to fill gaps in the knowledge of the distribution of these plants in the Venice Lagoon. The recorded data have been summarised in a thematic map realised by the Consorzio Venezia Nuova (referenced together with the papers).

The above work is not only the most recent study on distribution of marine phanerogams but is also the sole monitoring done on them. Some areas of the Lagoon were found to be covered by different species (mixed populations) and others had monospecific (pure) populations. When compared with earlier information, these data clearly show that the area covered by phanerogams is smaller than it was some years ago. This phenomenon, observed by some authors (Sfriso, 1987), is related to environmental degradation due to pollution and the widespread development of opportunistic macroalgae (Caniglia *et al.*, 1992; Sfriso *et al.*, 1990).

Based on recent data, the Venice Lagoon can be divided into three different zones with respect to phanerogam populations (Caniglia *et al.*, 1992). The *northern area* is characterised by scattered, small-size populations (localised particularly around the margins of channels and almost completely absent from shallow waters). The *central area* is almost devoid of them; only a few populations were discovered around the Malamocco inlet and near the littoral. The *southern area* has a high abundance of pure and mixed populations.

In terms of individual species, the same results can be described as follows (Caniglia *et al.*, 1992):

Zostera noltii is the most common species, with pure and mixed populations spread over an extended area (4235 ha). It has a preference for silty mud substrates, inhabits both intertidal and submerged zones and withstands extreme conditions such as prolonged periods of emersion and salinity variations.

Zostera marina occupies 3635 ha with pure and mixed populations. It inhabits different sediment types, with little clay and variable percentages of sand, silt and mud.

Cymodocea nodosa is less abundant and is found in the southern area. It inhabits zones with good water exchange and sandy sediments.

The seagrass populations are, in general, abundant in areas with good water exchange, high transparency, a prevalence of siliceous and sandy fractions in the sediments and a low abundance of macroalgae (Consorzio Venezia Nuova, 1991; Rismondo and Scarton, 1992). Some of these characteristics differentiate the southern basin of the Lagoon from other zones. Availability of quantitative data for calibration and validation is fundamental in developing any model. In this case, the data collected in the study by Caniglia *et al.* (1992) and published in Consorzio Venezia Nuova (1991) were the only data available to us. These included changes in total biomass for all the three seagrasses (Figure 27.1) monitored in an area near the Chioggia inlet in the southern part of the Lagoon.

MODELLING EELGRASS GROWTH: LITERATURE AND THE PROCESSES CONSIDERED

There are few papers on modelling eelgrass growth and they all deal with *Zostera marina* (Verhagen and Nienhuis, 1983; Bach, 1993).

Seagrass dynamics are a function of seasonal variations of the biomass of shoots and rhizomes. The forcing functions that influence eelgrass production

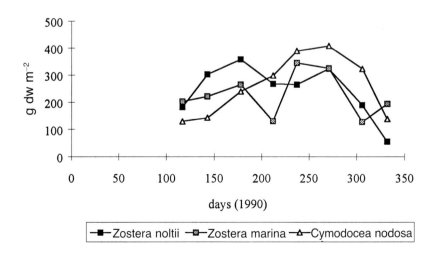

Figure 27.1 Changes in the biomass of the three phanerogams in 1990 near the Chioggia inlet (source: Consorzio Venezia Nuova, 1991)

457

dynamics are mainly light and water temperature. Nutrient dependence of production has recently become a subject of study. Marine phanerogams can take up nutrients both from the water (by leaves) and from the sediment (by roots). Some authors (Verhagen and Nielnhuis, 1983) do not believe that a relationship between nutrients and growth is important (in a model); the high amount of nutrients in interstitial waters of the sediments throughout the year would ensure that there is no deficiency. More recently, Bach (1993) proposed nutrient-dependent growth in *Zostera* for a dynamic model (Bach, 1993). A few investigations have focused on the possible nutrient limitation of marine phanerogams (Orth, 1987; Dennison *et al.*, 1987).

A relationship between the growth rates of shoots and rhizomes is normally taken into account because of the existence of translocation phenomena between aboveground and belowground biomass: carbohydrates synthesised during photosynthesis are transported from leaves to rhizomes and roots. Accordingly, the increase of belowground biomass is modelled with a direct dependence on the growth of leaves. A relationship between growth of shoots and below-ground biomass, in view of transport of nutrients, may also be considered.

Some authors (Bach, 1993) have introduced into the production equation a term for variations of daylength in the year: the longer the day, the higher the production. Others (Verhagen and Nienhuis, 1983) consider the influence of an age factor on growth and decomposition rates and space-limitation of shoot growth.

The decomposition of eelgrass biomass is a metabolic function but it is also influenced, when considered from a macroscopic point of view, by sloughing of shoots; this effect is related to wave impact and becomes evident in shallow waters.

MODELLING *ZOSTERA NOLTII* GROWTH

The present model was constructed as a very simple version of a *Zostera* model that can be considered representative of the situation in the Venice Lagoon. Almost all the information on the physiological mechanisms of growth available in the literature concern *Zostera marina*. These data were extended to *Zostera noltii* and used to model its dynamics.

The model was calibrated with data on the biomass of *Zostera noltii* (see above). The available data were taken as representative of a typical annual cycle of *Zostera noltii* biomass in the Venice Lagoon but the real situation could be different.

In this model only vegetative shoots are considered because sexual reproduction appears to be a rare event in the Venice Lagoon (Comune di Venezia, 1991). The state variables considered are:

S   Biomass of *shoots* [g dw m^{-2}]
R   Biomass of *rhizomes and roots* [g dw m^{-2}]

Their dynamics are represented by the following equations:

$$\frac{dS}{dt} = (\mu - \omega_S) \cdot S$$

$$\frac{dR}{dt} = (K \cdot \mu \cdot f_6(R) - \omega_R) \cdot R$$

where: μ is the specific growth rate of aboveground biomass [day^{-1}],
ω_S is the specific decay rate of aboveground biomass [day^{-1}]
and ω_R is the specific decay rate of belowground biomass [day^{-1}].

Specific growth rate of aboveground biomass is:

$\mu = \mu_{MAX} \, f_1(L) \, f_2(T) \, f_3(S) \, f_4(R)$
where: μ_{MAX} = maximum growth rate [day^{-1}]
$\qquad$ $f_1(L)$ $\quad$ = light limitation function
$\qquad$ $f_1(L)$ $\quad$ = 0 $\qquad\qquad\qquad\qquad$ if L < LCOMP

$$f_1(L) = \frac{L - LCOMP}{LSAT - LCOMP} \qquad \text{if } LCOMP \le L < LSAT$$

$f_1(L) = 1$ $\qquad\qquad\qquad\qquad\qquad$ if L ≥ LSAT

with: L = total daily incident light,
LCOMP = compensation light intensity,
and LSAT = saturation light intensity, all in units of [kcal m^{-2} day^{-1}].

Information about light dependence of photosynthesis in *Zostera* sp. was taken from literature (Drew, 1979; Dennison, 1987; Olesen and Sand-Jensen, 1993).

$f_2(T)$ = temperature growth limitation function

$$f_2(T) = \exp\left(-\left(\frac{T - TOPT}{SIGMAT}\right)^2\right)$$

with: T = water temperature [°C],
TOPT = optimal water temperature [°C]
and SIGMAT = temperature growth coefficient [°C].

In spite of what is usually used in models for eelgrass, an optimal temperature was considered in this work since it was assumed that high summer temperatures attained in the Venice Lagoon can affect the growth of marine phanerogams. Such an effect was observed in larger plants like *Zostera marina* and *Cymodocea nodosa* (Den Hartog, 1970). Data on optimal temperature and temperature dependence of growth in *Zostera* are reported by Drew (1979) and Bulthuis (1987). The quantitative relationships of photosynthesis and growth in *Zostera* sp. with temperature are still unclear and need further investigation.

$f_3(S)$ = space limitation function for shoot growth

$$f_3(S) = 1 - \exp\left(-\frac{(S - SMAX)^2}{SLIMS}\right) \qquad \text{if } S < SMAX$$

$$f_3(S) = 1 \qquad \text{if } S \geq SMAX$$

SMAX = maximum shoot biomass [g dw m^{-2}], and
SLIMS = space limitation coefficient [g dw m^{-2}].

This term sets a limit to the growth of aboveground biomass, i.e. the maximum number of shoots that can grow on a given quantity of rhizomes. Data on the highest values attained by shoot biomass under several environmental conditions can be found in Jacobs (1984).

$f_4(R)$ = rhizomes limitation function

$$f_4(R) = \frac{R}{R + KR}$$

KR = rhizome limitation coefficient [g dw m^{-2}].

This term links shoot growth with rhizome biomass and takes into account the transfer of nutrients from the roots to the leaves.

Specific decay rate for aboveground biomass is:

$$\omega_S = OMEGAMAXS \; f_5(T)$$

where: OMEGAMAXS = maximum shoot decay rate [day^{-1}],
$\qquad f_5(T) \qquad$ = temperature decay limitation function,
$\qquad f_5(T) \qquad$ = $KMORT^{(T-TOPT)}$
$\qquad$ KMORT $\qquad$ = temperature decay coefficient

$K, f_6(R)$ and ω_R, together with μ, determine the dynamics of rhizome biomass.

K = the fraction of photosynthetic products fixed by aboveground biomass and translocated to belowground biomass.

$f_6(R)$ = space limitation function for rhizome growth

$$f_6(R) = 1 - \exp\left(-\frac{(R - RMAX)^2}{SLIMR}\right) \qquad \text{if } R < RMAX$$

$$f_6(R) = 1 \qquad \text{if } R < RMAX$$

RMAX = maximum rhizome biomass [g dw m^{-2}],
and SLIMR = space limitation coefficient [g dw m^{-2}].

Data on the highest values attained by rhizome and root biomass can be found in Jacobs (1984).

Decay constant of belowground biomass is:

$\omega_R = 0.005$ [day^{-1}]

Since no literature data on the respiration of belowground biomass are available, a low respiration rate was fixed by calibration and, assuming that the sediment temperature is constant, was taken to be temperature-independent.

ANALYSIS AND CONCLUSIONS

Forcing functions of light and temperature calculated by interpolating typical Lagoon values were used for the simulation. Daily incident solar radiation data for the ten-year period 1981–1990 are presented and analysed in Consorzio Venezia Nuova (1991). These data were compared with those recorded at Ente Zona Industriale di Porto Marghera and used for the model of *Ulva* by Coffaro and Sfriso in Chapter 28. Data on water temperature, recorded for more than one year at different sites in the Lagoon, can be found in the literature (Sfriso *et al.*, 1990; Sfriso and Pavoni, 1993). Initial condition values of the model were set at 10 g dw m^{-2} for shoot biomass and 50 g dw m^{-2} for rhizome biomass. Based on general knowledge of the life cycle of marine phanerogams, it was assumed that most of the winter biomass was in the belowground fraction.

With this model, a simulation over several years was needed to achieve steady state conditions. A 10-year simulation was therefore carried out and the simulated curve for the tenth year was calibrated with observed data. The only data available to us for calibration were the total biomass data from *Consorzio Venezia Nuova*. The sum of the values computed by the model for shoot and rhizome biomass was calibrated with these data. Figure 27.2 shows the calculated and observed values for total biomass. Figure 27.3 shows above- and belowground biomass, calculated separately.

Acquisition of still more data is necessary if an improvement and strengthening of the model is envisaged. These include time series data of above- and belowground biomass, net growth rates of shoots and rhizomes, light and temperature dependence of growth, influence of nutrients (both in sediments and in water) on growth, time series data of internal nutrient quota and measurements of growth rates with different levels of internal and external nutrients.

This model represents a first attempt to introduce marine phanerogams among the subjects for study of the status and development of the Venice Lagoon. The basic driving forces behind this idea are:

(1) Marine phanerogams are not only important as primary producers but also as constituents of seagrass ecosystems that are generally characterised by a high biodiversity;

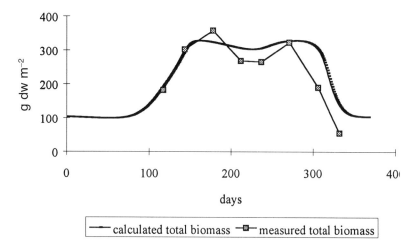

Figure 27.2 Modelled and observed total (above- and belowground) biomass of *Zostera noltii*

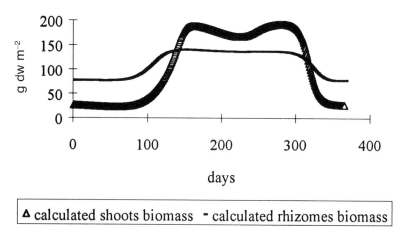

Figure 27.3 Above- and belowground biomass of *Zostera noltii* calculated from the model

(2) Their sensitivity to water quality makes them good indicators of the health of Lagoon ecosystems;

(3) Their interaction with macroalgae is of considerable ecological interest; and

(4) Transplantation of marine phanerogams to denuded areas, as a means of eco-restoration, is gaining importance at present.

The model presented here, when improved and supported with field data, can be used as a tool for long-term simulations under different environmental conditions.

ACKNOWLEDGEMENTS

This work was partially supported by the EU - MAS2-CT - 92-0036 contract.

REFERENCES

Back, H. J. (1993). A dynamic model describing the seasonal variations in growth and the distribution of eelgrass (*Zostera marina* L.). I. Model theory. *Ecol. Modelling,* **65**: 31–50

Bulthuis, D. A. (1987). Effects of temperature on photosynthesis and growth of seagrass. *Aquat. Bot.,* **27**: 27–40

Caniglia, G., Borella, S., Curiel, D., Nascimbeni, P., Paloschi, A. F., Rismondo, A., Scarton, F., Tagliapietra, D. and Zanella, L. (1992). Distribuzione delle fanerogame marine [*Zostera marina* L., *Zostera noltii* Hornem, *Cymodocea nodosa* (Ucria) Asch.] in Laguna di Venezia. *Lavori–Soc. Ven. Sc. Nat.,* **17**: 137–50

Comune di Venezia (1991). Le alghe della laguna di Venezia. *Arsenale Editrice, Venezia,* Vol. I

Consorzio Venezia Nuova (1991). *Rilievi sui popolamenti delle barene ed aree circostanti e sulla vegetazione dei bassifondi.* Studio A-3.16: Rapporto finale.

Consorzio Venezia Nuova (1991). *Interventi per l'arresto del degrado connesso alla proliferazione delle macroalghe in Laguna di venezia–3° ciclo.* Analisi statistica dati meteo

Consorzio Venezia Nuova–Sevizio Informativo. Marine phanerogams map – Scale 1:50000. Carta distributiva delle fanerogame marine

Den Hartog, C. (1970). *The Seagrasses of the World.* North-Holland Publishing, Amsterdam

Dennison, W. C. (1987). Effects of light on seagrass photosynthesis, growth and depth distribution. *Aq. Bot.,* **27**: 15–26

Dennison, W. C., Aller, R. C. and Alberte, R. S. (1987). Sediment ammonium availability and eelgrass (*Zostera marina*) growth. *Mar. Biol.,* **94**: 469–77

Drew, E. A. (1979). Physiological aspects of primary production in seagrass. *Aquat. Bot.,* **7**: 139–50

Jacobs, R. P. W. M. (1984). Biomass potential of eelgrass (*Zostera marina*) beds. *Mar. Biol.,* **66**: 59–65

Nielsen, S. N. (1992). Strategies for structural–dynamic modelling. *Ecol. Modelling,* **63**: 91–101

Olesen, B. and Sand-Jensen, K. (1993). Seasonal acclimatation of eelgrass *Zostera marina* growth to light. *Mar. Ecol. Prog. Ser.,* **94**: 91–9

Orth, R. J. (1977). Effect of nutrient enrichment on growth of the eelgrass *Zostera marina* in the Chesapeake Bay, Virginia, USA. *Mar. Biol.,* **44**: 187–94

Rismondo, A. and Scarton, F. (1992). Praterie a fanerogame marine in Laguna di Venezia–Distribuzione, importanza e proposte per il loro ripristino. *Ambiente, risorse e salute*, 4/1992

Sfriso, A. (1987). Flora and vertical distribution of macroalgae in the Lagoon of Venice: a comparison with previous studies. *Giornale Botanico Italiano*, **121**, (1-1): 69–85

Sfriso A., Marcomini A., Pavoni, B. and Orio, A. A. (1990). Eutrofizzazione e macroalghe. La Laguna di Venezia come caso esemplare. *Inquinamento*, **4**: 62–78

Sfriso, A. and Pavoni, B. (1994). Macroalgae and phytoplankton competition in the central Venice Lagoon. *Environmental Technology*, **15**: 1–14

Verhagen, J. H. G. and Nienhuis, P. H. (1983). A simulation model of production, seasonal changes in biomass and distribution of eelgrass (*Zostera marina*) in Lake Grevelingen. *Mar. Ecol.*, **10**: 187–95

CHAPTER 28

CALIBRATION AND VALIDATION OF A MODEL FOR ULVA RIGIDA GROWTH IN THE VENICE LAGOON

G. Coffaro and A. Sfriso

SUMMARY

The model presented here demonstrates that it is possible to simulate the life cycle of *Ulva rigida* C.Ag. in the Venice Lagoon with a mathematical description of considerable accuracy. The model not only accounts for physiological processes (photosynthesis, nutrient limitation, temperature effects and decomposition) but also highlights the importance of generic environmental disturbances, represented here by water turbulence and resuspension of sediments induced by strong winds (> 3 m s^{-1}). More knowledge on environmental disturbance is needed in order to better understand the influence such factors can have on growth and distribution of *Ulva rigida*.

INTRODUCTION

The marked expansion of industrial activities after the Second World War, the increased input of municipal sewage from Mestre and the surrounding hinterland, the growth of tourism in the Venice area and drainage from an intensively cultivated watershed (1800 km^2) have together led to a large flux of nutrients into the Lagoon. At present, more than 6900 tons of nitrogen and 827 tons of phosphorus are discharged every year into the Lagoon from civil, agricultural and industrial sources (Bendoricchio *et al.*, 1993).

The increased nutrient availability, along with hydrological changes caused by digging of large (80–150 m) and deep (12–17 m) commercial canals such as Malamocco–Marghera and Vittorio Emanuele II, induced marked shifts in the composition and distribution of macrophytes in the central Lagoon and favoured proliferation of monospecific populations of the green alga, *Ulva rigida*. From the 1970s, this species showed prolific growth, covering vast areas of the Lagoon and sustaining a biomass that at times was higher than 20 kg wet weight m^{-2}. Sfriso and Marcomini (1994) estimated net and gross production of *Ulva rigida* at ~1.6 and ~10.2 million tons until 1990 in the central part of the Lagoon (132/432 km^2). This production was equivalent to a macroalgal recycling of ~1900, ~21,800 and ~210,000 tons of phosphorus, nitrogen and carbon.

CONCEPTUAL MODEL

The initial *Ulva* growth in March–April is supported by increasing temperature, high nutrient availability and optimal light. The nutrients needed for spring bloom are provided, and temporarily buffered, by the nutrient stock stored in surface sediments and/or accumulated in fronds of the algae that survived the previous winter. Spring temperature and light trigger a rapid increase of biomass. *Ulva* biomass during the blooms becomes markedly stratified, restricting photosynthesis to the upper layers. Farther down the canopy, *Ulva* does not receive enough light for photosynthesis, and respiration and (or) decomposition prevails. When the upper layers also become unproductive owing to: (a) exposure to strong insolation culminating even in desiccation, (b) inhibition of growth due to resuspension of sediments, and (c) insufficient nutrient availability following high demand (up to 33, 3.4 and 0.3 g m^{-2} day^{-1} of C, N and P, respectively) and losses of C and N through respiration and denitrification, a total collapse of the bloom can occur. The decay process results in a high deposition of organic matter on the bottom and, coupled with reduced photosynthetic oxygen production, increases the oxygen deficit. The high temperatures of late spring accentuate this process and lead to total anoxia which, in a few days, extends through the whole water column. Under these conditions, also characterised by high H$_2$S concentrations, all aerobic life forms disappear. In late summer, nutrients regenerated from the decomposed biomass are usually sufficient to support a second growth of *Ulva rigida*. Because of the inhibiting effect of temperature (25–28 °C) and the high wind-induced water turbulence that enhances sediment resuspension (Sfriso *et al.*, 1992), in late summer compared with spring, growth is markedly low in intensity.

DATA DESCRIPTION

Calibration and validation of the model were done with time series data collected at quasi-weekly intervals on changes in *Ulva rigida* biomass, concentrations of dissolved nutrients and water temperature in the Venice Lagoon (Sfriso *et al.*, 1993). The set of forcing variables was completed with radiation and wind data recorded by Ente Zona Industriale Porto Marghera for the same period.

The sites studied were Lido (station 1 in Figure 28.1; from March 1985 to June 1986), Sacca Sessola (station 2 in Figure 28.1; from April 1990 to June 1991), San Giuliano (station 3 in Figure 28.1; from April 1990 to June 1991) and Palude della Rosa (station 4 in Figure 28.1; from April 1992 to May 1993). All these sites are shallow water areas (0.4–1.2 m) characterised by different hydrodynamic conditions. The model was calibrated with data from the first three sites and validated with data from the last.

466

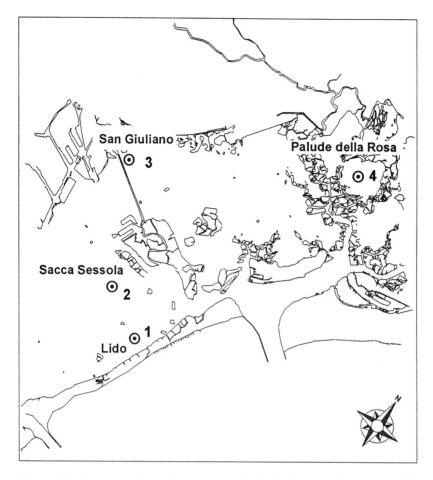

Figure 28.1 Location of sampling stations in the Venice Lagoon where data used in the model were collected

MODELLING *ULVA RIGIDA* GROWTH

In the model presented here, the state variables are represented by biomass (B) and internal nitrogen (N) and phosphorus (P) concentrations. Light intensity, temperature, external nutrients (ammonium, nitrate and phosphate) and wind are the forcing functions. More details on mathematical physiological description are reported by Coffaro (1993).

The dynamics of *Ulva rigida* biomass can be taken to be governed by the following processes: growth, non-predatory mortality (respiration and decomposition), and export due to environmental disturbance. A general equation for biomass (B) which includes these processes would be:

$$\frac{\partial B}{\partial t} = (\mu - \Omega) \cdot B - D \cdot B \tag{28.1}$$

where: μ = specific growth rate (l/day),
Ω = specific decomposition rate (l/day),
D = rate of decrease due to environmental disturbance (l/day), and
t = time (day).

Growth rates of algae are regulated by temperature, light and nutrient availability. Thus, algal growth rate can be written in general functional form as:

$$\mu = \mu_{max}(T_{ref}) \cdot f(T) \cdot f(L) \cdot f(C,N,P) \cdot RD(t) \tag{28.2}$$

where: μ = algal growth rate (l/day);
$\mu_{max}(T_{ref})$ = maximum growth rate at a particular reference temperature (T_{ref}) under conditions of saturating light intensity and excess nutrients (l/day);
f(T) = temperature function for growth, where T stands for temperature in °C and f(L);
f(N,P,C) = growth limiting functions for light and nutrients, where L stands for light intensity, and N, P, C stand for available inorganic nutrient (nitrogen, phosphorus, carbon) concentrations in mg l^{-1}; and
RD(t) = relative day length as a fraction.

Temperature is the most important factor controlling seasonal and annual growth patterns. *Ulva rigida* appears to have a temperature optimum within the range of 15–20 °C (Steffensen, 1976; Duke *et al.*, 1989). Of particular interest is the study by Steffensen (1976) where he tested the growth of *Ulva lactuca* at a temperature range from 6–25 °C, with specimens from two localities with different water temperature optima. The optimum temperatures for growth were 16 and 19 °C, respectively, for attached and drifting plants. The floating plants were older and from warmer waters than the attached ones. In both the cases, the growth rate decreased when temperature exceeded 20 °C and was completely arrested at 25 °C.

From these results, it is obvious that the effect of temperature on the growth rate of *Ulva rigida* should include a term for adaptation to optimum temperature and an allowance for the reduction in growth rate with ageing (30% reduction in the maximum growth rate for drift plants).

Different expressions can be used to describe the temperature dependence of growth rate. Thornton and Lessem (1978) developed a temperature optimum curve by combining two logistic equations, one for the ascending part of the curve and the other for the descending part, giving an asymmetric pattern with respect to optimum temperature, as observed in Steffensen's data (Steffensen, 1976):

$$f(T) = K_a \cdot K_b \tag{28.3}$$

with:

$$_a = \frac{K1 \cdot \exp(\gamma_1 \cdot (T - Tmin))}{1 + K1 \cdot (\exp(\gamma_1 \cdot (T - Tmin)) - 1)} \qquad (28.4)$$

$$_b = \frac{K4 \cdot \exp(\gamma_2 \cdot (Tmax - T))}{1 + K4 \cdot (\exp(\gamma_2 \cdot (Tmax - T)) - 1)} \qquad (28.5)$$

where:

$$\gamma_1 = \frac{1}{Topt - T\,min} \cdot \ln\left(\frac{0.98 \cdot (1 - K1)}{0.02 \cdot K1}\right) \qquad (28.6)$$

$$\gamma_2 = \frac{1}{T\,max - Topt} \cdot \ln\left(\frac{0.98 \cdot (1 - K4)}{0.02 \cdot K4}\right) \qquad (28.7)$$

Tmax = maximum temperature, set to 30 °C,
Tmin = minimum temperature, set to 8 °C;
T*opt* = optimal temperature (°C); and
K1 and K4 = rate multipliers.

Taking into account adaptation to temperature, the optimum temperature can be considered as a function of time or, as suggested by some authors, of the past temperature changes experienced by the algae. In our model, T*opt* has been modified to include the effects of past temperature changes the algae were subjected to, as suggested by O' Neill *et al.* (1972):

$$T_{opt} = T_{opt} + T_{sht} \qquad (28.8)$$

with:

$$Tsht = Ac \cdot (1 - \exp(- KT \cdot (Tavg - Topt)) \qquad (28.9)$$

where:

T*sht* = magnitude of acclimation (°C),
Ac = maximum magnitude of acclimation (°C),
KT = acclimation rate coefficient, and
Tavg = average temperature (°C) of previous two weeks.

The ageing effect on temperature dependence of growth rate was linked to adaptation to optimal temperature (the higher the *Topt*, the older the algae in a seasonal cycle) as follows:

$$gx = 1 - T_{rid} \cdot \frac{Topt - Opt_{min}}{Opt_{max} - Opt_{min}} \qquad (28.10)$$

where:

gx = reduction factor due to ageing of the algae,
T_{rid} = maximum value of reduction factor, and
Opt_{max}, Opt_{min} = maximum and minimum optimal temperature (°C).

This reduction factor, multiplied by the earlier formulation for f(T), reduces its value according to algal ageing:

$$f(T) = K_a \cdot K_b \cdot gx \tag{28.11}$$

Such temperature limitation is the most important factor in correctly formulating the model to describe the observed seasonal pattern.

In all models, the attenuation of light with depth is defined essentially by the Beer–Lambert law:

$$(Z) = I(0) \cdot \exp(-K \cdot z) \tag{28.12}$$

where:

$$= K_0 + UlvExt \cdot \frac{B}{z} \tag{28.13}$$

I(z) = light intensity at depth z below the surface (MJ/m²/day),
I(0) = light intensity at the surface (MJ/m²/day),
Ko = light extinction coefficient (l/m),
UlvExt = biomass-specific light extinction coefficient,
B = biomass (gfw/m²), and
z = depth (m).

Parameters Ko and z, different for each site, were set to:

Ko = 0.4, z = 1 for station 1;
Ko = 0.4, z = 0.6 for station 2;
Ko = 2, z = 0.7 for station 3, and
Ko = 0.4, z = 1 for station 4.

Experimental data reported for *Ulva rigida* in the literature (Keith and Murray, 1980; Brocca and Felicini, 1981) show that light saturation occurs at about 20 klux. The light limitation factor adopted in the model is given by the average over the depth (z) of a saturation-type curve analytically integrated on depth:

$$f(L) = \frac{1}{EXT \cdot z} \cdot \ln\left(\frac{KL + I(0)}{KL + I(0) \cdot \exp(-EXT \cdot z)}\right) \tag{28.14}$$

with:

$$XT = K_0 + UlvExt \cdot \frac{B}{z} \tag{28.15}$$

I(0) = light intensity at the surface (MJ/m²/day),
KL = half saturation coefficient of light intensity (MJ/m²/day),
Ko = light extinction coefficient (l/m),
UlvExt = biomass specific light extinction coefficient,
B = biomass (gfw/m²), and
z = depth (m).

It has been shown experimentally in several studies (Fujita,1985; Lundberg *et al.*, 1989; Lavery and McComb, 1991) that the past history and the consequent nutritional state of *Ulva rigida* can affect its nutrient uptake rate. These observations support the theoretical implications of an approach of variable internal quota of stored nutrients. This approach requires the definition of a limitation factor and uptake kinetics regulated through a feedback mechanism by internal nutrient content.

Usually, N and P are considered the major nutrients required for growth in seaweeds. For *Ulva rigida* in shallow water environments, however, N is more likely to be the limiting factor. This could be explained by the fact that the P requirement in *Ulva rigida* is proportionally smaller than that for N: the maximum tissue P content (2.2 mg/g dry weight) was almost 9 times the saturation level (0.25 mg/g dry weight) while the maximum tissue N content (40 mg/g dry weight) was only 50% of the saturation level (20 mg/g dry weight).

The minimum among the following expressions, at each time interval, was used to determine the growth-limiting factors as a function of internal quota of nitrogen f(N) and phosphorus f(P):

$$f(N) = 1 - \frac{N - QNM}{QNX - QNM} \tag{28.16}$$

$$f(P) = 1 - \frac{P - QPM}{QPX - QPM} \tag{28.17}$$

where:

N = internal quota of nitrogen (mg N/g fw),
P = internal quota of phosphorus (mg P/g fw),
QNM, QNX = respectively, minimum and maximum internal quota for N (mg N/gfw), and
QPM, QPX = respectively, minimum and maximum internal quota for P (mg P/g fw).

The similarity of the growth curves for *Ulva lactuca* with ammonium and nitrate as substrates does not indicate a preference for either of the two N compounds. (Waite and Mitchell, 1972; Steffensen, 1976) suggesting that total inorganic nitrogen can be used in a model, instead of the two separate forms.

Rates of nutrient uptake by *Ulva rigida* reported in the literature (Waite and Mitchell, 1972; Fujita, 1985; Lavery and McComb, 1991) vary widely because they were often measured in short laboratory experiments and/or without controls on internal quota of nutrients. However, the model proposed here is not sensitive to different values for N and P uptake rates.

The equation used for nutrient uptake is the classic Michaelis–Menten kinetics:

$$up = VNmax \frac{TIN}{KN + TIN} \qquad (28.18)$$

$$up = VPmax \frac{RP}{KP + RP} \qquad (28.19)$$

where:

Nup = N uptake (mg N/g fw),
Pup = P uptake (mg P/g fw),
VNmax = maximum N uptake rate (mg N/g fw),
VPmax = maximum P uptake rate (mg P/g fw),
TIN = total dissolved inorganic N (mg/l),
RP = dissolved phosphate (mg/l), and
KN, KP = half saturation constants (mg/l) for N and P uptake, respectively.

In general, uptake rate increases with external nutrient supply and at the same time decreases as internal nutrient levels approach saturation values. Uptake rate becomes zero when either external nutrients are depleted or internal nutrients are at saturation.

The equations used for N and P feedback control were set to:

$$fb = \frac{QNX - N}{QNX - QNM} \qquad (28.20)$$

$$fb = \frac{QPX - P}{QPX - QPM} \qquad (28.21)$$

In summary, the differential equations used to describe internal nutrient dynamics are:

$$\frac{\partial N}{\partial t} = Nupt \cdot Nfb - \mu \cdot N \qquad (28.22)$$

$$\frac{\partial P}{\partial t} = Pupt \cdot Pfb - \mu \cdot P \qquad (28.23)$$

The variations of an internal nutrient quota are obtained by balancing uptake from the environment with depletion due to growth. The decrease in internal nutrient content represents metabolic utilisation of nutrients for growth which is usually quantified by a linear relationship with the computed growth rate.

The decomposition kinetics of *Ulva rigida* as function of environmental conditions has not been studied earlier. Unfortunately, in the shallow waters of the Venice Lagoon, where *Ulva* biomass reaches very high levels, decomposition dynamics are an important process which, coupled with the resultant nutrient recycling, can reasonably explain the development trend of such seaweed populations.

The most important component of the non-predatory mortality term is respiration. The effects of an increase in temperature manifest in a reduction of oxygen saturation, an increase in respiration and decomposition rates and, consequently, an increase in organic matter flux toward the bottom, and an increase in bacterial activity. All these effects contribute to oxygen depletion and persistent anoxia lasting some days is usually observed when decomposition is at its maximum. However, without more direct evidence, it is difficult to state whether oxygen depletion is a consequence of decomposition of algae or its cause.

All the sharp decreases in biomass in our observations occurred at ~27 °C. High temperature, therefore, can be an important factor in decomposition kinetics and its influence was set as:

$$\Omega = \text{MaxMor} \cdot \exp(T - 29) \tag{28.24}$$

where:

Ω = mortality rate (l/day),
MaxMor = maximum mortality rate (assumed to be at 29 °C) (l/day), and
T = temperature (°C).

The rapid and strong biomass variations can be explained only by considering a physical disturbance (in terms of growth inhibition and biomass loss, independent of natural processes). Standing stock often increases with increasing distance from channels (Comune di Venezia, 1991) and floating algae are absent from areas where water currents are normally high. Because of this, free-floating algae like *Ulva rigida* invade confined shallow water areas where water velocity and tidal currents are weak. In such areas, like those used for calibration and validation of the models, wind effects can occasionally increase water turbulence and be responsible for export of biomass, high resuspension of sediments and deposition of a layer of particulate material on the algal fronds, inhibiting their growth (Sfriso and Pavoni, 1994).

This hypothesis of occasional disturbance due to strong winds is supported, at least in a qualitative way, by the fine correlation between the decline of biomass and wind events. This is shown in Figure 28.2 where changes of standing stocks of *Ulva rigida* and strong (> 3 m/s) winds (vertical arrows) for the four stations are plotted against time. Obviously, the magnitude of the influence of such a physical disturbance is dependent on morphology and location of stations (i.e. sensitivity to wind from a given direction, proximity to channels and emerged areas, sensitivity to resuspension of sediments). The relationship between loss of biomass and strong wind events has been set to:

$$D = \text{Sens} \cdot \exp(0.12 \cdot (B - 7000)) \tag{28.25}$$

where:

D = loss rate due to environmental disturbance (l/day),
Sens = site sensitivity to disturbance, and
B = biomass (g fw/m^2).

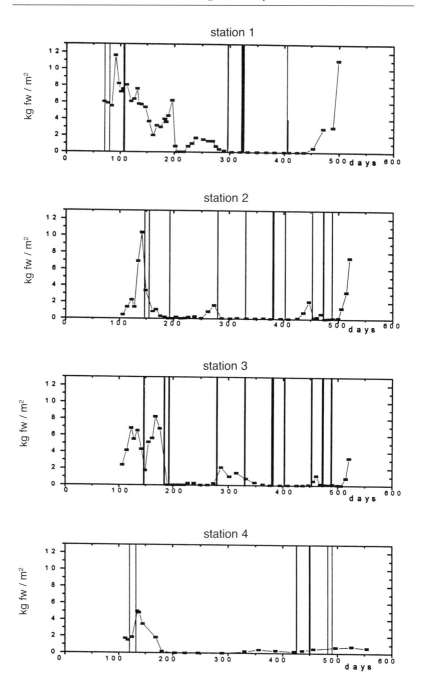

Figure 28.2 Biomass of *Ulva rigida* (as kg fresh weight/m²) observed at each site. Vertical lines indicate the time of occurrence of a strong wind event. Days are numbered starting from 1st January

In this formulation, B depends on whether the biomass is floating and aggregated or not, while *Sens* accounts for the site characteristics that might favour export of algae. *Sens* values for each site were tuned with calibration and set to:

-0.2 for station 1;
-0.8 for station 2;
-0.3 for station 3;
-0.4 for station 4.

Understandably, this formulation is empirical and relies on some qualitative hypothesis but is very important to improve the usefulness of the model.

RESULTS OF THE MODEL

Simulation of the growth of algae required a suitable set of values for the parameters in the model. A calibration of the model was needed to account for simplifications introduced by the mathematical formulations.

An extensive literature search was undertaken to obtain a range of suitable values for each parameters (see Table 28.1). For some parameters (i.e. maximum uptake and growth rates, half saturation constants and internal quota of nutrients) data were found easily, while for others a value was obtained by fitting experimental data reported in the literature to equations used in the model. For some others, such as mortality and decomposition of *Ulva rigida*, parameter values were obtained while calibrating the model. The calibration was carried out using an algorithm employing the *downhill simplex* method that reduces the difference between the calculated and observed values for biomass. The best fit of the model was obtained with the parameter values in Table 28.1.

Figure 28.3 shows the simulated biomass values (solid line) along with the standard deviation (error bars) of the observed values (stations 1, 2 and 3) used for calibration of the model. Notwithstanding the difficulties of removing the noise from the experimental data (see above), the model reproduces the main features of the *Ulva* life cycle, with statistically significant correlations between observed and calculated values ($r^2 = 0.82$ for station 1, $r^2 = 0.73$ for station 2, $r^2 = 0.8$ for station 3).

The model simulations are, however, too smooth with respect to changes of the standing crop measured, and this could be due to wind-induced inhibition of algal growth. The term for disturbance included in the model is effective but it cannot account for several other factors that act concomitantly on the spatial distribution of algae. The importance of the disturbance term is also shown in Figure 28.3 by a comparison of the biomass computed by the model with (solid line) and without (dotted line) this term.

The results of validation of the model with data from station 4 (Figure 28.4) also gave a good correlation ($r^2 = 0.75$) between observed and computed biomass values. In validating the model, a calibration of the parameter *Sens*,

Table 28.1 List of parameters used in the model

Symbol	Description	Unit	Literature range	Value used	Sources
μmax	Growth max rate	d^{-1}	0.36–0.50	0.4	Lapointe and Tenore, 1981; Rosenberg and Ramus, 1982; Henley and Ramus, 1989
VNmax	Uptake max rate for ammonium and nitrate as N	mg g dw^{-1}h^{-1}	0.56–5.2	0.56	Fujita, 1985; Lavery and McComb, 1991; Lapointe and Tenore, 1981; Menesguen, 1988
VPmax	Uptake max rate for phosphate	mg g dw^{-1} h^{-1}	0.23–1.09	0.23	Dion, 1988; Lavery and McComb, 1991
KN	Half sat. constant for total inorganic nitrogen	mg l^{-1}	0.2–0.6	0.25	Fujita, 1985; Lavery and McComb, 1991
KP	Half sat. constant for phosphate	mg l^{-1}	0.025–0.113	0.025	Dion, 1988; Lavery and McComb, 1991
QNX	Maximum intracellular quota for nitrogen	mg g dw^{-1}	40–60	40	Ho, 1981; Fujita, 1985; Bjornsater and Wheeler, 1990
QNM	Minimum intracellular quota for nitrogen	mg gdw^{-1}		10	Ho, 1981; Fujita, 1985; Bjornsater and Wheeler, 1990
QPX	Maximum intracellular quota for phosphorus	mg g dw^{-1}		3.9	Sfriso et al., 1990

Continued on next page

Table 28.1 *Continued*

Symbol	Description	Unit	Literature range	Value used	Sources
QPM	Minimum intracellular quota for phosphorus	mg g dw^{-1}		1.1	Sfriso *et al.*, 1990
KL	Half sat. constant for photosynthesis	kJ m^{-2} day^{-1}		1000	Brocca and Felicini, 1981 Henley and Ramus, 1989 Rosenberg and Ramus, 1982
UlvExt	Algal light extinction			0.001	Calibration
K1	Temper. adj. coeff.			0.3	"
K4	Temper. adj. coeff.			0.01	"
Trid	Ageing limitation factor			0.5	"
Ac	Opt. temperature acclimatation rate	°C day^{-1}		0.3	"
KT	Opt. temp. coefficient			0.003	"
MaxMor	Death max rate	d^{-1}		0.8	"

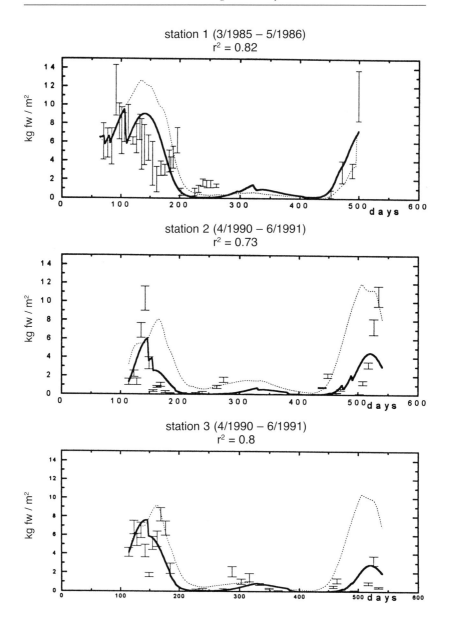

Figure 28.3 Biomass values obtained with the model after calibration (solid line). Dotted line shows model simulation without disturbance term. Vertical bars represent the standard deviation of the experimental values. Days are numbered starting from 1st January

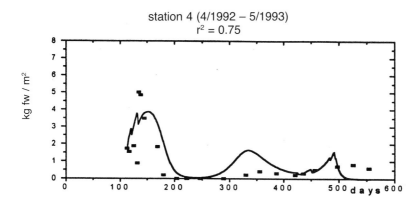

Figure 28.4   The results of the validation of the model. The solid line represents biomass computed with the model while the symbols represent the measured biomass. Days are numbered starting from 1st January

which accounts for site sensitivity to disturbance induced by wind, was necessary.

ACKNOWLEDGEMENTS

This work was carried out in the framework of the UNESCO Venice Lagoon Ecosystem Project which is part of the Italian Ministry for Scientific Research Project *Sistema Lagunare Veneziano*, from which it received a partial financial contribution.

REFERENCES

Bendoricchio, G., Di Luzio, M., Baschieri, P. and Capodaglio, A. (1993). Diffuse pollution in the lagoon of Venice. *Wat. Sci. Tech.*, **3–5**: 69–78

Bjornsater, B. R. and Wheeler, P. A. (1990). Effect of nitrogen and phosphorus supply on growth and tissue composition of *Ulva fenestrata* and *Enteromorpha intestinalis* (Ulvales, Chlorophyta). *J. Phycol.*, **26**: 603–11

Brocca, M. and Felicini, G. P. (1981). Autoecologia di *Ulva rigida*: 1. Influenza dell'intensità luminosa e della temperatura sulla produzione di ossigeno. *Giorn. Bot. Ital.*,. **115**: 285–90

Comune di Venezia (1991). Le alghe della laguna di Venezia. *Arsenale Editrice*, pp. 118

Coffaro, G. (1993). Model for *Ulva rigida* growth in the lagoon of Venice. *Technical Report to UNESCO*, Contract n. SC 218.354.1

Dion, F. (1988). Evaluation du role potentiel du phosphore dans la proliferation des *Ulvales* en baie de St. Brieuc. *Rapport de Synthèse C.E.V.A*

Duke, C. S., Litaker, W. and Ramus, J. (1989). Effects of temperature, nitrogen supply, and tissue nitrogen on ammonium uptake rates of the chlorophyte seaweeds *Ulva curvata* and *Codium decorticatum*. *J. Phycol.* **25**: 113–20

Fujita, R. M. (1985). The role of nitrogen status in regulating transient ammonium uptake and nitrogen storage by macroalgae. *J. Exp. Mar. Biol. Ecol.*, **92**: 283–301

Keith, E. A. and Murray, N. M. (1980). Relationship between irradiance and photosynthesis for marine benthic green algae (Chlorophyta) of differing morphologies. *J. Exp. Mar. Biol. Ecol.*, **43**: 183–92

Harlin, M. M. and Thorne-Miller, B. (1981). Nutrient enrichment of seagrass beds in a Rhode Island coastal lagoon. *Mar. Biol.*, **65**: 221–9

Henley, W. J. and Ramus, J. (1989). Photoacclimation of *Ulva rotundata* (Chlorophyta) under natural irradiance. *Mar. Biol.*, **103**: 261–6

Ho, Y. B. (1981). Mineral element content in *Ulva lactuca* with reference to eutrophication in Hong Kong coastal water. *Hydrobiol.*, **77**: 43–7

Lapointe, B. E. and Tenore, K. R. (1981). Experimental outdoor studies with *Ulva fasciata* Delile I. Interaction of light and nitrogen on nutrient uptake, growth and biochemical composition. *J. Exp. Mar. Biol. Ecol.*, **92**: 135–52

Lavery, P. S. and McComb, A. J. (1991). The nutritional ecophysiology of *Chaetomorpha linum* and *Ulva rigida* in Peel Inlet, Western Australia. *Bot. Mar.,* **34**: 251–60

Lundberg, P., Weich, R. G., Jensen, P. and Vogel, H. J. (1989). Phosphorus-31 and nitrogen-14 NMR studies of the uptake of phosphorus and nitrogen compounds in the marine macroalgae *Ulva lactuca*. *Plant Physiol.*, **89**: 1380–7

Menesguen, A. and Salomon, J. C. (1988). Eutrophication modelling as a tool for fighting against *Ulva* coastal mass blooms. *Comp. Model. Ocean Eng. Model.*, **2**: 147–65

O'Neill, R. V., Goldstein, R. A., Shugart, H. H. and Mankin, J. B. (1972). Terrestrial ecosystem Energy Model. *Eastern Decidous Forest Biome. Memo Report*, 72–9

Rosenberg, G. and Ramus, J. (1982). Ecological growth strategies in the seaweeds *Gracilaria foliifera* and *Ulva* sp.: photosynthesis and antenna composition. *Mar. Ecol. Prog. Ser.*, **8**: 233–41

Sfriso, A., Pavoni, B., Marcomini, A. and Orio, A. (1988). Annual variations of nutrients in the lagoon of Venice. *Mar. Pol. Bull.,* **19**: 54–60

Sfriso, A., Marcomini, A., Pavoni, B. and Orio, A. (1990). Eutrofizzazione e macroalghe: la laguna di Venezia come caso esemplare. *Inquinamento,* **4**: 62–78

Sfriso, A., Pavoni, B., Marcomini, A. and Orio, A. A. (1992). Macroalgae, nutrient cycles and pollutants in the Lagoon of Venice. *Estuaries,* **15**: 517–28

Sfriso, A., Pavoni, B., Marcomini, A. and Orio, A. (1993). Macroalgae, nutrient cycles and pollutants in the lagoon of Venice. *Estuaries*, **4**: 517–28

Sfriso, A. and Marcomini, A. (1994). Gross primary production and nutrient behaviour in a shallow coastal environment. *Biores. Tech.*, **47**: 59–66

Sfriso, A. and Pavoni, B. (1994). Macroalgae and phytoplankton competition in the central Venice lagoon. *Environ. Tech.*, **15**: 1–14

Steffensen, D. A. (1976). The effect of nutrient enrichment and temperature on the growth in culture of *Ulva lactuca*. *Aqu. Bot.*, **2**: 337–51

Thornotn, K. W. and Lessem, A. S. (1978). A temperature algorithm for modifying biological rates. *Trans. Am. Fish. Soc.*, **107**: 284–7

Vermaat, J. E. and Sand-Jensen, K. (1987). Survival, metabolism and growth of *Ulva lactuca* under winter conditions: a laboratory study of bottlenecks in the life cycle. *Mar. Biol.*, **95**: 55–61

Waite, T. and Mitchell, R. (1972). The effect of nutrient fertilization on the benthic alga *Ulva lactuca*. *Bot Mar.*, **15**: 151–6

CHAPTER 29

FAR-FROM-EQUILIBRIUM THERMODYNAMIC MODELLING OF AQUATIC SHALLOW ENVIRONMENTS

S. N. Lvov, R. Pastres, A. Sfriso and A. Marcomini

SUMMARY

Prigogine's far-from-equilibrium thermodynamics was applied for modelling an open shallow aquatic ecosystem in the Venice Lagoon. A thermodynamic stability analysis of the ecosystem characterised by the processes of gas diffusion, primary production and respiration, as well as degradation and sedimentation of algal biomass, was performed based on mass conservation and entropy balance principles. Excess entropy production of the system, as a Lyapounov function, was analysed and a stability criterion for sustainable biomass growth was found. It suggests that, in addition to the well-known limiting factors of biomass production (temperature, light, concentrations of N and P), carbon dioxide scarcity can also act as a limiting factor explaining the observed fast decline of macroalgal biomass. The calculated saturation factor of aqueous carbon dioxide was found to be highly correlated with the experimentally measured saturation factor of aqueous oxygen, which is easily explained by the simplest photosynthesis–respiration reaction. The calculated saturation factor of calcium carbonate suggests the possibility of high-rate precipitation of calcium carbonate over the whole year and, therefore, a significant flux of $CaCO_3$ inside and possibly into the ecosystem.

INTRODUCTION

Understanding the behaviour of aquatic shallow coastal systems such as the Venice Lagoon is a complicated but very important task because these environments are heavily affected by anthropogenic activities, which can be a cause of dystrophic crises. To describe and interpret the functioning of these systems, what is needed is not only the development of high quality experimental and observational monitoring methodologies, but also the elaboration of an adequate theoretical framework capable of describing and interpreting the physical, chemical, and biological processes occurring simultaneously within these systems. As pointed out by several authors (Mauersberger, 1978, 1979, 1983; Straskraba and Gnauck, 1985; Jorgensen *et al.*, 1992; Ortoleva, 1994; Lvov *et al.*, 1996), thermodynamics of far-from-equilibrium open systems, developed mainly by Prigogine and co-workers (Glansdorff and Prigogine, 1971; Nicolis and Prigogine, 1977; Prigogine, 1980), appear to provide a suitable frame for understanding the global behaviour of complicated systems.

From the thermodynamic point of view, aquatic shallow coastal environments are open, non-equilibrium systems, exchanging energy and matter with their surroundings. The most significant exchange processes between an aquatic ecosystem and external environments include the transfer of energy and mass across the air–water, water–biomass and water–sediment interfaces. The most important processes, such as physicochemical reactions, primary production and respiration, degradation of biomass, etc., should be taken into account and could be connected by applying the second law of thermodynamics. In addition, all the processes are deeply influenced by external variables, such as air temperature, light, input and output flows of chemicals, etc. An opportunity to unify and to link the physical, chemical and biological processes of the ecosystem is the key feature of the thermodynamic approach.

The entropy production could serve as a master variable in thermodynamic modelling, as it combines many functional relationships through a few basic equations. If the system is near equilibrium and its evolution can be described by the Prigogine theorem, which says that the total entropy production is positive whereas its rate of change is negative.

In the non-linear range of irreversible processes there are no 'global' master variables which could help to understand the ecosystem evolution (Mauersberger, 1983), but excess entropy production $\delta_x P$ (Nicolis and Prigogine, 1977), which is connected with a Lyapounov stability function L (Lyapunov, 1972), may serve as a general criterion of local stability of an ecosystem outside the linear region.

An aquatic shallow coastal ecosystem, such as the Venice Lagoon, is maintained in more or less stable state by extraordinary fluxes of energy and matter inside the system and through its boundaries and can evolve towards some dissipative structures, which are not necessary stable. A self-organising system can be maintained if there are non-linear couplings and competition between the entities constituting the ecosystem (Jorgensen *et al.*, 1992). The persistence of a self-organising system depends on resources necessary for substantial survival of the biological components, but can also be threatened when nutrients are supplied at an excessive rate. The result of such actions may be fatal at least for some subsystems and the elimination of some of the entities, or appearance of new species, is likely to occur. Some significant internal fluctuations of the physical, chemical, or biological parameters can lead to the transition of the system to a new type of organisation and new nonpredictable behaviour (Prigogine, 1980).

To calculate the excess entropy production $\delta_x P$ correctly the problem of speciation of major seawater components has to be solved, but virtually all chemical speciation models constructed for seawater are appropriate only within relatively narrow limits of physical chemistry parameters, such as temperature, salinity, pH and nutrient concentrations (Millero, 1990). However, in shallow coastal water basins such as the Venice Lagoon, salinity varies between 20 and 40‰, pH between 6 and 10, temperature between 0 ° and 40 °C and concentrations of nutrients, such as nitrogen and phosphorus compounds, can reach hundreds of μmol per kg of seawater.

The scarcity of appropriate thermodynamic data in a wide range of temperature and salinity conditions has inhibited the development of seawater speciation models which can be applied to shallow coastal environments, but the relatively sufficient information actually existing gives us a chance to solve the speciation problem, at least for the major chemical elements of seawater and nutrients such as N and P compounds.

THERMODYNAMIC MODEL OF THE SYSTEM AND ITS STABILITY

The first step of any modelling research is the choice of its boundaries and of its state variables. Because the analysis of a very complicated system could produce, at this stage, misleading results, it was decided to consider the simplest possible thermodynamic model.

Figure 29.1 is a schematic representation of our open-to-air reservoir (g) system which has two main compartments: water column (w) and upper sediments (s). Each of them is made up of multicomponent and multiphase sub-systems, where there is no equilibrium between biotic and abiotic components, but some partial equilibrium could be found between some inorganic species. An intense exchange of matter takes place between compartments, because the internal processes create strong gradients at their interface. As a first, rough and simplified approximation we can suppose that there are no

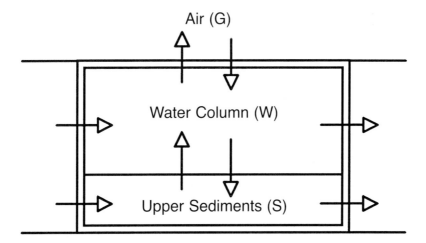

Figure 29.1 Diagram of the thermodynamic system under consideration

gradients in each phase of each compartment of the system, but in a higher approximation the diffusion processes of the chemical and biotic components should also be taken into account.

In our approach we assume that absorption of solar radiation constitutes the main energy supply for the system and also the processes of heating and cooling are significant for the energy balance. Because of the small depth, we presume that there is no temperature gradient inside the system, but its temperature can change over time. In other words, it is as if our system were controlled by a thermostat, whose temperature changes over time, and slowly enough to ensure that the temperature of the system equals the temperature of the thermostat within 1–2 °C.

Intensive fluxes of matter of each component from and to the system take place and the significant fluxes of several gases, such as CO_2, O_2 and possibly some others, through the water column–air interface must be also taken into consideration.

If our system is in the domain of validity of local equilibrium thermo-dynamics, two key expressions are employed for understanding its behaviour. The principle of mass conservation together with Gibbs equation (De-Groot, 1951) leads to the expressions of the time derivatives of mass density of the j-th component ρ_j and the entropy density s:

$$\frac{d\rho_j}{dt} = -div\mathbf{J}_s - div\mathbf{J}_j + \sum_i v_{ji} w_i \qquad (29.1)$$

$$\frac{ds}{dt} = div\mathbf{J}_s - \sum_{ji} v_{ji} \mu_j w_i \qquad (29.2)$$

where $\mathbf{J}_j$ is the mass flux of the j-th component by diffusion,
$\mathbf{J}_s$ is the entropy flux in and out of the system because of the heat flow,
μ_j is the chemical potential per unit mass of the j-th constituent,
w_i is the rate of the i-th bio-physicochemical process, and
v_{ji} is the stoichiometric coefficient for the j-th constituent in the i-th reaction.

It is significant to note that equation (29.1) is needed to calculate temporal and spatial distribution of any j-th constituent, while equation (29.2) should be applied to appreciate the evolution and stability of the system.

Stability is a concept that has been applied in many different ways and there is no possibility to devise a universally applicable definition of the term (Himmelblau and Bischoff, 1968). Generally speaking, a system is stable if it returns to its original state on removal of a disturbance applied on it. We will analyse the stability of the non-equilibrium thermodynamic steady states and, therefore, we are interested in finding a criterion of thermodynamic stability of the system under consideration. For this purpose, the theory of stability initiated by Lyapounov's classic work (Lyapunov, 1949) can be applied to understand the thermodynamic stability of the aquatic shallow ecosystem.

In the range of linear irreversible thermodynamics the total entropy production was found to be a Lyapounov function which can serve as a stability criterion of a steady state. The phenomenon of thermal diffusion is a classic example of the application of linear irreversible thermodynamics (Prigogine, 1962) and the entropy production plays a key role in formulating a steady-state solution of a non-isothermal aquatic system (Lvov *et al.*, 1993a).

Another Lyapounov function was introduced by Glansdorff and Prigogine (Glansdorff and Prigogine, 1971) for far-from-equilibrium systems. This function is the time derivative of the second-order variation of the total entropy of the system, which is the excess entropy production $\delta_\chi P$:

$$\delta_\chi P = \frac{1}{2} \int_V \frac{d(\delta^2 s)}{dt} \, dV \tag{29.3}$$

where V is total volume of the system.

It was found (Nicolis and Prigogine, 1977) that if, for all t larger than a fixed time t_0, which could be the starting time of the perturbation, then

$$\delta_\chi P \geq 0 \tag{29.4}$$

the stability of the system is ensured and the steady state is asymptotically stable; otherwise if the excess entropy production $\delta_\chi P$ is negative, then

$$\delta_\chi P < 0 \tag{29.5}$$

we have an unstable steady state.

The excess entropy production $\delta_\chi P$ depends only on the characteristics of the system, namely on a set of parameters, such as chemical affinities, kinetic constants, etc., which are a quantitative estimate of the distance from a steady state. This set can be modified to an extent that at some critical parameters, let's say λ_c, the sign of equation (29.4) will be inverted and the steady state will lose its stability. This critical situation corresponds to a 'bifurcation point' (Hale and Koçak, 1991) and in this case the excess entropy production is defined by

$$\delta_\chi P(\lambda_c) = 0 \tag{29.6}$$

Equation (29.6) is very significant in the thermodynamic theory of stability because it allows the calculation of the values of the constraints capable of inducing an instability of the previously stable thermodynamic branch (Prigogine, 1980). Furthermore, the new thermodynamic branch can give, through the fluctuations, a new space–time structure which was called by Prigogine a 'dissipative structure' and which makes a system self-organised.

Analysing equations (29.1–29.3), we can conclude that, even for the simplest thermodynamic system, the values w_i for all bio-physicochemical processes as well as the values ρ_j and μ_j for all constituents, must be inferred.

The chemical potential and speciation treatment

The chemical potential μ_j is a key value for the calculation of excess entropy production and both experimental or theoretical investigations of the speciation problem need to be provided to apply the thermodynamic theory of stability for an aquatic shallow ecosystem (Lvov *et al.*, 1993b). If some species are not in equilibrium with their surroundings, ρ_j should be inferred by applying a dynamic model of the process, or it must be measured experimentally. If some species are in a local thermodynamic equilibrium we could solve the problem of speciation theoretically and calculate ρ_j by minimisation of the Gibbs energy G (Stumm and Morgan, 1981):

$$G = \sum_{j=1}^{q} \mu_j n_j \rightarrow \min \tag{29.7}$$

where n_j is the number of moles of j-th species in the aqueous phase of the water column or the interstitial water.

We have investigated the hypertrophic seawater composition in an ecosystem of the Venice Lagoon (Sacca Sessola station) by using experimental, high-quality monitoring data (Sfriso *et al.*, 1992, 1993, 1994) and proposed a method for the speciation treatment. Short-range, ion–water interaction has been taken into account by applying the models developed by Helgeson and co-workers (Helgeson, 1992; Johnson *et al.*, 1992). Ion–ion interactions have been calculated by the Millero approach (Millero and Schreiber, 1982; Millero, 1990; Millero and Hawke, 1992) which is, in fact, an extension of the well-known Pitzer's aqueous solution theory (Pitzer, 1973) for aquatic marine environments.

Equation (29.7) was solved by applying a linear programming approach and an iterative algorithm to take into account the ion–ion interaction. The validation of the proposed chemical speciation model has been tested by comparing the simulation of complex electrolyte solutions with actual measured values, as well as with highly reliable calculated literature data.

The experimental values of temperature, salinity, pH, as well as concentrations of $NH_3(aq) + NH_4^+(aq)$, $NO_2^-(aq)$, NO_3^- (aq) and total inorganic P(aq) (Sfriso *et al.*, 1992, 1993, 1994), reported in Table 29.1, were taken as input data. The amounts of 11 chemical elements (H, B, C, O, Na, Mg, S, Cl, K, Ca, Si) were calculated by using conventional seawater composition with a correction to account for salinity change from 35‰ to any other experimentally measured value.

The concentrations of 45 species were calculated with the proposed approach for a set of 45 field measurements acquired over 430 days, starting from April 23, 1990 and lasting until June 25, 1991 at the Sacca Sessola station in the Venice Lagoon. The location of the sample collection is shown in Figure 29.2.

Table 29.1 Physicochemical parameters in the water column at the Sacca Sessola station of the Venice Lagoon (Sfriso *et al.*, 1992, 1993, 1994)

Date 1990/91	Temp. °C	pH	Eh mV	Salin. ‰	O_2 %	tot.P	NH_4+NH_3 μmol/l	NO_2	NO_3
24-Apr	15.8	8.81	403	27.6	254	1.10	14.27	0.54	2.70
03-May	21.8	9.56	331	30.2	218	0.95	4.83	0.40	3.46
08-May	23.0	8.76	378	30.0	220	0.90	5.29	0.46	1.34
15-May	23.8	8.78	334	30.7	120	0.95	6.86	0.37	2.95
22-May	23.8	8.53	351	28.6	145	0.95	3.13	0.40	2.39
29-May	20.8	9.04	381	31.9	154	1.35	31.67	0.84	3.17
05-Jun	21.2	8.72	370	31.2	114	0.55	9.21	0.30	0.89
12-Jun	21.0	8.41	383	28.7	131	1.05	12.99	0.63	4.69
18-Jun	23.8	8.43	250	30.0	153	0.90	10.06	0.63	1.23
26-Jun	26.9	8.41	342	29.3	136	1.80	9.66	1.23	1.90
04-Jul	23.2	7.60	293	31.2	83	1.35	16.39	0.35	2.38
10-Jul	25.2	8.20	360	30.3	60	2.55	8.10	0.51	3.39
18-Jul	25.8	8.34	432	31.9	113	1.00	8.10	0.61	6.64
25-Jul	25.8	8.25	338	31.6	106	0.60	8.10	0.58	1.02
02-Aug	27.0	8.29	385	32.3	180	0.90	7.25	0.61	1.46
09-Aug	23.0	8.11	340	31.6	129	0.80	21.29	0.91	6.22
16-Aug	26.8	8.59	339	33.2	191	0.90	10.25	0.44	3.07
27-Aug	25.5	8.30	346	33.6	159	0.80	11.95	0.89	5.17
10-Sep	20.8	8.10	325	32.5	135	0.75	13.97	1.56	2.09
20-Sep	21.2	8.33	335	32.0	117	1.00	9.99	1.10	5.33
01-Oct	19.5	8.52	375	32.2	138	0.80	9.34	0.84	0.81
15-Oct	21.4	8.56	365	29.4	204	1.25	7.71	0.63	0.87
30-Oct	15.0	8.40	313	28.8	108	1.50	13.58	2.19	16.74
12-Nov	10.4	8.64	301	29.9	160	0.78	12.02	1.00	8.18
29-Nov	11.8	8.38	269	29.1	138	0.70	4.96	1.19	23.73
15-Dec	5.8	8.54	387	27.6	108	1.10	5.22	2.38	17.08
30-Dec	4.3	8.48	368	27.8	106	0.90	4.88	2.12	22.13
12-Jan	3.8	8.32	398	30.1	104	0.70	4.11	1.89	29.67
01-Feb	3.0	8.40	378	32.2	127	0.48	6.99	1.33	27.16
15-Feb	2.0	8.30	411	32.0	135	0.23	4.24	1.40	44.74
04-Mar	10.4	8.12	392	30.8	142	0.40	4.31	1.12	10.00
15-Mar	11.9	8.13	388	30.7	106	0.85	4.11	0.70	4.45
25-Mar	13.0	8.13	396	31.3	127	0.60	2.94	0.40	1.75
03-Apr	14.8	8.19	370	27.8	230	0.45	4.90	2.84	1.68
09-Apr	18.2	8.61	379	29.9	170	0.38	4.31	1.14	2.37
16-Apr	17.8	8.35	393	30.5	158	0.35	5.75	0.47	2.92
23-Apr	14.0	8.22	355	31.7	103	0.35	10.06	0.89	8.72
30-Apr	14.2	8.12	348	31.3	97	0.41	15.13	2.21	18.25
08-May	14.8	7.93	337	23.5	85	3.00	25.79	3.38	32.18
17-May	15.0	8.07	373	27.4	99	0.20	12.86	2.80	12.98
23-May	23.2	8.18	385	30.9	144	0.25	14.63	1.05	7.79
01-Jun	22.2	8.35	400	30.4	135	0.20	9.43	0.90	6.54
10-Jun	21.0	8.29	406	29.4	114	0.25	5.03	0.77	5.04
18-Jun	25.2	8.40	320	29.0	163	0.05	6.59	0.51	0.69
25-Jun	26.6	8.19	363	29.8	146	0.45	3.40	0.26	4.09

Figure 29.3 shows the saturation factor of $CO_2(aq)$, Ω_{CO_2}, calculated by:

$$\Omega_{CO_2} = m_{CO_{2(aq)}} / m^*_{CO_{2(aq)}} \tag{29.8}$$

where $m_{CO_{2(aq)}}$ and $m^*_{CO_{2(aq)}}$ are molal concentrations of $CO_2(aq)$ in a definite time interval and in a case of equilibrium between $CO_2(aq)$ and $CO_2(gas)$, respectively. It is inferred from figure 3 that Ω_{CO_2} appears to be inversely correlated with the saturation factor of aqueous oxygen $O_2(aq)$, Ω_{O_2} which has been measured experimentally (see Table 29.1). This correlation corresponds to the general photosynthetic reaction according to which a minimum of Ω_{CO_2} should correspond to a maximum of Ω_{O_2} and vice versa.

The low level of Ω_{CO_2} over the year, probably due to the enormous biological activity of macroalgae at the Sacca Sessola station of the Venice Lagoon, suggests that CO_2 plays a significant role as a possible limiting factor of biomass production when other well-known limiting factors (temperature, light, concentrations of N and P) are not so significant.

Figure 29.4 shows the saturation factor of calcium carbonate Ω_{CaCO_3} calculated by:

$$\Omega_{CaCO_3} = m_{Ca^{2+}} \cdot m_{CO_3^{2-}} \gamma_{Ca^{2+}} \gamma_{CO_3^{2-}} / K_{CaCO_3\,(s)} \tag{29.9}$$

where $K_{CaCO_3\,(s)}$ is the solubility equilibrium constant, $m_{Ca^{2+}}$ and $m_{CO_3^{2-}}$ are molal concentrations of $Ca^{2+}(aq)$ and $CO_3^{2-}(aq)$ ions, respectively, and $\gamma_{Ca^{2+}}$ and $\gamma_{CO_3^{2-}}$ are the activity coefficients of these ions.

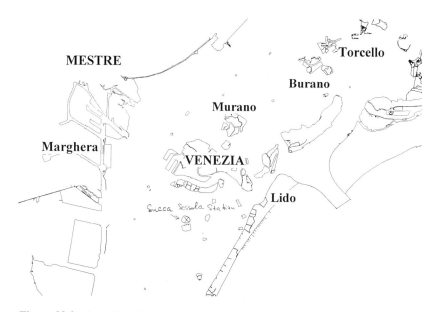

Figure 29.2 Location of the sampling stations in the Venice Lagoon

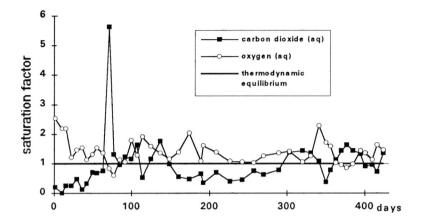

Figure 29.3 Variation of the saturation factors of carbon dioxide and oxygen in the water column at the Sacca Sessola station of the Venice Lagoon

Because of the high level of Ω_{CaCO_3} over the year, we can expect a high rate of calcium carbonate precipitation and some significant fluxes of $CaCO_3(aq)$ inside and possibly into the ecosystem.

Thermodynamic stability analysis of the simple shallow water ecosystem

We have conducted a thermodynamic stability analysis for a very simple aquatic ecosystem, named 'Photosynthesator' (Lvov *et al.*, 1993b), which takes into account just three chemical elements: C, H and O, producing the following minimum list of chemical species: $O_2(g)$ (atmospheric oxygen), $CO_2(g)$ (atmospheric carbon dioxide), $O_2(w)$ (dissolved oxygen), $CO_2(w)$ (dissolved carbon dioxide), $CH_2O(w)$ (living biomass), $H_2O(w)$ (water), $CH_2O(s)$ (dead biomass). A short list of the main physical, chemical and biological processes which occur in the system could consist of:

(1) Diffusion of $CO_2(g)$ through the air-water interface,

$$CO_2(g) \overset{w_1}{\underset{w_{-1}}{\longleftrightarrow}} CO_2(w)$$

(2) Diffusion of $O_2(g)$ through the air–water interface,

$$O_2(g) \overset{w_2}{\underset{w_{-2}}{\longleftrightarrow}} O_2(w)$$

(3) Primary production (photosynthesis) of the biomass $CH_2O(w)$ in the water column,

$$CO_2(w) + H_2O(w) + CH_2O(w) \overset{w_3}{\longrightarrow} 2CH_2O(w) + O_2(w)$$

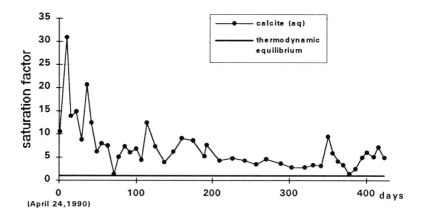

Figure 29.4 Variation of the saturation factor of calcite in the water column at the Sacca Sessola station of the Venice Lagoon

(4) Respiration of $CH_2O(w)$ in the water column,

$$CH_2O(w) + O_2(w) \overset{w_4}{\rightarrow} CO_2(w) + H_2O(w)$$

(5) Degradation (mortality of biomass) and sedimentation of $CH_2O(w)$,

$$CH_2O(w) \overset{w_5}{\rightarrow} CO_2O(s)$$

A significant feature of our approach is that the photosynthetic biomass production process is presented as an autocatalytic reaction and it tends to give thermodynamic instability (Glansdorff and Prigogine, 1971), destabilising the aquatic ecosystem if such a process is part of an open system.

By taking into account the results of ecological observations (Lvov *et al.*, 1993b) we can express the following set of kinetic equations for the five processes listed above:

$$w_1 = k_1 p_{CO_2(g)} - k_{-1} p_{CO_2(w)} \tag{29.10}$$

$$w_2 = k_2 p_{O_2(g)} - k_{-2} p_{O_2(w)} \tag{29.11}$$

$$w_3 = f k_3 p_{CO_2(w)} p_{CH_2O(w)} \tag{29.12}$$

$$w_4 = k_4 p_{CH_2O(w)} \tag{29.13}$$

$$w_5 = k_5 p^2_{CH_2O(w)} \tag{29.14}$$

where $p_{CO_2(g)}$ and $p_{O_2(g)}$ are partial pressures of carbon dioxide and oxygen in the atmosphere;

k_1, k_2, k_3, k_4, k_5 are, respectively, the rate constants of $CO_2(g)$ diffusion, $O_2(g)$ diffusion, primary production (photosynthesis), respiration and mortality of biomass; k_{-1} and k_{-2} are the backward rate constants of $CO_2(g)$ and $O_2(g)$ diffusion, respectively;

f is a daily-averaged limiting function of the external factors such as solar radiation, temperature, concentrations of nutrients, etc., which can vary between 0 and 1 depending on the season of the year and some particular climatic and hydrodynamic conditions.

If we suppose now that there is no convective flow in the system, the set of the mass balance equation (29.1) will be reduced to:

$$\frac{d\rho_{CO_2(w)}}{dt} = k_1 P_{CO_2(g)} - k_{-1}\rho_{CO_2(w)} - f k_3 \rho_{CO_2(w)}\rho_{CH_2O(w)} + k_4 \rho_{CH_2O(w)} \qquad (29.15)$$

$$\frac{d\rho_{CH_2O(w)}}{dt} = f k_3 \rho_{CO_2(w)}\rho_{CH_2O(w)} - k_4 \rho_{CH_2O(w)} - k_5 \rho^2_{CH_2O(w)} \qquad (29.16)$$

$$\frac{d\rho_{O_2(w)}}{dt} = k_1 P_{O_2(g)} - k_{-1}\rho_{O_2(w)} + f k_3 \rho_{CO_2(w)}\rho_{CH_2O(w)} - k_4 \rho_{CH_2O(w)} \qquad (29.17)$$

$$\frac{d\rho_{CH_2O(s)}}{dt} = k_5 \rho^2_{CH_2O(w)} \qquad (29.18)$$

which explicitly describes the density variation of the main four species of the system.

Two steady-state solutions that have a physical meaning for all four species can be analytically inferred from equations (29.15–29.18). The first steady state, which corresponds to zero biomass production, is defined by:

$$\rho^*_{CH_2O(w)}(1) = 0 \qquad (29.19)$$

$$\rho^*_{CO_2(w)}(1) = K_{CO_2}P_{CO_2} \qquad (29.20)$$

$$\rho^*_{O_2(w)}(1) = K_{O_2}P_{O_2} \qquad (29.21)$$

$$\rho^*_{CH_2O(s)}(1) = 0 \qquad (29.22)$$

The solution of the second steady state, which corresponds to maximum biomass production, can also be provided as an analytical solution (Lvov *et al.*, 1993b).

Because there are two possible steady states, the system can evolve towards one of them depending, first, on the numerical value of the parameters $k_1, k_{-1}, k_2, k_{-2}, k_3, k_4, k_5, P_{CO_2(g)}, P_{O_2(g)}, f$ in the set of differential equations (29.15–29.18) and, second, on the initial values of the variables.

In order to understand the system behaviour as a whole, a criterion for its development needs to be found. This problem can be solved by applying the second law of thermodynamics on the basis of thermodynamic stability theory. By substituting both equations (29.10–29.14) for the rates and corresponding expressions for the chemical potentials in equations

(29.2–29.4) we have found the following stability criterion for the first steady state, given by (29.19–29.22), of the Photosyntesator:

$$\alpha = \frac{f\,k_3 K_{CO_2} P_{CO_2}}{k_4} \leq 1 \qquad (29.23)$$

the validity of which has been confirmed by both a mathematical theoretical analysis as well as a numerical evaluation of equations (29.15–29.18) (Lvov *et al.*, 1993b).

The stability criterion for the second steady state can be given by:

$$\alpha > 1 \qquad (29.24)$$

The critical value of the stability criterion $\alpha_c = 1$ is a bifurcation point of the Photosyntesator so that in the region $0 < \alpha \leq$ the first steady state (zero biomass production) takes place, while in the region $\alpha > 1$ the second steady state (maximum biomass production) is stable.

The so-called bifurcation diagram (Hale and Koçak, 1991) of the Photosyntesator's behaviour (Figure 29.5) was drawn for the main variables in the convenient scaled variables (Lvov *et al.*, 1993b), which are the dimensionless steady state density of the dissolved water column carbon dioxide $\rho^*_{CO_2(w)} f\,k_3/k_4$ (curve (a)) and the dimensionless steady state density of the living biomass $\rho_{CH_2O(w)} f\,k_3/k_{-1}$ (curve (b)). The stable steady states are drawn in solid lines and the unstable states in dashed lines. As soon as α reaches its bifurcation value equal to 1, the first steady state becomes unstable and another thermodynamic branch corresponding to non-zero biomass appears. At this point, the steady state density of $CO_2(w)$ suddenly changes its direction with a consequent increase of the water column pH.

Expressions (29.23) and (29.24) are the main result of our thermodynamic stability analysis of the Photosyntesator and can be employed for description of a shallow water ecosystem and for understanding its behaviour. It follows from equations (29.23–29.24) that, in addition to the well-known limiting factors of biomass production (temperature, light, concentrations of N and P), the scarcity of CO_2 can also act as a limiting factor explaining the fast decline of macroalgal biomass.

The stability criteria (29.23) and (29.24) corresponding to the two possible steady states can be used to understand the behaviour of the Photosyntesator without numerical analysis of the system. This great advantage of the thermodynamic stability analysis will be illustrated by the following simple example. If the rate constants of the biomass primary production k_3 and respiration k_4 for some macroalga species, e.g. *Ulva Rigida*, quite common in the Venice Lagoon, are assumed to be equal to 40,000 (day/M) and 0.216 (day), respectively (Solidoro *et al.*, 1993), and making the Henry's constant K_{CO_2} and

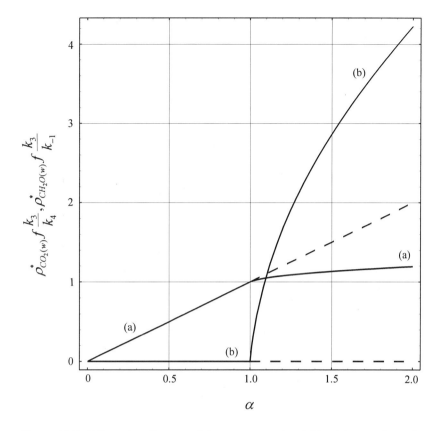

Figure 29.5 Bifurcation diagram of the ecosystem: the stable states are drawn in solid lines and the unstable states in dashed lines; (a) and (b) are scaled steady-state densities of $CO_2(w)$ and $CH_2O(w)$, respectively

partial pressure p_{CO_2} of carbon dioxide equal to 0.03067 M/bar, and 0.00033 bar, respectively, (Skirrow, 1975), we can calculate the average limiting function f by applying the equality of equation (29.23), which leads to $f = 0.5335$. Now we can conclude that, if $f \le 0.5335$, the system would evolve towards the first steady state (i.e. zero biomass production) and, if $f > 0.5335$, macroalgal growth will take place. Therefore, we can draw a conclusion about system stability under some particular conditions without the need of numerical simulations based on linear stability analysis.

CONCLUSIONS

In this work we have developed far-from-equilibrium thermodynamic analysis for an ecosystem in the Venice Lagoon by taking into account a minimum of bio-physicochemical processes, namely gas diffusion, primary production and respiration of biomass, as well as its degradation and sedimentation. This simplification offers the advantage to show clearly how we can employ Prigogine's far-from-equilibrium thermodynamics for coupling bio-physico-chemical processes in open systems and for understanding the system stability without carrying out numerical evaluation of any set of mass conservation differential equations.

From our consideration of the aquatic shallow water ecosystem, we have found that in addition to the well-known limiting factors of biomass production, such as solar radiation, temperature and concentrations of N and P, the thermodynamics and kinetics of CO_2 fluxes through the air–water and water–sediment interfaces must be taken into consideration to understand the ecosystem behaviour correctly. Therefore, in addition to the common limiting factors, CO_2 scarcity can also act as a limiting factor explaining the rapid decline of macroalgal biomass.

To be closer to the real aquatic ecosystem, however, some additional significant bio-physicochemical processes in the upper sediment compartment should be taken into account which may lead to the discovery of new instabilities. Also, temperature and salinity changes, as well as the periodical flow of the water, will give rise to other instabilities of the system. Investigation of these problems is the focus of our current research and requires serious efforts in both theoretical analysis and experimental investigations of the most significant processes occurring in shallow aquatic environments.

The composition of the Venice Lagoon hypertrophic seawater (Sacca Sessola station) has been investigated using experimental, high-quality monitoring data and the proposed method of speciation treatment. The calculated saturation factor of aqueous carbon dioxide Ω_{CO_2} was found to be highly correlated with the experimentally measured saturation factor of aqueous oxygen Ω_{O_2}, which is easily explained by the simplest photosynthesis–respiration reaction. The calculated saturation factor of calcium carbonate Ω_{CaCO_3} suggests the possibility of high-rate precipitation of calcium carbonate over the whole year and, therefore, a permanent flux of $CaCO_3$ inside and possibly into the ecosystem. These calculations also clearly demonstrate that an ecosystem such as the Venice Lagoon can never be in thermodynamic equilibrium with its surroundings and that intense inner and external fluxes of energy and matter must be provided for its survival.

ACKNOWLEDGEMENTS

S.N.L. wishes to acknowledge the support of the Division of UNESCO Basic Sciences (Project Venice Lagoon Ecosystem) through grants SC 233.130.2 and SC 233.251.2. Part of this work was presented at the European Research Conferences on Natural Waters and Water Technology held in Italy (1993).

REFERENCES

De-Groot, S. R. (1951). *Thermodynamics of Irreversible Processes.* North-Holland, Amsterdam

Glansdorff, P. and Prigogine, I. (1971). *Thermodynamic Theory of Structure, Stability and Fluctuations.* Wiley, London

Hale, J. and Koçak, U. (1991). *Dynamics and Bifurcations.* Springer Verlag, New York

Helgeson, H. C. (1992). Effect of complex formation in flowing fluids on the hydrothermal solubilities of minerals as a function of fluid pressure and temperature in the critical and supercritical regions of the system. *Geochim. Cosmochim. Acta,* **56**: 3191–207

Himmelblau, D. M. and Bischoff, K. B. (1968). *Process Analysis and Simulation: Deterministic Systems.* Wiley, New York

Johnson, J. W., Oelkers, E. H. and Helgeson, H. C. (1992). A software package for calculating the standard molal thermodynamic properties of minerals, gases, aqueous species, and reactions from 1 to 5000 bars and 0 to 1000 °C. *Computers and Geosciences,* **18**: 899–947

Jorgensen, S. E., Patten, B. C. and Straskraba, M. (1992). Ecosystems emerging: toward an ecology of complex systems in a complex future. *Ecol. Modelling,* **62:** 1–27

Lvov, S. N., Marcomini, A. and Suprun, M. M. (1993a). Thermodynamics of non-isothermal electrolyte solutions. *Proc. 4th Int. Symp. on Hydrothermal Reactions,* 135–8

Lvov, S. N., Pastres, R. and Marcomini, A. (1993b). Irreversible thermodynamic approach for coupling bio-physicochemical processes in aquatic shallow environments. *Proc. of European Research Conferences on Natural Waters and Water Technology, Acquafredda di Maratea, Italy,* **22**

Lvov, S. N., Pastres, R. and Marcomini, A. (1996). Thermodynamic stability analysis of the carbon biogeochemical cycle in aquatic shallow environments. *Geochim. Cosmochim, Acta.,* **60**: 3569–79

Lyapunov, A. A. (1949). Problème gènèral de la stabilitè de mouvement. *Annals of Math. Studies, Princeton University, NJ.,* 17

Lyapunov, A. A. (1972). *Cybernetic Problems in Biology.* Nauka, Moscow. (in Russian)

Mauersberger, P. (1978). On the theoretical basis of modelling the quality of surface and subsurface waters. *IAHS-AISH Publication,* **125**: 14–23

Mauersberger, P. (1979). On the role of entropy in water quality modelling. *Ecol. Modelling,* **7**: 191–9

Mauersberger, P. (1983). General principles in deterministic water quality modelling. In Orlob, G. T. (ed.). *Mathematical Modelling of Water Quality: Streams, Lakes, and Reservoirs.* Wiley, New York., Chap. 3, 43–115

Millero, F. J. and Schreiber, D. R. (1982). Use of the ion pairing model to estimate activity coefficients of the ionic components of natural waters. *Am. J. Sci.,* **282**: 1508–40

Millero, F. J. (1990). Marine solution chemistry and ionic interactions. *Mar. Chem.,* **30**: 205–29

Millero, F. J. and Hawke, D. J. (1992). Ionic interactions of divalent metals in natural waters. *Mar. Chem.,* **40**: 19–48

Nicolis, G. and Prigogine, I. (1977). *Self-Organisation in Nonequilibrium Systems, from Dissipative Structures to Order through Fluctuations.* Wiley, New York

Ortoleva, P. (1994). *Geochemical Self-organization: Oxford Monographs on Geology and Geophysics 23*. Oxford University Press, Oxford

Pitzer, K. S. (1973). Thermodynamics of electrolytes. I. Theoretical basis and general equations. *J. Phys. Chem.*, **77**: 268–77

Prigogine, I. (1962). *Introduction to Nonequilibrium Thermodynamics*. Wiley, New York

Prigogine, I. (1980). *From Being to Becoming*. Freeman, California

Sfriso, A., Pavoni, B., Marcomini, A. and Orio, A. A. (1992). Macroalgae, nutrient cycles and pollutants in the lagoon of Venice. *Estuaries*, **15**: 517–28

Sfriso, A., Marcomini, A., Pavoni, B. and Orio, A. A. (1993). Species composition, biomass and net primary production in coastal shallow waters: the Venice Lagoon. *Bioresource Technology*, **44**: 235–50

Sfriso, A., Pavoni, B. and Orio, A. A. (1994). Flora and macroalgal biomass production in different nutrient-enriched areas of the Venice Lagoon. *Final Report of BEST (Benthic Studies) Programme*

Skirrow, D. (1975). *The Dissolved Gases – Carbon Dioxide, in: Chemical Oceanography*. 2nd Edition Riley, J. P. and Skirrow, G. (eds), Academic Press, London., Vol. 2, Chap. 9, 1–192

Solidoro, C., Deiak, C., Franco, D., Pastres, R. and Pecenik, G. (1993). Modelling macroalgae and phytoplankton growth and competition in the Venice Lagoon. *Proc. Int. Congr. on Modelling and Simulation*, Vol. 4, 1371–6

Straskraba, M. and Gnauck, A. H. (1985). *Freshwater Ecosystems: Modelling and Simulation*. Elsevier, Amsterdam

Stumm, W. and Morgan, J. J. (1981). *Aquatic Chemistry*. Wiley, New York

INDEX